Studies
in the History of Mathematics and Physical Sciences

11

Editor
G. J. Toomer

Advisory Board
R. P. Boas   P. J. Davis   T. Hawkins
M. J. Klein   A. E. Shapiro   D. Whiteside

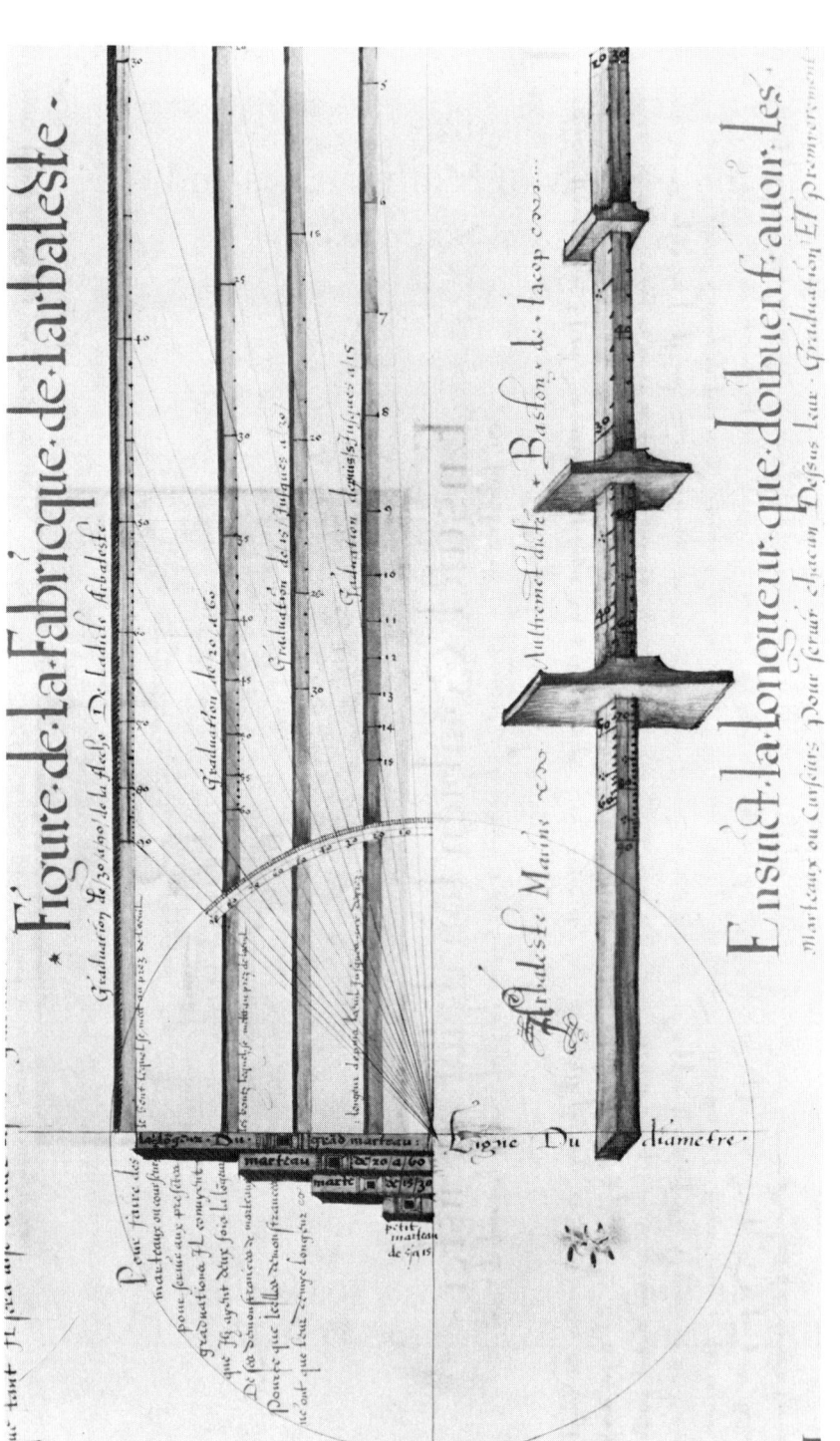

Bibliothèque Nationale, Paris, MS français 150, folio 16: *Les Premières Euvres de Jacques de Vaulx, pillote en la marine, au Havre* (1583). In this navigational manual, we are shown a Jacob Staff, an astronomical instrument invented by Levi ben Gerson, and a method for graduating it. (Phot. Bibl. Nat., Paris.)

Bernard R. Goldstein

The Astronomy of Levi ben Gerson (1288–1344)

A Critical Edition of Chapters 1–20 with Translation and Commentary

With 68 Figures

Springer-Verlag
New York   Berlin   Heidelberg   Tokyo

Bernard R. Goldstein
Jewish Studies Program
University of Pittsburgh
2604 Cathedral of Learning
Pittsburgh, PA 15260
U.S.A.

AMS Classification: 01A30, 01A35, 85-03

Library of Congress Cataloging in Publication Data
Goldstein, Bernard R.
  The astronomy of Levi ben Gerson.
  (Studies in the history of mathematics and physical sciences; 11)
  Bibliography: p.
  Includes index.
  1. Astronomy—Early works to 1800. 2. Astronomy, Jewish. 3. Levi ben Gershom, 1288–1344. I. Title.
II. Series.
QB41.G477   1985   520'.92'4   85-2751

© 1985 by Springer-Verlag New York Inc.
All rights reserved. No part of this book may be translated or reproduced in any form without written permission from Springer-Verlag, 175 Fifth Avenue, New York, New York 10010, U.S.A.

Typeset by Asco Trade Typesetting Ltd., Hong Kong.
Printed and bound by R.R. Donnelley and Sons, Harrisonburg, Virginia.
Printed in the United States of America.

9 8 7 6 5 4 3 2 1

ISBN 0-387-96132-1 Springer-Verlag New York Berlin Heidelberg Tokyo
ISBN 3-540-96132-1 Springer-Verlag Berlin Heidelberg New York Tokyo

To O. Neugebauer
My teacher, friend, and guide

Preface

It would seem that S. Munk was the first modern scholar to draw attention to the significance of Levi ben Gerson's *Astronomy*, surely the most original work on astronomy written in Hebrew in the Middle Ages. Munk (1859, p. 500) called for a specialist to undertake a serious study of this work, but there was little response to his plea in the succeeding century. Indeed, this is the first edition of the Hebrew text of any part of Levi's *Astronomy* but for the table of contents (Renan, 1893, pp. 624–32), and the poems celebrating the invention of the Jacob Staff that appear in chapter 9 (Carlebach, 1910a, pp. 152–53).

The text of Levi's *Astronomy* is written in a ponderous Hebrew style but the content sparkles with originality. The Ptolemaic tradition is subjected to a profound critique based on the idea that the planetary models must conform both to Levi's own observations as well as those of the ancients, and the claim that astronomical theory must be philosophically sound. The enduring vigor of the Ptolemaic tradition has been characterized by O. Neugebauer as follows: "There is no better way to convince oneself of the inner coherence of ancient and medieval astronomy than to place side by side the *Almagest*, al-Battānī's *Opus astronomicum* and Copernicus's *De Revolutionibus*. Chapter by chapter, theorem by theorem, table by table, these works run parallel" (1957, pp. 205–6).

It is surprising, therefore, to find that Levi departs from this well-trodden path, and opens his *magnum opus* with a discussion of instruments he himself invented, of models not previously considered, and of observations to test the theories presented. In addition to the Jacob Staff and the camera obscura, Levi describes a new instrument for finding the Moon's elongation from the Sun (chapter 10) and a modification of the astrolabe for purposes of observation (chapter 12).

Levi tells us that his *Astronomy* was completed in 1328, but it has long been noted that observations made as late as 1340 are included in it. The text is quite long (over 250 folios) and so it seemed appropriate to publish a part of it, with others to come. Indeed, my own study of it began in 1968 and the fruits of this research have appeared in a series of publications beginning in 1969. Among aspects of Levi's work that do not appear in the first 20 chapters are his analysis of precession (Goldstein, 1975), his solar and lunar tables (Goldstein, 1974a), his observations and analyses of solar and lunar eclipses (Goldstein, 1979), his lunar models (Goldstein, 1974a, pp. 53–74; 1974b), his views on cosmic distances (Goldstein, 1969a), and his criticisms of al-Biṭrūjī (Goldstein, 1971, vol. 1, pp. 40–43). His planetary theories that are introduced beginning in chapter 21, and the observations on which they are based, remain to be analyzed, a difficult task because of lacunae that afflict all the manuscripts (both in the original Hebrew and in the medieval Latin translation).

In the Introduction an attempt is made to place Levi in the astronomical tradition. Since he mentions few of his predecessors and none of his contemporaries, considerable uncertainty remains. It is equally difficult to assess his influence on subsequent generations of astronomers, for references to Levi are rare both in Hebrew and Latin, and one is often forced to rely on indirect evidence.

Parts of this edition have been adapted from previous publications and appear here in revised form (Goldstein, 1974a, 1977, 1980). In order to facilitate references to the text, sentence numbers have been added, and they are used in the translation and the commentary as well as in the critical apparatus. The Latin translation has not been consulted systematically; it seems to be a quite literal rendering of the Hebrew.

It is a pleasure to thank the institutions that have supported this research in various ways over the years: Yale University, The Hebrew University, Brown University, University of Pittsburgh, The Institute for Advanced Study (Princeton), Centre national de la recherche scientifique, Observatoire de Paris, The Museum for History of Science (Oxford), Istituto e Museo di Storia della Scienza (Florence), The American Philosophical Society, The American Council of Learned Societies, The John Simon Guggenheim Memorial Foundation, The National Science Foundation, and The National Endowment for the Humanities. The cooperation of librarians in many countries was kindly forthcoming, and I am especially grateful to M. Garel (Bibliothèque Nationale, Paris), M. Schmelzer (The Jewish Theological Seminary of America, New York), and B. Richler (National Library, Jerusalem).

Numerous colleagues have assisted in this research and their suggestions have always been welcome. In particular, I wish to thank G. Beaujouan, B. Blumenkranz, J. Lay, G. Nahon, E. Poulle, R. Rashed, and J.-P. Verdet (Paris); F. Maddison and J. Roche (Oxford); R. Scheindlin and N. Siraisi (New York); P. Galluzzi (Florence); and A. Aaboe (New Haven). A special debt of thanks is due to G. J. Toomer, the editor of this series, for his careful reading of the completed draft of this book, and his many detailed and illuminating comments.

Photographs appear courtesy of the Bibliothèque Nationale, the Adler Planetarium (Chicago), and the Istituto e Museo di Storia della Scienza (Florence). The Hebrew text was typed by Mrs. Karen Rosenzweig whose assistance is gratefully acknowledged.

Paris
December 1984

Bernard R. Goldstein

Manuscript Sigla

N. Naples, Biblioteca Nazionale, heb. III F.9 (undated), with marginal notes by Mordecai Finzi (not in his hand).
P. Paris, Bibliothèque Nationale, heb. 724 (dated 1397).
Q. Paris, Bibliothèque Nationale, heb. 725 (dated 1510).

For more information on the Hebrew and Latin manuscripts of Levi ben Gerson's *Astronomy*, see Goldstein, 1974a, pp. 74–83; and Goldstein, 1981, p. 250.

Notation

Paragraphs, sentences, and sentence numbers in the text have been added by the editor. Sentence numbers and folio numbers for all three manuscripts may be found in the margin of the edited text: a vertical stroke in the text indicates the place where a new folio begins. The supplementary pagination for the Hebrew text appears inside square brackets. In the translation sentence numbers appear in square brackets, and in the commentary they are preceded by the symbol §. Moreover, words added by the editor to facilitate reading the translation have been put inside square brackets.

Contents

Preface	vii
Manuscript Sigla and Notation	x
Introduction	1
Translation	17
Commentary	131
References	198
Index	207
Hebrew Text	215

Introduction

Levi ben Gerson (1288–1344), sometimes called Gersonides or Leo de Balneolis, is well known as a philosopher, biblical exegete, mathematician, and astronomer (cf. Touati, 1973; and Feldman, 1984). He lived in Orange and occasionally visited Avignon where his brother Solomon was physician to Pope Clement VI. Although the family name was de Balneolis, there is no evidence that he himself was born or ever lived in Bagnols (cf. Shatzmiller, 1972, 1974). Levi did not cite any contemporaries and little is known of his life. More surprisingly, his scientific work was rarely cited in the subsequent literature, though a number of manuscript copies survive both in the original Hebrew and in Latin translations. The works on which Levi depended were all available in Hebrew and there is no reason to believe he read astronomical texts in Latin or in Arabic. A recent article has brought attention to a list of the books in Levi's library that had not been properly identified in the manuscript catalogue (Weil, 1980, p. 590; cf. Loewinger and Weinryb, 1965, p. 32). Many of the 140 items concerned astronomy and, judging from the incomplete published list, they were all in Hebrew.

Levi's *Astronomy* forms Book V, Part 1, of the *Wars of the Lord*, his *magnum opus* on religious philosophy. It survives in separate manuscript copies from the remainder of this treatise, and was omitted from the printed editions (1569, 1866). The *Astronomy* is quite long—136 chapters that fill 257 folios in MS P—and, of the Hebrew text, only the table of contents has previously been published (Renan, 1893, pp. 278–86). Recently, there appeared a translation and commentary on the section concerning creation (*Wars*, VI: 2.1–8) in which Levi alluded to some of his astronomical views (Staub, 1982: for a general survey of the literature on Levi, see Kellner, 1979).

In his astronomy, Levi was deeply indebted to Ptolemy whom he frequently cited, and his profound criticisms of the Ptolemaic tradition are very refreshing in comparison to the repetitive character of so much of medieval astronomical writings. The arrangement of his *Astronomy* does not follow that of Ptolemy's *Almagest*, and a wider range of planetary models were considered. Of special interest in the twenty chapters edited and translated here are his descriptions of new observational instruments, notably the Jacob Staff. His introduction of an "experiment" for finding the center of vision within the eye (chapters 6, 7), and his careful examination of optical phenomena relating to the camera obscura (chapters 5, 9) suggest that his understanding of physical phenomena and physical reasoning was as advanced as his understanding of astronomy. He also discussed instrumental error and suggested methods for measuring lines and arcs by means of fine graduations along transversals (chapters 7, 12).

It remains to consider Levi's place in the history of astronomy. I have divided this discussion into two sections.

Section 1: Levi and His Predecessors

1. In order to assess Levi ben Gerson's contributions to astronomy, it is well to review some of the characteristic features of this discipline in antiquity and the Middle Ages. One of the major issues that confronted Greek and medieval astronomers was the relationship between physics and mathematics in astronomy. The goal of mathematical astronomy was the computation of the planetary positions at any time, past or future and, by appealing to geometric models, rather than say, the arithmetic schemes of the Babylonians, the astronomers seemed to suggest that these geometric models were an accurate representation of physical reality. However, when we examine the relevant discussions in ancient and medieval literature, it becomes clear that they did not uniformly incline to this view.

2. The earliest unambiguous example of a geometric model for planetary motion is due to Eudoxus (flourished ca. 350 B.C.), but it is most probable that his homocentric spheres were only intended to describe planetary motion qualitatively (cf. Goldstein and Bowen, 1983). This scheme was modified by Aristotle and as such became the foundation of most ancient and medieval scientific cosmologies in which it was generally agreed that motion descends from the Prime Mover to the planetary spheres, and that no empty space intervenes between these spheres. Some of the difficulties that a homocentric model presented were already noted in antiquity, and Simplicius reports from Sosigenes (second century A.D.) that some planets approach and recede from us as is seen from the variation in their apparent sizes, an insight that is probably due to Autolycus of Pitane (ca. 300 B.C.: cf. Dreyer, 1953, p. 141). This settled the matter in antiquity and no new homocentric models are known to have been invented until the problem arose again in the Middle Ages.

3. It is no longer doubted that both epicyclic and eccentric models were known before Apollonius (ca. 200 B.C.), but we cannot be much more specific about their development prior to him (cf. Neugebauer, 1975, pp. 805 ff;

Neugebauer, 1972; Aaboe, 1963). From the *Almagest* it is clear that he was able to demonstrate their exact equivalence, and in fact the elegant methods for deriving the parameters for the solar model and the simple lunar model from specific dated observations may also be due to him, rather than Hipparchus (ca. 150 B.C.) as is usually assumed (cf. Neugebauer, 1975, p. 265). In planetary theory Apollonius proved a theorem on stationary points that satisfactorily explained retrograde motion, but it would seem that this entire discussion was a matter of pure mathematics (Neugebauer, 1975, pp. 270 f). The application of these models to "save the phenomena" was one of the principal contributions of Hipparchus, and in Ptolemy's *Almagest* we learn the extent of his success. But many difficulties remained, notably Hipparchus could not account for the observed planetary motions. Nevertheless, Ptolemy mentioned no improved models for planetary motion between Hipparchus's time and his own (cf. *Almagest*, IX, 2; trans. Toomer, 1984, p. 422).

4. In the *Almagest* we find a new and complicated model to account for lunar motion, and an equant model for the planets (for a brief description of these models, see Neugebauer, 1957, pp. 191 ff). Ptolemy was aware of Aristotle's principles for astronomy as well as the tension between them and the models he presented. Essentially his response was that in a conflict between "principles" and accounting for the observations, the latter must take precedence even if this requires us to modify our notions of simplicity in nature. In a passage concerning the theory of planetary latitudes, Ptolemy says the following (*Almagest* XIII, 2):

> Let no one object to these hypotheses on account of the great complication of our system. For it is not fitting to draw comparisons between human and divine matters nor to seek proofs for such things as these from totally dissimilar examples. For what is more dissimilar than eternally unchanging things to the ever-changing, or than things altered by every kind of exterior influence to things that never alter even from causes emanating from themselves? We should try to find the simplest hypothesis we can by which we can harmonize the heavenly motions, and failing this be content with what is possible ... It is better not to judge of the simplicity of heavenly matters from the things that appear simple to us especially as even among us the same thing is not always similarly simple for everyone.

5. This argument did not satisfy his successors because it avoided an important issue, namely, Ptolemy first set out the principle of uniform circular motion and then proceeded to violate it by introducing the equant, i.e., a model in which a point on the circumference of a circle no longer moves uniformly about the center of that circle. This in turn raises the question: did Ptolemy intend his models to be considered physically real or merely computational devices? In the *Almagest* true planetary distances are not discussed, and one may be tempted to believe that only the computation of

planetary positions in longitude and latitude is involved. However, in the *Planetary Hypotheses* Ptolemy elaborated his models and constructed a cosmological scheme (similar to what was later known as the Ptolemaic System) which leaves no doubt that his ultimate intention was to present physical models, and this sytem introduced additional tension with Aristotle's system of homocentric spheres.

6. There was also a tradition that the domains of physics and astronomy differed, but this could not be invoked to reduce the tension between the views of these two major ancient authorities. According to this view physical inquiry was concerned with such things as the substance, force, quality, shape, and perhaps even the arrangement of the heavens, whereas astronomy had to deal with the movements, sizes, and distances of the heavenly bodies by means of arithmetic and geometry. But the roles of the physicist and the astronomer are not equal, for the latter is subordinate to the former. We learn from Simplicius (Heath, 1913, pp. 275 f):

> It is no part of the business of an astronomer to know what is by nature suited to a position of rest, and what sort of bodies are apt to move, but he introduces hypotheses under which some bodies remain fixed, while others move, and then considers to which hypotheses the phenomena actually observed in the heaven will correspond. But he must go to the physicist for his first principles, namely that the movements of the stars are simple, uniform and ordered, and by means of these principles he will then prove that the rhythmic motion of all alike is in circles, some being turned in parallel circles, others in oblique circles. Such is the account given by Geminus, or Posidonius in Geminus, of the distinction between physics and astronomy, wherein the commentator is inspired by the views of Aristotle.

Thus Ptolemy could not be excused, for example, from adhering to the principle of uniform circular motion.

7. The earliest serious critique of Ptolemy's astronomy on physical grounds was presented by Ibn al-Haytham (d. ca. 1039), better known today for his work in optics than for his contributions to astronomy (cf. Sabra, 1972). In his *Doubts on Ptolemy* he attacked the inconsistencies in Ptolemy's arguments, e.g., he noted that Ptolemy set himself the task of accounting for the planetary phenomena by means of uniform circular motions (*Almagest* IX, 2), and then introduced an equant model that violated this principle (*Almagest* IX, 6). Moreover, Ibn al-Haytham objected to Ptolemy's complex lunar model (*Almagest* V, 5) because, among other reasons, an imaginary point opposite the center of the deferent controlled the position of the apogee of the lunar epicycle, and this seemed physically impossible (Sabra and Shehaby, 1971, pp. 15 ff; see also Pines, 1964a). He further argued that astronomy must deal with real bodies and not imaginary ones, and that by introducing such imaginary points and circles Ptolemy had not succeeded in

"saving the phenomena." On the other hand, Ibn al-Haytham is not known to have proposed any alternative models, and his most influential work on planetary theory was his adaptation of Ptolemy's models in the *Planetary Hypotheses*. Indeed, this system of nested spheres was the Ptolemaic System that dominated European thinking until the seventeenth century (Goldstein, 1967a; Aiton, 1981; Dreyer, 1953, pp. 258 ff).

8. Some of Ibn al-Haytham's Muslim successors, notably Naṣīr al-Dīn al-Ṭūsī (thirteenth century) and Ibn al-Shāṭir (fourteenth century), came up with alternatives that were meant to satisfy some of Ibn al-Haytham's objections (Kennedy, 1966, especially p. 365, n. 1). They were able to depend on a new mathematical tool, now called the "Ṭusi-couple", by which non-uniform motion can be produced by the vector sum of uniform circular motions. Thus they were able to introduce devices that were effectively equivalent to the equant.

9. In Islamic Spain a different aproach was taken. Maimonides (d. 1204) informs us that Ibn Bājja (Avempace in Latin: d. ca. 1139) rejected the epicyclic theory because a point on an epicycle does not move toward the center, away from the center, or around it, and these are the only motions permitted (Maimonides, trans. Friedländer, 1956, pp. 196 f). According to Maimonides the eccentric model is also contrary to Aristotelian principles because a point on the eccentric does not move around the center of the Universe, but around an imaginary point removed from the center. Maimonides adds that these centers are not all below the sphere of the Moon, i.e., they must participate in the motions of the planetary spheres. For example, he says that the center for Jupiter lies between the spheres of Venus and Mercury: "consider how improbable all this appears according to the laws of Natural Science" (Maimonides, trans Friedländer, 1956, p. 196). Maimonides indicates that his source for determining the planetary distances and eccentricities in terms of terrestrial radii was al-Qabīṣī's *Treatise on Distances*, a work thought to be lost, but in fact it survives in an Istanbul manuscript: MS Aya Sofya 4832, 88b–94a (cf. Sezgin, 1974, p. 312). I have examined that text and it contains the information that Maimonides ascribed to it, e.g., the center for the deferent of Jupiter is computed to be 525 terrestrial radii away from the center of the earth, and that distance is greater than the Ptolemaic value for the maximum distance of Mercury and less than the Ptolemaic value for the maximum distance of Venus (MS Aya Sofya 4832, 93a : 20). However, al-Qabīṣī was only interested in computing these distances; he drew no philosophical or cosmological conclusions as did Maimonides. The planetary eccentricities could easily have been computed in terms of terrestrial radii by anyone familiar with Ptolemy's *Planetary Hypotheses* from Ptolemy on, but it seems that al-Qabīṣī (tenth century) was the first to do so.

10. Maimonides described other philosophical and mechanical difficulties

that arise in Ptolemy's theory and clearly he was familiar with all its technical features. He referred to Ptolemy's attempt to deal with these difficulties in the *Almagest* XIII, 2 (cited above), but found that approach unsatisfactory. On the other hand he was equally unhappy with the Aristotelian rejection of Ptolemy's models because calculations based on these models agree so well with observations. His solution was a radical reformulation of Geminus's argument (see the quotation from Simplicius, above):

> These difficulties do not concern the astronomer; for he does not profess to tell us the existing properties of spheres, but to suggest, whether correctly or not, a theory in which the motion of the planets is circular and uniform and yet in agreement with observation ... Man's faculties are too deficient to comprehend even the general proof the heavens contain for the existence of Him who sets them in motion. It is in fact ignorance or a kind of madness to weary our minds with finding out things which are beyond our reach without the means of approaching them (Maimonides, trans. Friedländer, 1956, p. 198).

We may note that Maimonides was not concerned with the problem introduced by the equant. Moreover, one may ask: if physical principles are not required to guide the astronomer, why must he adhere to the principle of uniform circular motion? In effect, Maimonides has compartmentalized physics and astronomy, such that the astronomer is seemingly free to base his calculations on any mathematical model he invents that can produce agreement with the observations. This is somewhat reminiscent of the position expressed centuries later by Cardinal Bellarmine in response to the Copernican hypothesis: the mathematician is to provide models that can account for the observed planetary motions, but he may not ascribe physical reality to them. Bellarmine wrote to Foscarini in 1615 (de Santillana, 1955, p. 99):

> It seems to me that your Reverence and Signor Galileo act prudently when you content yourselves with speaking hypothetically and not absolutely, as I have always understood that Copernicus spoke. To say that on the supposition of the Earth's movement and the Sun's quiescence all the celestial appearances are explained better than by the theory of eccentrics and epicycles is to speak with excellent good sense and to run no risk whatever. Such a manner of speaking is enough for a mathematician.

11. Another member of the Aristotelian School that flourished in twelfth century Islamic Spain to deal with astronomy was al-Biṭrūjī, who alone among the members of that school presented alternative models to replace those of Ptolemy (Goldstein, 1971). He was equally critical of epicycles and eccentrics on philosophical grounds and, like Maimonides, he did not mention the equant specifically. Indeed, he somewhat innocently assumed that Ptolemy's models could be placed on the surface of the planet's sphere, rather than in the plane of the ecliptic, without disturbing the accuracy of the com-

putations based on the models. In this way he felt he had reconciled Aristotelian principles with the Ptolemaic models. Although he succeeded to a limited extent in some cases, he was not sufficiently expert in astronomy and mathematics to recognize the great shortcomings of his models. Moreover, it was soon noted that his homocentric theory could not explain the observed variable brightness of Mars, the same problem that afflicted Eudoxus's homocentric models (cf. Dreyer, 1961, p. 251).

12. Levi ben Gerson was one of al-Biṭrūjī's most severe critics. After dismissing al-Biṭrūjī's models he went on to introduce his own new theories to account for planetary motion (cf. Goldstein, 1971, vol. 1, pp. 40 ff). Although critical of Ptolemy as well, Levi accepted the equant model and even explored additional configurations that Ptolemy did not describe (see chapter 20, below). However, in his lunar model Levi did not follow Ptolemy, preferring a far more complicated arrangement, and he argued that this model agreed with his own observations far better than Ptolemy's model did (see Goldstein, 1974a, pp. 52–74). Levi did not attempt to justify his models mechanically or physically, and therefore he might be viewed as a disciple of Maimonides in separating mathematical from physical astronomy. But despite his lack of concern for mechanical considerations, Levi was intent on harmonizing physics and astronomy, as he makes clear at the beginning of his treatise (chap. 1:13):

> In its perfection this investigation [astronomy] belongs to both sciences—to mathematics because of the geometric proofs, and to natural philosophy because of the physical and philosophical proofs.

Levi's view is clearly that natural philosophy and mathematics are coordinate in astronomy, in opposition both to the Aristotelians and Maimonides. Ibn al-Haytham expressed a similar opinion some centuries earlier (Sabra, 1976, p. 444: quoted from Ibn al-Haytham's *Optics*):

> Our inquiry combines the natural and the mathematical sciences. It is dependent on the natural sciences because vision is one of the senses and these belong to the natural things. It is dependent on the mathematical sciences because sight perceives shape, position, magnitude, movement and rest, in addition to its being especially concerned with straight lines. Since it is the mathematical sciences that investigate these things, the inquiry into our subject truly combines the natural and the mathematical sciences.

It may be noted that Averroes rejected Ibn al-Haytham's assertion and preferred to retain Aristotle's view of subordination (Sabra, 1976, p. 449).

13. Levi sought to shift the emphasis in astronomy by introducing observations of the apparent sizes of the planets and the fixed stars as part of the data that the models had to satisfy (see chapters 17, 18, below). Ptolemy

had discussed these apparent sizes in the *Planetary Hypotheses*, but they did not influence his choice of models or his determination of the parameters for his models. Ptolemy's attitude persisted among astronomers throughout the Middle Ages, with perhaps the unique exception of Levi. Once apparent sizes were taken seriously for astronomy, Levi felt he had to investigate certain optical problems, notably those associated with the camera obscura (see, for example, chapter 5).

14. A physical consideration that Ibn al-Haytham and Levi both discussed is the apparent shape of the planets (for Ibn al-Haytham, see ᶜArafat and Winter, 1971; for Levi, see chap. 54 [unpublished]: MS P 103b–104a). If the planets receive their light from the Sun, they should often appear crescent-shaped like the Moon: since this is inconsistent with observation, they must be self-luminous (Ptolemy and Copernicus were both silent on the issue of planetary phases). Indeed, the phases of Venus were only observed after the invention of the telescope, and Galileo correctly remarked that one set of phases would be seen if the orb of Venus lay entirely below the Sun (the Ptolemaic view), and another set if the orb of Venus surrounded the Sun (the Copernican view: cf. Drake, 1957, pp. 74 f, 93 ff; Rosen, 1965, pp. 67 ff). Moreover, Levi's rejection of the epicyclic model was based on a physical argument not introduced by any of his predecessors, namely its incompatibility with the fact that under all circumstances only one side of the Moon faces the earth (see chap. 20:38–41; Goldstein, 1972).

15. Levi might have defended his new models by invoking Ptolemy's argument, cited above, that simplicity cannot overrule agreement between theory and observation. Implicit in this argument is that no two models have identical consequences, and hence one can always choose between them on the basis of the observed phenomena. This became explicit in Kepler (Westman, 1977, p. 58: quoted from Kepler's *Apologia Tychonis contra Ursum*):

> If in their geometrical conclusions two hypotheses coincide, nevertheless in physics each will have its own peculiar consequence.

The view existed in the Middle Ages that two or more models might be equivalent in the sense that they produced equally close agreement in position with the phenomena. For example, Naṣīr al-Dīn al-Ṭūsī invented a lunar model that succeeded in producing positions and distances that were observationally indistinguishable from those produced by Ptolemy's model (cf. Hartner, 1969) and, according to al-Bayhaqī, Ib al-Haytham envisaged the possibility of different astronomical theories that would all satisfactorily explain the observational data and so be equally valid (cf. Pines, 1964a, p. 549). But Ṭusi's model adhered to the principle of uniform circular motion whereas Ptolemy's did not, and thus a philosophical principle was the decisive factor for those who preferred Ṭusi's model. Levi, on the other hand, correctly affirmed that his lunar model was more successful in accounting for the phenomena than that of Ptolemy. Further, he was virtually alone in the medieval period in insisting that his own observations, rather than the observations of the ancients, serve as the crucial test for the verification of his theories.

Section 2: Levi's Influence on Subsequent Astronomers

1. Levi was cited by a number of astronomers who wrote in Hebrew in the fourteenth through sixteenth centuries, but none of them alluded to the most original aspects of his work. Most noteworthy in the fourteenth century was Immanuel ben Jacob Bonfils of Tarascon (fl. ca. 1360); nevertheless, his astronomical tables are based on those of al-Battānī rather than those of Levi (cf. Goldstein, 1978, p. 46). In addition to a table for the motion of the solar apogee based on Levi's parameter (MS Rome, Casanatense, heb. 204, 115b), Bonfils apparently followed Levi in setting up a table for the astrological houses with respect to the ascendant (MS Florence, Laur., heb. Pl. 88/30, 34a : 8; cf. Goldstein, 1974a, pp. 81–82, concerning Levi and tables for the astrological houses). Another fourteenth century Hebrew author to cite Levi was the astronomer Jacob ben Yomtov Poel (cf. MS Vatican, heb. 368, 99a). An anonymous Hebrew text of the fourteenth or fifteenth century cites Levi's work as follows: "Levi ben Gerson wrote down the positions of some of the fixed stars in longitude and latitude for the year 1325 according to the tables of al-Battanī ..." (MS London, Montefiore Library, heb. 425, 8a; cf. Hirschfeld, 1904, p. 126). I have recently discovered that this star list also appears in Levi's *Ḥug ha-shamayim* (MS Mantua, heb. 10, folios 12a–13a) with minor textual variants and that it was computed from Abraham Bar Ḥiyya's star list of 1104 which in turn depends on al-Battānī's star catalogue (Goldstein, 1985d). This text, *Ḥug ha-shamayim*, describes an armillary sphere, and not a Jacob Staff as was once thought (Langermann, 1985; cf. Touati, 1973, p. 55).

2. Profet Duran, or Isaac ben Moses ha-Levi, known as Ephod or Ephodi (fl. ca. 1400) referred to Levi's *Astronomy* in at least three places: on Duran's astronomical works, see Steinschneider, 1964, pp. 180–2; and Renan 1893, p. 744.

   (a) MS Oxford, Can. Misc. 334, 100b: in a note appended to the end of an astronomical work by Joseph ibn Naḥmias (fourteenth century, Spain) in which a modification of al-Biṭrūjī's homocentric models was proposed, Duran remarks that this new homocentric system also fails to account for the phenomena because, among other reasons, Levi ben Gerson had shown decisively that the solar orb is eccentric, based on observations of the Sun's diameter at 0° and 180° of solar anomaly. Here Duran correctly presents the arguments found in Levi's *Astronomy*, chapter 56 (cf. Goldstein, 1974a, p. 93).

(b) A second passage is found in a marginal note (MS Paris, heb. 1026, 1a) where the text concerns Duran's comments on Averroes's *Epitome of the Almagest*. In the note Duran explicitly quotes from Levi's *Astronomy*, chapter 51, to argue for the sphericity of the heavens. The quotation is slightly garbled, but the text can easily be restored from the corresponding passage in Levi's *Astronomy*, chapter 51: MS Paris, heb. 724, 101a : 9–13.

(c) In a text concerning the equation of time (MS Leyden, Warner 43, beginning on folio 30), Duran explicitly refers to Levi's *Astronomy*, chapter 62, section 9, for the view that at Aquarius 20° the equation of time is 4;2° (fol. 35: cf. Goldstein, 1974a, pp. 102–3).

3. These citations strongly suggest that Duran had read extensively in Levi's work, although there is no indication that he appreciated the more profound aspects of it, e.g., the revised lunar theory presented by Levi in chapter 71 (cf. Goldstein, 1974a, pp. 53–74).

4. In the fifteenth century Mordecai Finzi (Mantua, d. 1473) cited Levi in a number of places: e.g., MS Oxford, heb. Mich. 350 (presumably in the hand of Finzi: cf. 61b), 8a, 25a, etc. Finzi also wrote comments on the text of Levi's *Astronomy*: see below, especially the comments on chapter 4 : 33 as found in MS N, 16b and 17a. Moses Farissol Botarel (Avignon, fl. ca. 1465) wrote a commentary on Levi's astronomical tables, but he only described their use and not their construction (MS Oxford, heb. Regio 14, 44–56: cf. Steinschneider, 1893, p. 648; another copy has recently come to light in the private collection of Mr. Michael Meer, New York).

5. Abraham Zakkut (d. after 1521) referred to Levi in the introduction to his astronomical tables (radix 1473 for Salamanca, Spain): see Cantera Burgos, 1931, pp. 111, 207, 244 n. 94. Moreover, Zakkut mentioned Levi in his 1513 tables for Jerusalem and, presumably on the basis of this reference, Levi was cited by Simon ben Jonah Mizraḥi (Baghdad, fl. ca. 1696): see Goldstein, 1981, pp. 239, 245, 248.

6. Another reference to Levi's *Astronomy* is found in a sixteenth century Hebrew text, probably by al-Faji of Istanbul (MS London, British Library, Add. 15,454, 3b; cf. Margoliouth, 1915, p. 346). It should also be noted that MS P of Levi's *Astronomy* was owned by Kaleb Afendopolo (d. after 1499) in Istanbul, and that MS Q was copied in Istanbul in 1510 (cf. Goldstein, 1974a, pp. 74–76).

7. Levi's astronomical text is cited in a number of Latin and vernacular texts in the fifteenth and sixteenth centuries. We will also present some evidence to suggest other possible influences of his work.

8. Most striking is the way George of Trebizond (1395–1472), who migrated from Crete to Italy at about age 20, referred to Levi (see Monfasani, 1976, pp. 4, 196, 234). There follows a paraphrase of two passages from his works.

(a) In each discipline there is one ancient scholar of distinction whom we should follow as, for example, Aristotle for laying bare the secrets of nature, Euclid for the elements of geometry, Homer for Greek poetry, Virgil for Latin poetry, Demosthenes and Cicero for oratory, and surely one person for astronomy, namely, Ptolemy. Now, among the later commentators, some have made [new] instruments for themselves, as we learn from their writings, that have led them—and those who follow them—into great errors. Indeed, a certain Jew, Leo, describes the positions of both planets and fixed stars by means of his own instruments ... (Monfasani, 1984, p. 250: from the preface to George's commentary on the *Almagest*).

(b) But Leo and others descended to inept demonstrations trying to save the appearances ... seeking glory most basely by false and deceitful detraction of divine men. (Monfasani, 1984, p. 681: from George's commentary on the *Almagest* I, 8).

9. It seems certain that the Leo of this passage is Levi ben Gerson, known in Latin as Magister Leo de Balneolis (cf. Renan, 1893, p. 621). George also censured Levi (Leo Judaeus) in a brief work entitled: *Liber de antisciis et cur astrologorum iudicia plerumque fallant*, dated 1456, "for positing a ninth sphere and for attributing precession to a 'lag' of the eighth sphere" (Monfasani, 1984, pp. 678, 696 [where the versions of 1454 and 1456 are distinguished]; cf. Thorndike, 1934, vol. 4, pp. 395–6). What is striking in these texts is that Levi is given such prominence as a dissenter from the views of Ptolemy. It is generally agreed that George was not among the great astronomers of his age—indeed, Regiomontanus severely criticized George's commentary on the *Almagest* (cf. Rose, 1975, p. 102)—and so it may be that George was simply opposing the positive evaluation of Levi's contributions by his colleagues (cf. Pico's views cited below). George was well connected with other humanists in Italy (though not always on good terms with them), notably those associated with Cardinal Bessarion (1403–1472: Rose, 1975, pp. 44–46, 94). Another member of that circle was Regiomontanus (d. 1476): there is only one reference to Levi in his works, as far as I know, and it is in his *Defensio Theonis contra Trapezuntium* (Murr, 1801, pp. 11–12; cf. Monfasani, 1976, p. 296). Moreover, a copy of the Latin translation of Levi's description of the Jacob Staff was among the possessions of Bernhard Walther, Regiomontanus's disciple, and it has been assumed that Walther acquired this copy from his teacher who was also interested in making observations with this instrument (Petz, 1888, p. 260; cf. Rose, 1975, p. 107).

10. An important treatise against astrology by Giovanni Pico della Mirandola (d. 1494) was published in 1496: in this work Levi is cited three times. The first is in connection with the order of the celestial spheres (*Disputationes adversus astrologiam*, VIII, 1, ed. Garin, 1952, vol. 2, p. 232). In the second reference Levi's observational instrument is mentioned: "Leo Hebraeus, vir

insignis et celeber mathematicus, quasi veteribus parus fidens, excogitavit novum instrumentum ..." (*Disputationes adversus astrologiam*, IX, 9, ed. Garin, 1952, vol. 2, p. 324). In contrast to George of Trebizond, Pico referred to Levi in laudatory terms as an eminent and famous mathematician. Pico continued with a description of Levi's instrument that is remarkably similar to a passage in George of Trebizond's commentary on the *Almagest*. Pico wrote: "... verumtamen eo duas non erraticas bis eodem anno cum observasset, discrepasse priorem altera observationem duobus gradibus dicit, quod in aeris statum putat referendum" (*loc. cit.*). George expressed the same idea as follows: "Leo enim Iudeus aliquis, non planetarum dumtaxat, sed etiam stellarum loca instrumentis persequens suis, duas fixas eodem instrumento bis anno eodem observavit, differentiamque observationum graduum duorum ait fuisse, et tanti erroris causam non instrumentis suis nec sibi ipsi, qui forsan non recte instrumentis utebatur, sed dispositioni aeris attribuit" (Monfasani, 1984, p. 250). In both texts we are told that Levi observed the distance between two fixed stars twice within one year and ascribed the discrepancy to the condition of the air (i.e., atmospheric refraction): for Pico this was praiseworthy, whereas for George it was blameworthy. Since the wording of the two texts differs, I do not think that Pico used George's text, and it is possible that Pico learned of Levi's work from Jewish informants or from his own reading of the Hebrew text (rather than from the Latin translation). On Pico's relations with Jewish scholars and his study of Hebrew see, for example, Novak, 1982, p. 129.

11. Pico's third reference to Levi is in connection with the motion of the eighth sphere, i.e., the precession of the equinoxes, and the motion of the solar apogee: "... recentiores volunt, quos retaxat Hebraeus Leo, veram sententiam Ptolemaei defendens" (Pico, *Disputationes adversus astrologiam*, IX, 11, ed. Garin, vol. 2, p. 348). Just prior to this passage (Garin, 1952, vol. 2, p. 346) two Jewish astronomers of the fourteenth century are named: Isaac Israeli (fl. ca. 1310), and Immanuel Bonfils of Tarascon (fl. ca. 1360: here called "Emanuelis ... Baal Cuaphaim", i.e., Immanuel, author of the [six] wings [read *knaphaim* instead of *cuaphaim*]). Also mentioned is Angelus Mantuanus who can be identified as Mordecai Finzi of Mantua, known in Italian as Angelo (cf. Steinschneider, 1964, p. 193; Roth, 1959, p. 231). In the same passage we find a reference to the (astronomical) tables of Oxford, i.e., the tables of Batecombe for 1348 that Finzi translated into Hebrew (MS Oxford, Lyell 96: cf. Goldstein, 1985a, 1985b; North, 1977). Since Finzi's activities are known to have begun as early as 1435 (Roth, 1959, p. 231), it may be that Finzi was the common source for both George of Trebizond and Pico. The Mantuan court under the Gonzagas was most supportive of the humanistic study of mathematics and achieved great prestige among humanists throughout Italy (Rose, 1975, pp. 16–17): it is highly likely that Finzi had

FIGURE 0.1. A Jacob Staff as illustrated in *The Mariner's New Calendar* by Nathaniel Colson (London, 1753), p. 64.

connections with the court (certainly through his friend Bartolomeo Manfredi who was an astrologer and clockmaker for the court: on Finzi and Manfredi, see Goldstein, 1985b).

12. A fifteenth century work in French concerning famous astrologers by Symon de Phares (d. after 1499: Thorndike, 1934, vol. 4, p. 557) mentions Levi briefly:

> 1336. Leo de Balneolis, Juif, florit en ce temps ès parties du Dauphiné vers Orenge, lequel en son temps fist plusieurs jugemens de la science de astrologie, dignes de memoire tant en prenosticacions, comme en interrogacions particulieres. Fist aussi une verifficacion des estoilles fixes, moult precise et bien ordonnée (Wickersheimer, 1929, pp. 214–15).

Of Levi's prognostications only one is known, and in it he predicted a calamity to take place shortly after the conjunction of Mars, Jupiter, and Saturn in 1345 (a year after Levi's death); this was later understood to be a prediction of the Black Death: it was already cited in 1350 by Simon de Covino (Touati, 1973, p. 43). The text has long been known on the basis of copies in Latin, and a fragment of the original Hebrew version has recently been identified (MS Cambridge, heb. Add. 1563, 105b–106b: cf. Touati, 1973, p. 58). At the end of the Latin version we are told that Levi died on 20 April 1344, and that the Hebrew text was translated by Petrus de Alexandria *ordinis fratrum Heremitarum sancti Augustini* with the aid of Levi's brother Solomon, who was physician to the Pope, Clement VI, in Avignon (cf. Thorndike, 1934, vol. 3, p. 310).

13. The Jacob Staff that Levi invented was used by a number of significant astronomers in the fifteenth century, notably Toscanelli (whose observations go back to 1433), Regiomontanus, and Walther (cf. Roche, 1981, p. 11). In the sixteenth century it was even more widely used and among those who proposed improvements was Peter Apian (in works published in 1524 and 1533). Indeed, it was still being used as late as the eighteenth century (see Fig.

0.1). A number of references to the Staff have been found in texts from sixteenth century England, including a treatise by Thomas Digges published in 1573 that Roche (1981, p. 19) has considered the best one to appear since that of Levi. However, direct references to Levi are rare in this literature, and even in the most extensive account of the Staff, published by Gemma Frisius in 1545, Levi is not mentioned (cf. Haasbroek, 1968; and Roche, 1981, pp. 16–18). Early in his career Tycho Brahe, the foremost observational astronomer of the sixteenth century, used a Staff constructed according to the design of Gemma Frisius, but later discarded it in favor of instruments he took to be more accurate (Roche, 1981, p. 18; Raeder, 1946, pp. 96–97). We learn that Ramus cited Levi in connection with the Staff (1559, p. 62), as did Commandino (1558, folios 60v–61r: cited in Roche, 1981, p. 8). Commandino apparently had access to a still extant manuscript of Levi's *Astronomy* (MS Vat. lat. 3380) that was in the library of Fulvio Orsini (Rose, 1975, p. 189; cf. Nolhac, 1887, pp. 249, 377).

14. There are a few examples of astronomical studies in the fifteenth and sixteenth centuries that may have been inspired by Levi's text. Amico (Italy, d. 1538: cf. Swerdlow, 1972) criticized the use of epicycles for the same reasons given by Levi (see chap. 20:38–41): he seems to be the only later astronomer to invoke this argument, and he did not cite Levi (cf. Dreyer, 1953, p. 302). In an early sixteenth century Latin work on the motion of the eighth sphere, Ricius presented views on the relationship of precession and the length of the tropical year that are similar to those of Levi. Ricius cited Levi explicitly in related passages, and it may be assumed that he depended on Levi in this regard (cf. Goldstein, 1975, p. 32).

15. A possible connection with astronomy in Florence was suggested by recent evidence on Toscanelli's meridian observations of the Sun inside the Florentine cathedral, Santa Maria del Fiore (Settle, 1978), and the subsequent meridian observations by lgnatio Danti (d. 1586) in another Florentine church (Righini Bonelli and Settle, 1979, p. 11). I know of no solar observations made on a meridian inside a church during the Middle Ages (for examples of other types of observations associated with churches, see Constable, 1975; and Beaujouan, 1974), and Levi's description of such interior observations, possibly made in a synagogue (see the commentary on chapters 13 and 15, below), may be the earliest in the literature of Western Europe. The sources on Toscanelli's astronomical activities are very meager (cf. Curtze, 1902, p. 264; Jervis, 1977), and so it is difficult to determine whether Levi influenced his work. It is clear, however, that Danti was aware of Toscanelli's observations of a century earlier, and that Danti made similar observations in a church in Bologna after he was forced to leave Florence in 1575 (Righini Bonelli and Settle, 1979, p. 7). It may be of some interest to note that in the seventeenth century Gian Domenico Cassini redid Danti's meridian observa-

tions in Bologna which helped to establish his reputation; he then continued the same kind of work at the Observatoire de Paris of which he was the first director (on Cassini, see Taton, 1971, especially pp. 100b–101a).

16. From the preceding examination of Levi's influence on subsequent astronomers, it may be said that most of his contributions went unnoticed, notably: his new lunar model (Goldstein, 1974a, pp. 53–74); his cosmological views and cosmic distances (Goldstein, 1974a, pp. 28–29); his observations of eclipses (Goldstein, 1979); his observations of the Sun, the Moon, and the planets that are cited in many places in his *Astronomy*; his discussion of the relationship of physics and mathematics (chapter 1); his discussion of planetary models (chapter 20); his instrument for observing the elongation of the Moon from the Sun (chapter 10); and so on. To be sure, Levi's Hebrew style is difficult and the Latin translation has many inadequacies, but I do not find these explanations sufficient to explain the absence of interest by successors who, at least in some cases, were aware of the significance of certain aspects of his work.

Translation

Wars of the Lord: Book V

[1]   Levi ben Gerson said: In previous books we alluded to explanations of some matters that will be explained in this book. [2]   Our intention in this book is to investigate how it is possible to establish models for the celestial bodies and their parameters (*mispar*) in such a way that their observed motions are perfectly represented. [3]   Moreover, they should produce agreement with the variation observed in the magnitudes of the planets as well as with physical principles. [4]   Later we shall investigate the motions that belong to these celestial bodies with respect to their speeding up and slowing down, their retrograde and direct motions, their inclinations to the north and to the south, as well as all their other properties. [5]   Then we shall investigate how the movers of the celestial bodies relate to one to another and how God relates to them, according to the ability of our mind.

[6]   Therefore we have divided this book into three parts: Part 1 is devoted to investigating the models and parameters of the celestial bodies; [7]   Part 2 is devoted to presenting the causes for the properties of the celestial bodies, according to what is possible for us; [8]   and Part 3 is devoted to investigating how the movers of the celestial bodies relate to one another and how God relates to them, according to what is possible for us.

PART 1. The investigation of the models and the parameters for the celestial bodies, and it is divided into 136 chapters.

CHAPTER 1: In it we shall explain the subject of the investigation in this book.

CHAPTER 2: In it we shall explain that the investigation of this subject is worthy of extensive treatment as befits the greatness of its nobility.

CHAPTER 3: In it we shall explain some of the difficulties that beset this subject. Moreover, our reasons for inventing a [new] instrument for making observations with the greatest ease and accuracy will become clear.

CHAPTER 4: In it we shall explain some useful theorems for investigating observations and for investigating the angles of correction in the models that will be introduced. It is divided into 5 sections:

SECTION 1: In it we shall explain the meaning of some terms used in this science.

SECTION 2: In it we shall present geometric proofs concerning arcs, sines, and versines, and they are most essential in this science.

SECTION 3: In it we shall present instructions for making tables to find sines and versines from their arcs, and vice versa.

SECTION 4: In it we shall make tables and indicate how to use them.

SECTION 5: In it we shall explain how to find angles and sides of triangles when some of them are known.

CHAPTER 5: In it we shall describe a useful method for finding the size of the radius of the Sun or Moon with respect to the circle of its revolution by means of the ray of its light that enters the window.

CHAPTER 6: In it we shall investigate the location of the center of vision in the eye of the observer in order to determine the [angular] distance between stars as observed with the instrument we invented for making observations.

CHAPTER 7: In it we shall mention the procedure for making this instrument and how to use it to determine the [angular] distances between stars with respect to the ecliptic.

CHAPTER 8: In it we shall present instructions for determining with this instrument the altitude of the Sun or any star with the greatest possible accuracy from which the time of day or night can be determined. Moreover, we shall indicate how to determine a star's distance from the equator from its meridian altitude.

CHAPTER 9: In it we shall explain that this instrument helps us to determine the size of the diameter of a star with respect to its circle of revolution [i.e., its angular diameter].

CHAPTER 10: In it we shall present instructions to use this instrument for determining the distance from the Moon to the Sun with respect to the ecliptic from which we shall be led to determine the positions of the fixed stars.

CHAPTER 11: In it we shall present instructions for using this instrument in such a way that errors will not be introduced into the observations.

CHAPTER 12: In it we shall present instructions for making an astrolabe with the greatest possible accuracy such that errors will not be introduced in the observations of stellar altitudes, and for dividing the degrees on it finely into minutes.

CHAPTER 13: In it we shall present instructions for finding the meridian at any location with the greatest possible accuracy.

CHAPTER 14: In it we shall explain the difficulties in determining the true positions of the fixed stars in longitude and latitude.

CHAPTER 15: In it we shall present instructions for finding the true position of the Sun at any time we wish.

CHAPTER 16: In it we shall present instructions for finding the positions of the fixed stars.

CHAPTER 17: In it the variation in the magnitudes of the planets [*lit.*: stars] will be used to demonstrate that their models cannot be those that Ptolemy set forth for them.

CHAPTER 18: In it we shall explain that this is not the place to prove that Ptolemy's models do not conform to the observations of planetary motions.

CHAPTER 19: In it we shall present as the first principles for this investigation the properties of the motions of the stars that were found by observation by us or by others, and they are required for the perfection of this investigation.

CHAPTER 20: In it we shall investigate all possible models by which we can account for a variation in the motion in longitude assuming this motion to be simple, and we shall mention their properties.

★ ★ ★

PART I. It is divided into 136 chapters as we mentioned.

CHAPTER 1

[1] None of the ancients whose works have reached us tried to investigate the science of astronomy (*hokhmat ha-tekhuna*) in its perfection, and thus some gaps remain in it. [2] For this reason we decided to investigate it here. [3] We found that these investigators, namely some of the mathematicians, decided it was sufficient to determine a model (*tekhuna*) from which there would follow that which is in close agreement with what is perceived by the senses, but they did not attempt to explain the model according to true principles. [4] Indeed the model they produced contains so many doubtful matters that it is altogether impossible that it be as they assumed. [5] Some natural philosophers (*ha-ṭiv'iim*) were aware of some of these doubtful points, and they decided that the model for the celestial orbs could not possibly be in that way, but they did not attempt to explain proper principles for the motions of the stars that agree with observation. [6] On the contrary they removed this burden from themselves, arguing that the mathematician is the proper person to investigate it, for it is not possible to investigate it in their capacity as natural philosophers. [7] The mathematician also said that in his capacity as mathematician he is unable to complete this investigation: it is sufficient for him to arrange a model in agreement with observation with regard to the motions of the planets in longitude and latitude, their stations, direct and retrograde motions, without any concern whether or not the model conforms to physical principles.

[8] Since this is so, a doubter may well ask: if this investigation is appropriate neither for the mathematician nor for the natural philosopher, for whom is it? [9] Is this an investigation for none of the masters of science (*'iyyun*)? [10] We say that this doubt may be resolved even though this investigation does not belong to the general science that is concerned with existing things insofar as they exist, but it is concerned with existing things that move with regard to their motion. [11] On the one hand, it does not belong to physics because the investigation of existing things with regard to their motion involves proofs that in almost all cases are drawn from geometry, and in general you explain the quantities of the motions that follow necessari-

ly from the model which you set down. [12] On the other hand, it does not belong entirely to mathematics for in that case the models might be thought to conform to the apparent motions without conforming to physical or philosophical principles. [13] In its perfection this investigation belongs to both sciences: to mathematics because of the geometric proofs, and to natural philosophy because of the physical and philosophical proofs. [14] Since this is so, this investigation cannot be split in such a way that part of it would belong to a master of one science and the remainder to a master of the other, because the second would not know what is missing from the investigation of the first unless he already knew what the first had explained in the matter under discussion in order to decide in this way what was required to perfect the investigation of the first. [15] But in this case the second is a master of both sciences, and this will be quite clear to anyone who studies this book; any further explanation would be superfluous.

[16] It follows that this investigation can only be undertaken in its perfection by one who is both a mathematician and a natural philosopher, for he can be aided by both of these sciences and take from them whatever is needed to perfect his work. [17] Since this treatise is of this character, i.e., it is not limited either to mathematics or to natural philosophy, it is clear that this investigation in its perfection will be appropriately treated in this book.

CHAPTER 2

[1]   Since we have explained that in this book this investigation ought to be perfect, we decided to expand it because this science is most valuable both for itself and for its relationship with the other sciences. [2]   Its value for itself is clear because the significance of the inquiry is proportional to the significance of the subject of the inquiry, and it is clear that the subject of this inquiry, i.e., celestial body, is the most noble of all natural bodies, and the form that moves it is the most noble of all natural forms, such that there is no relationship between them, for the term subject and form are predicated of these and other natural things only equivocally, and this is all explained in the science of physics. [3]   Moreover, it is most instructive for the other sciences, particularly physics and metaphysics, inasmuch as the study of its form [i.e., those of the spheres] and their relationship to the first form [i.e., God] is the ultimate goal of metaphysics, and it is owing to this investigation that this science [metaphysics] is characterized as divine. [4]   This investigation is also instructive for political philosophy in the way that Ptolemy explained in [Book] One of the *Almagest*.
   [5]   The prophets and those who spoke by virtue of the Holy Spirit made us aware that it is appropriate to expand this investigation because from it we are led to understand God, as will become evident in this study. [6]   Indeed these orbs and stars are created [bodies] by the word of God—as will be explained later in this treatise when we will discuss the substantiating power of the divine wisdom and the divine will—inasmuch as He brings these noble bodies into existence in this wondrous way by means of the [divine] wisdom. [7]   Diverse things emanate from them by which sublunar existence is perfected, whereas they are all of one nature separate from the qualities that emanate from them. [8]   He set them in such a way that they do not tire or get weary of the actions that God decreed for them on the day He created them.
   [9]   For this reason the prophet Isaiah said [Is. 40:26]: "Lift up your eyes on high, and see who created these, who brought forth the number of their hosts, and who gave them all names; so great is His strength, so mighty is He

in power, not one of them is missing." [10]   He meant by this: lift up your eyes on high to see the stars that you may understand who it is who created them, for this study leads you to recognize that they are created and to perceive who it is who created them, inasmuch as you will see the substantiating power of the [divine] wisdom and will in creating them which enables Him to give all of them names which distinguish one from another, in that God causes diverse actions (*pe^culot*) to emanate from them in this sublunary world. [11]   None of them is missing, i.e., none is lacking in strength or power to do what God has decreed on the day He created them. [12]   The meaning of "who brought forth the number of their hosts" is that He brought forth the host of the heavens from the eastern horizon in measured number in a way that teaches us that there is no weariness in their motion and that they always move with a unitary arrangement. [13]   He might also have meant by "who produces the number of their hosts" that He produces their hosts in the number appropriate for perfecting whatever is necessary. [14]   It follows from God's abundant power (*gevura*) that none of them lacks what is necessary for the perfection of existence. [15]   This is a kind of principle for a demonstration by which to explain the creation (*ḥiddush*) of these celestial bodies, as will become clear in what follows, God willing.

[16]   Accordingly, David said in describing the divine substantiating power [Ps. 8:4–5]: "When I behold Your heavens, the work of Your fingers, the Moon and the stars that You have established; what is man that You are mindful of him, etc." [17]   He meant that when I see Your heavens, it is revealed from them that they are the work of Your fingers because of the wonder and wisdom evident in their creation, and in particular from the models (*^cinyan*) for the Moon and the planets that You established, as will be clear from our remarks later on, God willing.

[18]   This matter is clearer as regards the Moon than it is as regards the other planets [*lit.*: stars]. [19]   Although the Moon and the planets all share the nature of celestial bodies, there is a great difference between the Moon and the other planets because the Moon is an opaque (*ḥashukh*) body that receives its light from elsewhere, and this is not the case with the other planets, as will be explained. [20]   Moreover, the parts of the Moon themselves differ from one another as is apparent from the shadow seen on the body of the Moon. [21]   Another marvelous thing seen in all of them is that although they all share the same nature, the color of each planet and the appearance of their rays differ, and this would all be impossible unless it is assumed that these celestial bodies are created, as will be explained in what follows. [22]   Since this is so, it is a most wondrous thing that despite the level of Your wondrous force that is revealed by studying these celestial bodies, You also remember and look after such a small and lowly creature, namely man. [23]   This is all among those things appropriate to rouse us to

expand this study because this marvelous investigation is that from which one reaches this marvelous fruit.

[24] Because of the high rank of this investigation, the small degree of understanding that we reached in it is more precious and more noble to us than the entire investigation of matters less noble and of lesser rank, and this is quite clear from the nature of human desires. [25] Accordingly, it is clear that if we could perfect this investigation fully without leaving any remaining doubt or confusion, it would be infinitely precious and noble, and therefore we ought to elaborate upon it, perfecting it in every way possible for us.

CHAPTER 3

[1] The difficulty of understanding this subject should not be overlooked, for it is difficult to observe what is needed by someone who wishes to complete this investigation. [2] All the more so in our time in which men have turned away from investigating this science with observational instruments, for they are satisfied with what the ancients said. [3] Indeed for this reason we did not find among our predecessors from Ptolemy to the present day observations that are helpful for this investigation except for our own, and they agree with what al-Battānī [derived] from his observations, but those observations are not known to us.

[4] Moreover, when we observed the planets [*lit.*: the stars], we found that their motions do not agree with what follows from Ptolemy's model. [5] Confusion seems to have fallen into them. [6] This error may come about because the apogee (*govah*) is not in the place that follows from the reckoning of Ptolemy or al-Battānī after him, or because the amount of the correction for the motion of the center does not agree with what follows from that reckoning, or because the amount of the correction for the motion in anomaly does not agree with what follows from that reckoning. [7] It may also come about because the amount of the correction for the inclination of the diameter of the motion in anomaly does not agree with what follows from that reckoning. [8] Or it may come about because the position of the planet in longitude does not agree with the position that follows from that reckoning. [9] Or because the position of the planet in anomaly does not agree with what follows from that reckoning. [10] Or it might be due to several of them or all of them, and that increases the confusion even more, for it is necessary to discern the reason for the error that entered. [11] This involves great difficulty, and requires many observations in order for it to be possible to discern this completely, as we will explain later on, God willing. [12] The numerous observations needed may require a very long time exceeding the human lifetime many times over.

[13] Even a single observation can only be gotten with difficulty because a true observation requires that a planet be observed in conjunction (*daveq*)

with one of the fixed stars whose position is known. [14] No doubt or confusion will enter such an observation, and no error should occur in it due to clouds or vapors that are in the air through which the observation takes place. [15] In this observation we do not need to use any observational instrument, and thus no error arises from that side. [16] But this observation only occurs rarely, and it is not possible to complete in this way the numerous [observations] required in order to discuss the types of error in the apparent positions of the planets according to Ptolemy's reckoning.

[17] You will find that all Ptolemy's predecessors preferred this type of observation because they realized that no error would enter them. [18] If we tried to take observations with the instrument that Ptolemy invented, namely the instrument of the rings, to determine the position of a planet at any time, a doubt which does not disappear would enter what is seen with such an instrument. [19] Indeed some error arises from the difficulty in manufacturing such an instrument, so much so that no artisan in our land knows how to make it with precision, and it is clear that the error in manufacture causes an error in the observations taken with it. [20] It may also happen that the instrument will be distorted after leaving the artisan, particularly some or all of the circles. [21] Therefore we do not trust such an instrument to yield the truth by which we can judge this matter. [22] Moreover, it is most difficult and virtually impossible to make observations with it through the holes. [23] This will be clear to you if you try to observe a star of small magnitude through the two holes of the alidade [*lit*.: table (*luaḥ*)] of the astrolabe. [24] Since this is so, it is clearly impossible to complete the observation of a fixed star with [this instrument] in the very short time after the observation of the other star [i.e., the sighting star] in such a way that the fixed star does not move a sensible amount between the two observations. [25] Therefore it is clearly impossible to use this instrument to determine the true distances between observed stars.

[26] If we grant that we can take observations in this way, or in the first way that we mentioned, we must first know the position of the [fixed] star with respect to which we seek the position of the planet. [27] But there is great doubt today on this point, just as there is concerning the position of the planets. [28] There is a grave doubt concerning the motion of the fixed stars, and the ancients disagreed greatly about it so that the positions of the fixed stars are doubtful. [29] We thought of determining the positions of the fixed stars with the astrolabe or quadrant from their altitudes as they cross the meridian, as is mentioned in the book, "The Explanation of the Astrolabe." [30] But many kinds of approximations could enter into such an observation. [31] The first kind is due to faulty manufacture, and a small error in finding the altitude produces a large error [in the resulting position]. [32] We already determined this with all the instruments that reached us,

and we found great errors in their construction reaching up to about 1°. [33] This happens either from faulty division into degrees or from the diameter of the instrument that inclines in its positioning (*maṣav*) to the left or to the right, or from the alidade (*luaḥ*) that has the two pierced vanes (*dappin*) on it. [34] This is the result if the holes are not set in such a way that the line from one to the other is parallel to the line of the alidade (*luaḥ*) which indicates the altitude of the star according to the place at which it intersects the degrees on the astrolabe. [35] It is possible for an error to arise from any of these causes, or from some or all of them, and this leads to a great error in the determination of the star's position.

[36] The second kind is due to the smallness of the degree [markings] because it is impossible to determine the subdivisions of the degrees except approximately.

[37] The third kind is due to approximation in the procedure for making observations with this instrument, so much so that the approximation can reach as high as 10°, especially if the star has [considerable] latitude in Gemini or Cancer—or in Sagittarius or Capricorn—because you will not notice any [difference in the] declination with these instruments at less than 10° or at more than 10°, i.e., from Gemini 20° to Cancer 10° you will not notice any [change in] declination. [38] Therefore a small error in the altitude of a star in this place brings about a large error in the longitude of the star.

[39] You should know that the declination of the stars is still not perfectly known, as has been made clear to us from many observations. [40] If we compute from the altitude of a star found by observation the corresponding declination according to Ptolemy's reckoning, a large error will occur. [41] Moreover, we cannot find their positions in longitude from their altitudes unless we first know the declination to the north or the south that corresponds to that longitude. [42] Therefore we must first know their positions in longitude before we can investigate their positions in longitude with these instruments, and this is improper (*megunneh*). [43] It is difficult to grasp this, as we explained, but it became clear to us on the basis of observations repeated many times that neither the motions of the stars nor their declinations conform to what follows from Ptolemy's models. [44] Therefore we tried to invent an instrument such that errors would arise neither in its construction, nor in its observations. [45] We began to make some observations with it that are very beneficial for this investigation that will lead us to the true model, according to which the motions and inclinations of the stars will agree with observations. [46] We did not wish to delay until all the observations necessary for this investigation were completed. [47] But we were satisfied with some observations from which with great effort we could discover the true model very nearly, for we feared lest this science suffer because of the difficulty and profundity of these procedures by which the truth was established such

that we would have to write a very long treatise. [48] But we did not wish to mention all the procedures by which we reached mechanical analogies (*heqqeshim taḥbuliim*) in the explanation for each star, for then this treatise would have been even longer. [49] With this there would also be such great profundity that the reader (*ha-meʿayyen*) would not have received any benefit, but would have tired of these matters because of their length and profundity. [50] Therefore we mentioned only a few of them so that a sound scholar could discover the way to construct these mechanical analogies from what we said, and from them he could understand our explanation for each star. [51] Moreover, we explained the true models for each star in such a way that no doubt remains, and such that no other model brings about agreement with the observations of the motions of the star [planet] in longitude and latitude. [52] This is sufficient for the reader (*ha-meʿayyen*), and there is no great benefit [to be derived from] the difficult investigations that led to the construction of the mechanical analogies needed for this subject except insofar as they help us to reach the truth. [53] But after this they have no benefit except for seeking honor.

[54] God willing—and may He decree [long] life so that we may complete the appropriate observations to explain these subjects more easily—we will return to this subject in a separate work to explain this from the observations, or in this book, if we complete it before this book is spread in the lands [i.e., published]. [55] Indeed, then we can expand the explanation of the procedures by which the mechanical analogies are taken without profundity and great length in these matters.

[56] Before we begin to discuss this instrument, we will present some useful remarks for our investigation of this instrument, and for the motions that follow for each model that can possibly be set down, in such a way that the motions of the stars, their swiftness and slowness, direct and retrograde motions will be apparent. [57] Moreover, we shall investigate the arcs and the chords [*or*: sines] for without them there is no way to enter this investigation. [58] Although what Ptolemy said in the *Almagest* is sufficient, we decided not to omit from this book anything needed for this investigation because this book might fall into the hands of someone who did not have access to the *Almagest*, and thus the benefit of this book would be withheld from him. [59] For this reason, we decided to describe some subjects which were treated satisfactorily by Ptolemy, such as the right and oblique ascensions of the zodiacal signs, and similar matters.

CHAPTER 4

[1] It is divided into 5 sections.

[2] SECTION 1. On the explanation of some of the terms which are used in this science. [3] You should know that astronomers are accustomed to divide the circumference of the circle into 360 parts each of which is called a degree, and each degree is divided into 60 parts called minutes or first parts, and each of these minutes is divided into 60 parts called seconds. [4] In this way seconds are divided into thirds, and thirds into fourths, and so on without limit, according to the desired precision.

[5] The orb of the zodiac is first divided into 12 equal parts each called a zodiacal sign. [6] Each zodiacal sign is divided into 30 degrees so that there are 360 degrees in the orb. [7] The diameter of a circle is divided into 120 degrees, although the amount of these degrees differ from the amount of the degrees on the circumference. [8] A part of the circumference is called an arc, and the straight line that joins the two ends of an arc is called the chord (*metar*) of that arc or of the remainder of the circle. [9] The line that joins the midpoint of the chord to the midpoint of that arc is called the sagitta (*heṣ*) of half the arc or the versine (*metar nazor*). [10] Half the chord is called the sine (*nokheḥut*) of half that arc or the half-chord. [11] Euclid explained[1] that the versine just described meets the sine at right angles, and it also lies on a diameter[2] of the circle, i.e., if we extend it to the other side of the circle it will pass through the center of the circle.

[12] We call the motion that Ptolemy ascribed to the epicycle the motion in anomaly (*tenuᶜat ha-ḥilluf*), and the place that Ptolemy called apogee (*govah*) for the planets we also call apogee simply for the sake of agreement on terminology, but the meaning is not the same because this place may not be farther from the earth than any other point on the orb.

---

[1] N mg. adds: in III.29 [Euclid, trans. Heath, III.30].
[2] N mg. adds: in III.1 of Euclid.

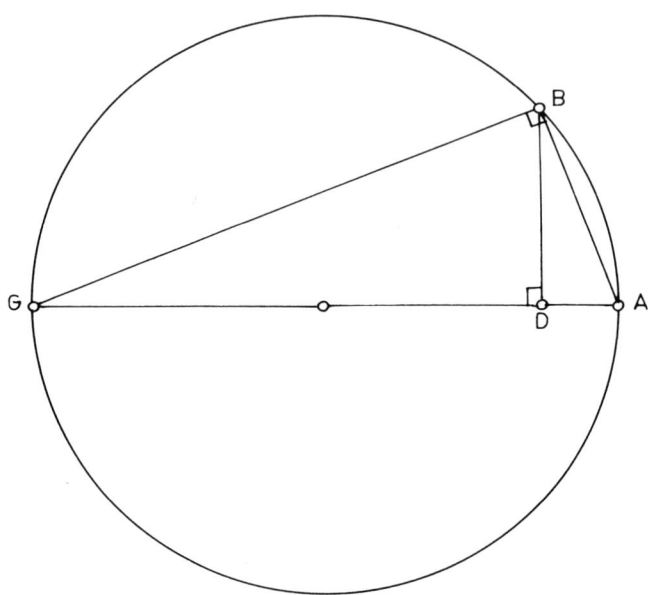

FIGURE 4.1. N 16b: complete circle; P 8b: semicircle only; Q 3a: semicircle only.

[13] SECTION 2. In it we will present some geometrical proofs to instruct us to find what is necessary for arcs and chords.

[14] The chord of an arc is also the chord of the remaining arc of the circle. [15] This is clear if you complete[3] the circle for that arc, for then you can see that this chord serves for both arcs. [16] From this it is clear that the half-chord of any arc is also the half-chord for the arc remaining in 180°.

[17] [Theorem]: The square on the chord of any arc is equal to the area formed at right angles by the versine of that arc and the diameter. [18] To illustrate this we consider arc AB whose chord is line AB, and the diameter of the circle is line AG, and line AD is the versine of arc AB [Fig. 4.1]. [19] I say that the product of line AB by itself is equal to the product of line AD by line AG. [20] Proof: We extend arc AB to G such that arc ABG is a semicircle, and we draw lines BG and BD. [21] It is clear that angle ABG is a right angle,[4] and angle BDA is a right angle because line BD is the sine of arc BA and the sine forms a right angle with the diameter. [22] I say that triangles

---

[3]N mg. adds: in III.24 of Euclid [trans. Heath, III.25].
[4]N mg. adds: in III.30 of Euclid [trans. Heath, III.31].

ADB and *ABG* are similar because angles *ADB* and *ABG* are right angles, and angle *A* is common to both triangles. [23] The remaining angle in triangle *ADB*, namely angle *ABD* is equal to angle *AGB* remaining in triangle *ABG*. [24] Since triangles *ADB* and *ABG* are similar, their sides which subtend equal angles are proportional. [25] Therefore, the ratio of line *AD* in triangle *ADB* to line *AB* in triangle *ABG* is equal to the ratio of line *AB* in triangle *ADB* to line *AG* in triangle *ABG*. [26] Since this is so, it follows that the product of the second by the third, i.e., the product of line *AB* by itself, is equal to the product of the first by the fourth, i.e., the product of line *AD* by line *AG*, and this is what we sought to prove.

[27] From this figure it is clear that if the amount of line *BA* is known, the amount of line *AD* is known, as follows. [28] Since the amount of line *BA* is known, its square is known, and when we divide it by line *GA* which is known because it is the diameter set at 120, the result is line *AD*. [29] It is clear from this figure that if the amount of *AD* is known, the amount of line *BA* is also known, as follows.[5] [30] When the amount of line *AD* is known, the product of line *AD* by line *AG* is known inasmuch as the amount of *AG* is already known. [31] The square root of this product is known, and it is the amount of line *AB*.

[32] I say further that when the chord or versine of any arc is known, the sine of that arc is also known, as follows. [33] For from one of them the other may be found, and from the two of them you can find the sine since the square of the chord is equal to the square of the versine plus the square of the sine.[6] [34] This is clear from the preceding figure (*ṣura*): since angle *BDA* is a right angle, the square of line *BA* is equal to the squares of lines *BD* and *DA*, and it is clear that line *BD* is the sine of arc *BA*. [35] Therefore, it is clear that if we subtract the square of line *AD*, which is known, from the square of line *AB*, the remainder is known, and when we take its square root, we find line *BD*.

---

[5] N mg. adds: If the versine is known, the chord is known.
[6] Note on N 16b, at the bottom of the page: M[ordecai] F[inzi] said: It is also clear that if the sine and the versine are known, the diameter is known, because from both of them we can find the chord, and its square. Since it is equal to the product of the versine by the diameter, it follows that if we divide it by the versine the result is the diameter (*alakson*), i.e., the diameter (*qoter*). The book does not mention this because it always sets the diameter at the same amount and that is appropriate for this investigation.

Note on N 17a, at the top of the page: M[ordecai] F[inzi] said: From the last theorem (*temuna*) in this section it is easily seen that if the versine is known, the sine is known because from it we know its distance from the center, i.e., the amount lacking from, or in excess of, 60. When we subtract the square of this diminution or excess from the square of 60 and take the square root, the remainder is what we sought to find.

[These notes may be found in the Appendix to the Hebrew text: p. 217.]

[36] This can be shown in another easier way. [37] We say that if the versine is known, the sine is known, i.e., that if in the preceding figure (*temuna*) the amount of line AD is known, the amount of line DB is known. [38] It has already been shown that triangles ABD and BDG are similar because angles ADB and BDG are right angles, and angle ABD is equal to angle G as before; it follows that angle A is equal to angle GBD. [39] Thus it is clear that the sides of these triangles which subtend equal angles are proportional. [40] Therefore, the ratio of line AD in triangle BDA to line BD in triangle BDG is equal to the ratio of line BD in triangle BDA to line GD in triangle GDB. [41] Since this is so, it is clear that the product of the first by the fourth is equal to the product of the second by the third, i.e., the product of line AD by line DG is equal to the product of DB by itself. [42] It follows that line GD is known since line AD is known and line AG, the diameter of the circle, is known, and it is also clear that line GD is known. [43] Since this is so, it is clear that the product of line AD by line DG is known, and when we take its square root, we find line BD; thus the amount of line BD is known, and this is what we sought to prove.

[44] From this it is clear that if the chord of any arc is known, the chord of twice that arc is known, and from it the sine of that arc may be found, and when we double it the result is the chord of twice the arc. [45] Moreover, I say that when the versine of any arc is known, the chord of its supplementary arc in 180° is known, as follows. [46] When we know line AD in the preceding figure, we know line GD which is the remainder of the diameter, and when we know the amount of line GD, we know the amount of chord BG, as before.

[47] [Theorem]: When we add the versine of any arc to the sine of the complementary arc in 90°, the sum is equal to the semidiameter. [48] To illustrate this, let us consider arc ABG as 90° about a center at point Z; the versine of arc AB is line AD, and the sine of arc BG is line BE [see Fig. 4.1a]. [49] I say that when we add the two lines AD and BE, the sum is equal to the semidiameter. [50] Proof: We draw line BD between points B and D, and it is clear that line BD is the sine of arc AB. [51] We extend line AD until it reaches point Z, the center of the circle, for the versine of the arc always lies on a diameter of the circle. [52] We also draw line GEZ, and it is clear that line GEZ is a single line because line GE is the versine of arc BG, and the versine must lie on a diameter of the circle, as was explained in Euclid.[7] [53] It is clear that angle AZG is a right angle because arc ABG is a quadrant. [54] Since angles ADB and GEB are also right angles, it is clear from Euclid that lines BD and EZ are parallel, and similarly it is clear that

---

[7]N mg. adds: in III.1 [of Euclid].

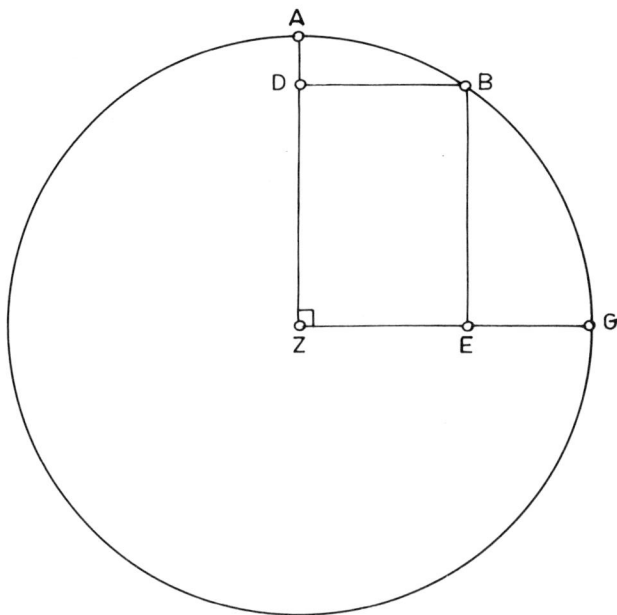

FIGURE 4.1a. N 17b: complete circle, *D* and *E* interchanged; P 9a: quadrant only; Q: 3b: quadrant only.

lines *BE* and *DZ* are parallel and equal; hence line *BE* equals line *DZ*. [55] Since [the sum of] lines *AD* and *DZ* is the semidiameter, it is clear that [the sum of] lines *AD* and *BE* equals the semidiameter, and this is what we sought to prove. [56] From this it is clear that if the versine of any arc is known, the sine of the complement in 90° is known, and vice versa, i.e., if the sine of any arc is known, the versine of the complement in 90° is known, and this is what we sought to prove.

[57] [Theorem]: We wish to prove that the versine of an arc greater than 90° is equal to the sum of the sine of 90° which is the semidiameter and the sine of the arc in excess of 90°. [58] To illustrate this, let arc *ABG* be greater than a quadrant about its center point *Z*, and let arc *AB* be 90° [see Fig. 4.2]. [59] The versine of arc *ABG* will be line *AZD*, and the sine of arc *BG* will be line *GE*. [60] I say that line *AZD* is equal to the semidiameter plus line *GE*. [61] Proof: We draw line *GD* which is the sine of arc *ABG*, and we draw line *BZ*. [62] It is clear that angles *GDZ* and *BEG* are both right angles, and therefore it is clear that lines *EG* and *ZD* are parallel and also equal. [63] Thus it is clear that line *AZD* is equal to [the sum of] the lines *AZ* and *EG*, and this is what we sought to prove.

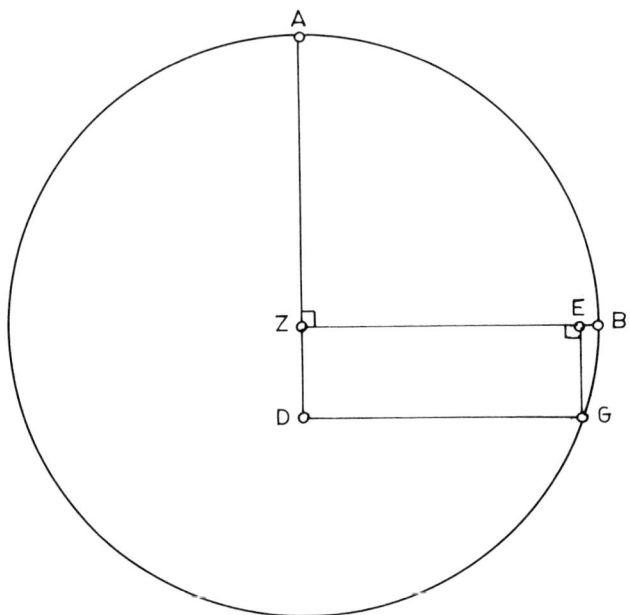

FIGURE 4.2. N 18a: the figure is rotated 90°; P 9a: point *D* is the center of the circle, points *E* and *B* coincide, and line *EG* meets line *DG* at a point inside the circle.

[64] [Theorem]: If the sines and versines of two different arcs are known, the chord of the arc that is equal to [the sum of] these two arcs is known and the chord of the arc of the excess of one arc over the second is known. [65] To illustrate this, let the two different arcs be arcs *AB* and *BG*; let arc *AB* be the greater arc and let line *AE* be the sine of arc *AB*, and *BE* its versine [Fig. 4.3]. [66] Let line *GZ* be the sine of arc *BG*, and its versine line *BZ*; it is clear that line *BZ* necessarily lies on line *BE* because both versines lie on the line from *B* to the center of the circle. [67] Let us consider arc *BGD* equal to arc *AB*, and extend line *AE* to point *D*; it is clear that it will reach point *D* because the chord of arc *ABD* is divided in half at point *E* such that line *AE* is the sine of arc *AB* and, therefore, it is clear that line *ED* is equal to line *AE*. [68] We draw line *GH* from *G* perpendicular to line *AD*, extended to point *H* if point *H* does not fall between points *A* and *D* as in the second figure [Fig. 4.4]. [69] We draw lines *AG* and *GD*; line *AG* is the chord of arcs *AB* and *BG* together, and line *GD* is the chord of the excess of arc *AB* over arc *BG* because arc *BGD* was set equal to arc *AB*. [70] I say that if the sines and versines of the two arcs *AB* and *BG* are known, namely lines *AE*, *GZ*, *BE*, and *BZ* in this example, the chord of arc *ABG* is known, namely line *AG* in this example, and the chord of the arc of the excess of arc

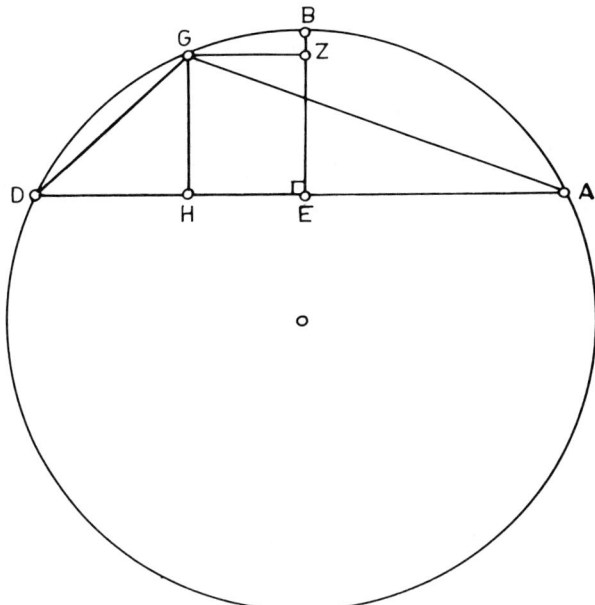

FIGURE 4.3. N 18a: complete circle; P 9a: arc *ABD* only; Q 4a; arc *ABD* only.

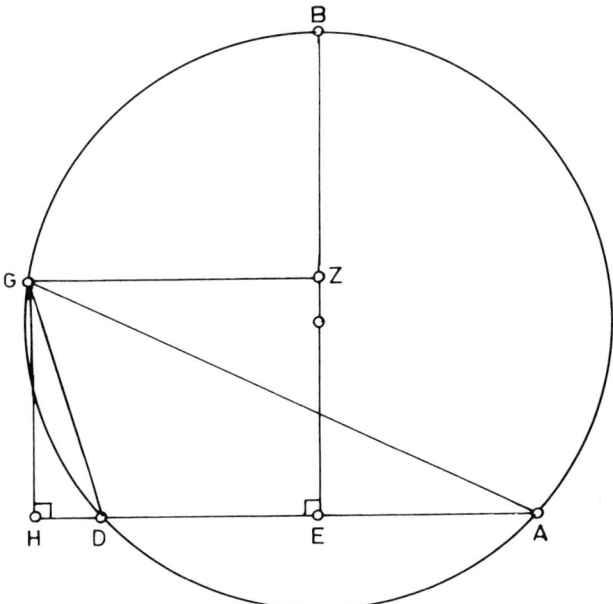

FIGURE 4.4. N 18b: complete circle; P 9a: arc *ABGD* only; Q 4a: arc *ABGD* only.

AB over arc *BG* is known, namely line *GD* in this example. [71] The chord of the sum of the two arcs squared is equal to the square of the line equal to the two sines of the two arcs plus the square of the excess of the greater versine over the lesser versine. [72] In this example line *AG* squared is equal to the squares of lines *AE* and *ZG* together plus the square of line *ZE*. [73] These squares are known because the lines which are their roots (*yesodot*) are known. [74] The chord of arc *GD* squared is equal to the square of the excess of the sine of one over [the sine of] the other plus the square of the excess of the versine of one over the versine of the other. [75] Since angles *Z*, *E*, and *H* are right angles, lines *GZ* and *HE* are parallel, and similarly lines *ZE* and *GH*; therefore they are also equal, i.e., line *HE* to line *GZ*, and line *GH* to line *ZE*. [76] Since angle *GHA* is a right angle, it is clear that line *GA* squared is equal to the sum of the squares of lines *GH* and *HA*. [77] But line *HA* is equal to the sum of the sines of arcs *AB* and *BG*, and line *GH* is equal to line *ZE*, the excess of one versine over the other.

[78] Therefore, it is clear that the chord of the sum of the two arcs squared is equal to the square of the line equal to the two sines of the two arcs plus the square of the excess of the greater versine over the lesser versine. [79] Similarly, it is clear from this figure that the square of the chord of the excess arc of the greater arc over the lesser arc is equal to the square of the excess of one sine over the other plus the square of the excess of one versine over the other. [80] It is clear in this example that line *GD* squared is equal to the sum of the squares of lines *GH* and *HD* where line *HD* is equal to the excess of one sine over the other and line *GH* is equal to the excess of one versine over the other, and this is what we sought to prove.

[81] [Theorem]: We now wish to show that if the sine of some arc is known, the versine of that arc is also known, for the distance of the sine from the center is known from which the versine may be determined. [82] To illustrate this, let the sine of arc *AB* be given, namely line *AE*, and let the center of the circle be point *D* [Fig. 4.5]. [83] We draw line *BE* on the diameter of the circle such that the center of the circle will lie on it between points *B* and *E* if arc *AB* is greater than a quadrant; in the first figure it is line *BDE*, and in the second figure [Fig. 4.6] it is line *BED* and the center comes after point *E* which is the case when arc *AB* is less than a quadrant. [84] The center will be at point *E* when arc *AB* is a quadrant. [85] If it is so arranged, the versine is known, namely line *EB*, because it is equal to the semidiameter of the circle.

[86] If point *E* does not lie at the center of the circle I say that the amount of line *ED* is determined because we can draw line *AD* which is known because it is the semidiameter, and when we subtract the square of line *AE* which is known from the square of line *AD* which is known, the square root of the remainder is equal to line *DE*, and so line *DE* is determined. [87] Since

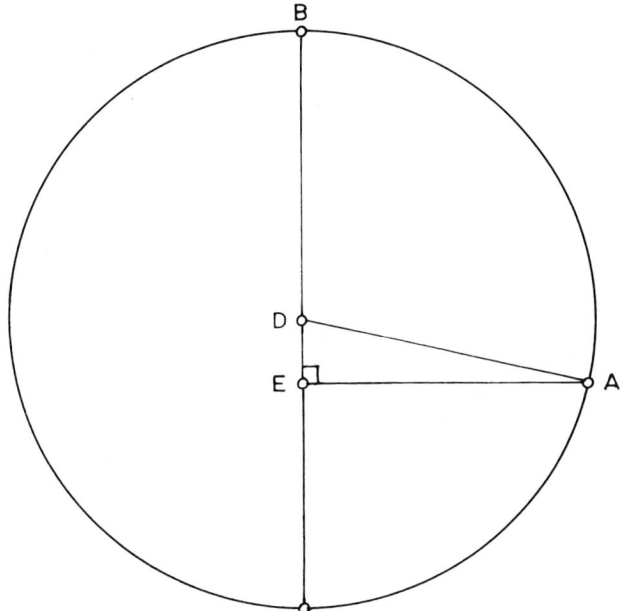

FIGURE 4.5. N 19a: complete circle; P 9b: arc *AB* only; Q 4b: arc *AB* only.

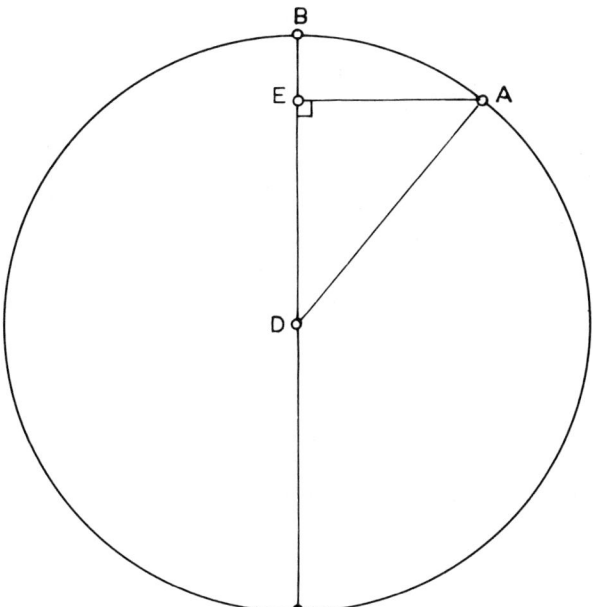

FIGURE 4.6. N 19a: complete circle; P 9b: arc *AB* only, *AE* is not drawn perpendicular to *BD*; Q 4b: arc *AB* only.

line *BD* is known, line *BE*, the remainder, is known as well. [88] It is clear that arc *AB* in the first figure is greater than a quadrant because angle *BDA* is greater than interior angle *E* in triangle *AED*. [89] Since angle *E* is a right angle, angle *BDA* is an obtuse angle from which it follows that arc *BA* is greater than a quadrant. [90] Similarly, it is clear that in the second figure angle *BDA* is acute from which it follows that arc *BA* is less than a quadrant. [91] From these figures it is also clear that the versine of an arc less than a quadrant is less than the semidiameter by the amount of line *DE* in our example, and that the versine of an arc greater than a quadrant is greater than the semidiameter by the amount of line *DE* in our example.

[92] [Theorem]: Since the chord of an arc can be found from its versine as we have shown already, it is clear from this figure that if the chord of twice some arc is known, the sine (*metar*: lit. chord) of that arc is also known. [93] [Proof]: If the sine of an arc is known the chord of twice that arc is also known because the sine is half that chord, and the chord of that arc is known from the sine, which is what we sought to prove.

[94] SECTION 3. We present instructions for making these tables from which the sines and versines of any arc may be found, and vice versa.[8]

[95] It has already been established that whoever knows the sines and the versines of any arc up to 45° knows all the sines and versines. [96] To illustrate this, consider as known the sine and versine of arc $\frac{1}{4}°$; I say that the versine and sine of $(90-\frac{1}{4})°$ are also known. [97] The versine of the second arc plus the sine of the first arc are equal to the semidiameter, as explained in the preceding section, and similarly for the sine of the second arc plus the versine of the first arc. [98] With the sine and versine of the second arc, the sine and versine of $(90+\frac{1}{4})°$ is known. [99] Its sine is the sine of the second arc and its versine when added to the versine of the second arc is equal to the diameter of the circle. [100] With the versine and sine of the first arc, the versine and sine of $(180-\frac{1}{4})°$ are known because its sine is the sine of the first arc, and its versine when added to the versine of the first arc is equal to the diameter of the circle. [101] Similarly, when the sines and versines are known up to 45°, the rest of the sines and versines are known.

[102] In the preceding section we have already explained how to find the unknowns from the knowns in all matters pertaining to sines and versines; now we will also assume that the values of 3 chords are known to the reader of this book. [103] The first chord is that of 180° because its amount is that of the diameter of the circle; from it we know the sine and versine of 90° both of which are equal to the semidiameter of the circle. [104] The second chord is

---

[8] P mg. adds: Constructing tables to find a chord (*metar*: sine?) [from its] arc, and vice versa.

that of 60° because its amount is that of the semidiameter of the circle, as is clear with a little thought. [105] Indeed, Euclid already proved this, and thus the sine of 30° is equal to a quarter of the diameter. [106] Its versine, as follows from what we already proved, is 8;2,18,30,46 very nearly. [107] The third chord is the side of a decagon and it is the chord of 36° about which Euclid proved that the square of the sum of the side of a decagon plus a quarter of a diameter is equal to the sum of the squares of the semidiameter and a quarter of the diameter.[9] [108] It follows that the chord of 36° is 37;4,55,20,30 very nearly, and that the sine of 18° is 18;32,27,40,15 very nearly. [109] Based on the preceding explanations, it follows that its versine is 2;56,11,45,58. [110] From the sine of 90° we can find the sines of 45°, $22\frac{1}{2}$°, $11\frac{1}{4}$°, and their versines. [111] From the sine of 30° we can find the sines of 15°, $7\frac{1}{2}$°, $(4-\frac{1}{4})$°, and their versines. [112] From the sine of 18° we can find the sines of 36°, 9°, $4\frac{1}{2}$°, $2\frac{1}{4}$° and their versines.

[113] It also follows from the preceding that from the sine of 30° and the sine of 18° and their versines we can find the sine of half their sum, namely 24°, from which we can find the sines of 12°, 6°, 3°, $1\frac{1}{2}$°, $\frac{3}{4}$°, and their versines. [114] With this procedure we can easily determine all sines of arcs at intervals of $\frac{3}{4}$°. [115] Indeed, from the sine of 15° and the sine of $1\frac{1}{2}$° and their versines we can determine the sine of $8\frac{1}{4}$° and from the sines of $8\frac{1}{4}$° and $1\frac{1}{2}$° and their versines we can find the sine of $(10-\frac{1}{4})$°, and similarly for the remaining [entries].

[116] We can find the sine of $\frac{1}{4}$° by a mechanical analogy (*heqqesh tahbuli*) as follows. [117] From the sine of $8\frac{1}{4}$° we can find the sine of $4\frac{1}{8}$° and in this way we can descend until we know the sine of $(\frac{1}{4}+\frac{1}{128})$°. [118] Similarly, from the sine of $(4-\frac{1}{4})$° we can proceed until we find the sine of $(\frac{1}{4}-\frac{1}{64})$°. [119] When we investigated this in this way we found that the ratio of the sine of $(\frac{1}{4}+\frac{1}{128})$° to the sine of $(\frac{1}{4}-\frac{1}{64})$° is very nearly equal to the ratio of one arc to the other arc without disturbing this ratio even to fourths. [120] Therefore, we established that sine $\frac{1}{4}$° is 0;15,42,28,32,7. [121] From this amount we can find all the remaining sines. [122] From sine $\frac{1}{4}$° and sine $\frac{3}{4}$° we can determine sine $\frac{1}{2}$°, 1°, 2°, 4°, 8°, 16°, and 32°. [123] From sine $\frac{1}{2}$° and sine 2° we can find sine $1\frac{1}{2}$°, $2\frac{1}{2}$°, 5°, 10°, 20°, and 40°. [124] By the same procedure we can easily find all sines and versines, very nearly, at intervals of $\frac{1}{4}$° up to 45° from which all the others follow, as explained above.

[125] We decided to determine these values at intervals of $\frac{1}{4}$°, not being satisfied with finding them at intervals of 1°, as is the case with many such tables, because we noticed that in some places the error reaches about 15 minutes of arc when we seek to find the arc corresponding to a [given] sine,

---

[9]N mg. adds: XIII.9,3; II.10 [of Euclid: trans. Heath, XIII.9,3; II.11].

especially when the arc is a little more or less than 90°. [126] To illustrate this, let the sine be 59;59,52. [127] According to the tables arranged at degree intervals, the corresponding arc is 89;45,27° if it is less than 90°, or 90;14,33° if it is greater than 90°. [128] It is clear from the preceding that if this were correct, 0;0,8 would be the versine of 0;14,33°. [129] But it follows from the preceding that when the amount of the versine is 0;0,8, the square of the sine of that arc is 0;15,59,58,56 for that is the result of multiplying 0;0,8 times the remainder of the diameter. [130] Therefore, the amount of the sine of that arc, 0;14,33°, is 0;30,59,0,53. [131] But this is false because this sine corresponds to an arc of 0;29,35°. [132] Therefore, it is clear that the arc corresponding to the aforementioned sine is greater or less than 90° by about 0;29,35°. [133] Thus, the error introduced by the other method is greater than 0;15°, and for this reason we decided to arrange these tables at intervals of $\frac{1}{4}°$ so that no perceptible error arises from linear interpolation [*lit.*: according to the ratio].

[134] SECTION 4. The construction of the tables to find the sine and the versine of any arc, and vice versa, and the way to use these tables.

[135] You ought to know that we have entered the arcs and their sines in these tables at intervals of $\frac{1}{4}°$; we have not entered the versines of these arcs or their chords. [136] But with these tables you may find the chord of any arc by first finding the sine of half that arc and doubling it. [137] Moreover, you may find the versine of any arc as follows. [138] If the arc whose versine is sought is less than 90°, look in these tables for the sine of the complement of that arc; the versine that is sought is the remainder of that amount in a semidiameter. [139] But if the arc is greater than 90°, look in these tables for the sine of the arc which exceeds 90° by that amount and add it to a semidiameter; the result is the versine that is sought.

[140] By the opposite procedure, you can find the arc corresponding to the versine, i.e., if it is less than 60, subtract it from 60 and look up the remainder in this table to see which arc corresponds to this sine; then subtract that arc from 90° and the result is the arc corresponding to that versine. [141] But if the versine is greater than 60, look up the excess to see which arc corresponds to that sine; add that arc to 90° and the result is the arc corresponding to that versine—all this is clear from what we mentioned in Section 2 of this chapter. [142] If you have an arc and wish to know the sine of that arc, look up that arc in these tables and its sine is found opposite it. [143] In this way you may also find the arc from the sine. [144] If you cannot find that arc in these tables, look up the two arcs closest to it, and take [as its sine an amount] between the two sines opposite them according to the ratio. [145] Proceed in this same way if you do not find the sine whose corresponding arc is sought. [146] Since the closer the sine approaches the semidiameter the greater the error in taking the corresponding arc according

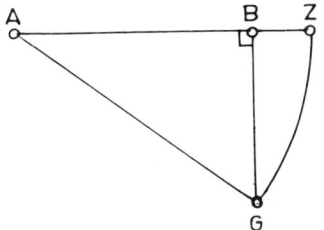

FIGURE 4.7. N 24a; P 12b; Q 7a.

to the ratio from the two sines closest to it, as is clear from what we mentioned with a little thought, you always ought to choose the arc that corresponds to the lesser sine whenever one of these arcs allows you to find (*modi*ᶜ*a*) the other. [147] You should note that if the sum [*lit.*: amount] of the two arcs of these sines is 90°, you can find the sine of the second arc from the first arc in which case you should choose the sine of the smaller arc.

[148] The form of these tables is as follows: we divide the table into 3 columns. [149] The first column is inscribed with arcs at intervals of $\frac{1}{4}°$ up to 90°. [150] The second column is inscribed with the supplement in 180° of the arc entered in column 1, row by row. [151] The third column is inscribed with the sines corresponding to these arcs, row by row, and this is the table.

[152] SECTION 5. The procedure for finding the angles and sides of a triangle when some of them are known.

[153] [Theorem]: If two sides of a right triangle are known, the remaining side and angles may be found. [154] [Proof]: Let *ABG* be a right triangle two sides of which are known; I say that remaining side and angles are also known [Fig. 4.7]. [155] Whichever two sides are known, the third side is known because the square of the side that is the hypotenuse is equal to the sum of the squares of the two remaining sides. [156] Thus, if they are known, it is known; and if it and one of the remaining sides are known, the third side is also known because its square is the difference between the square of the hypotenuse and the square of the other side.

[157] Let us assume that angle *ABG* is a right angle, and that sides *AG* and *BG* are known; I say that angle *BAG* is known, for if we consider point *A* as center, draw an arc *GZ* with *AG* as radius, and join line *ABZ*, it follows from the above that line *BG* is the sine of arc *GZ*. [158] Since lines *AG* and *BG* are known, it follows that line *BG* is known in the measure where line *AG* is the semidiameter of 60. [159] In this measure the arc corresponding to line *BG* considered as a sine can be looked up in the table of arcs and sines. [160] When arc *GZ* is found in this way, angle *BAG* is also known, as is

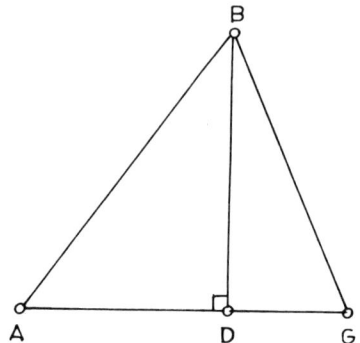

FIGURE 4.8. N 24b: *A* and *G* interchanged; P 12b; Q 7a.

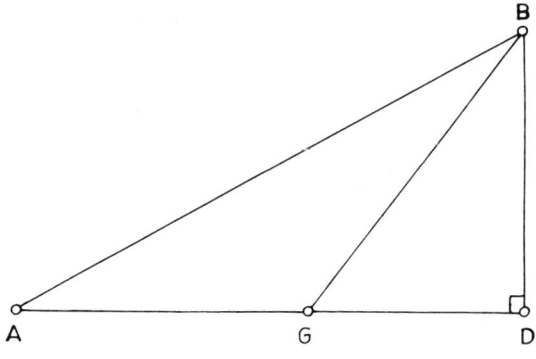

FIGURE 4.9. N 24b; P 12b; Q 7a.

clear from Euclid.[10] [161] From it angle *BGA* is also known because it is the complement in 90° inasmuch as angles *BAG* and *BGA* together are 90°.

[162] [Theorem]: If [all] sides of any triangle whatever are known, its angles are also known. [163] [Proof]: Let the sides of triangle *ABG* be known; I say that its angles are also known. [164] We drop perpendicular *BD* from point *B* to line *AG* extended if necessary—in the first figure point *D* falls within the triangle and in the second figure it falls outside the triangle; I say that the amount of *GD* is known. [165] When we take the excess of the squares of lines *GB* and *GA* over the square of line *AB* in the first figure [Fig. 4.8], or the excess of the square of line *AB* over the squares of lines *GB* and *GA* in the second figure [Fig. 4.9] and divide it by twice line *GA*, the result is

---

[10]N mg. adds: From the end of [Book] VI [i.e., Euclid, VI.32].

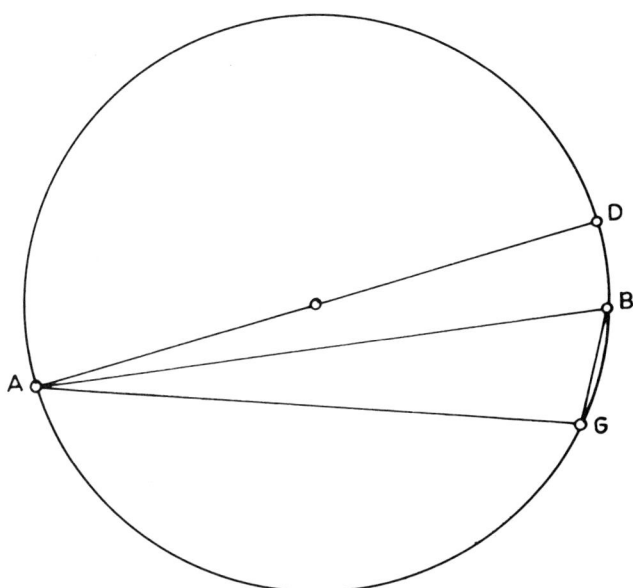

FIGURE 4.10 N 25a: lines are drawn from *B* and *G* to the center of the circle; P 12b; Q 7a.

equal to line *GD*—as will be clear with a little thought concerning Book II of Euclid—and thus the amount of line *GD* is known.[11] [166] Since the square of line *GB* which is known is greater than the square of line *BD* by [the amount of the square of line *GD*], the amount of line *BD* is known. [167] Therefore, it follows as before that all the angles of right triangle *BDG* are known. [168] In the first figure this yields angle *BGA* and one part of angle *GBA*, namely angle *GBD*, and in the second figure angle *BGA* is known because angle *BGD*, its supplement in two right angles, is known; also angle *GBD* is known. [169] Moreover, since lines *AG* and *GD* are known, the amount of line *AD* is known. [170] Thus all the sides of right triangle *BDA* are known, and therefore angle *BAG* is known in both figures. [171] The remaining angle *GBA* in the triangle is known because angle *GBD* is known and angle *DBA* is known from which it follows with a little thought that angle *GBA* is known in both figures. [172] Therefore it is clear that all sides and angles of triangle *ABG* are known, and this is what we sought to demonstrate.

---

[11]P mg. adds, in another hand, a brief note stating the theorems in Euclid, II.12,13.

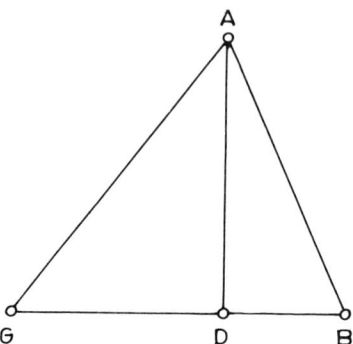

FIGURE 4.11. N 25b; P 13a; Q 7b.

[173] [Theorem]: If we know two sides of any triangle whatever and one angle such that one of the known sides subtends it, the other angles and the third side are known. [174] [Proof]: Let the two known sides be lines *AB* and *BG* in triangle *ABG*, and let angle *BAG* be known [Fig. 4.10]. [175] I say that line *AG* is known and that the remaining angles are known. [176] Let us circumscribe circle *BAG* about triangle *BAG*, and let us consider the diameter of the circle to be line *AD*. [177] Since angle *BAG* is known and we consider it as an inscribed angle, where two right angles are 360, arc *BG* is known.[12] [178] Therefore, the amount of the chord of this arc may be found from the table of arcs and sines in the measure where line *AD* is 120, and so the ratio of line *BG* to line *AD* is known. [179] Since the ratio of line *BG* to line *AB* is also known, the ratio of line *AB* to line *AD* is known. [180] It follows that line *AB* is known in the measure where line *AD* is 120, and thus arc *AB* may be found in the table of arcs and chords. [181] Since both arcs *BG* and *AB* are known, the remaining arc *GA* is known, from which angle *BGA* and line *GA* are known by the aforementioned procedures. [182] Thus it is clear that the sides and angles of triangle *ABG* are all known, and this is what we sought to demonstrate.

[183] You ought to understand from this explanation that if you consider the diameter of the circle to be 60 and the circumference to be 180° it is not necessary to compute the chords of these arcs in a table of arcs and chords because [the table of arcs and] sines serves [this purpose] inasmuch as the sine is half the chord of twice the arc. [184] The ratio of the sine of any arc to the semidiameter of the circle is equal to the ratio of the chord of twice that arc to

---

[12]N mg. adds: Its amount is known in the measure where two right angles are 180°; since angle *BAG* is an inscribed angle which is half the central angle, arc *BG* is known.

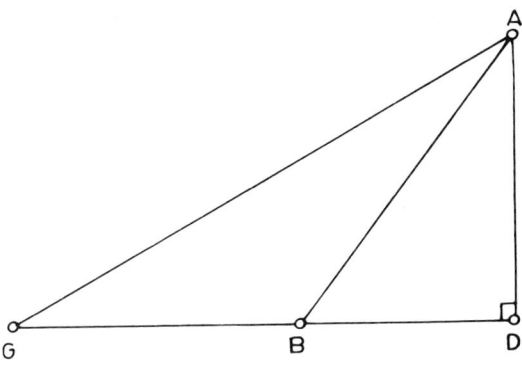

FIGURE 4.12. N 25b; P 13a; Q 7b.

the diameter of the circle. [185]   Therefore in this proof we chose the second alternative so that our remarks would be brief. [186]   We mention this here to avoid confusion in our subsequent proofs. [187]   From this theorem (*temuna*) it follows that in any triangle whose sides are straight lines, the ratio of one side to another is equal to ratio of the sines of the angles that they subtend. [188]   It also follows with a little thought that if the angles of a triangle with straight sides are known and one side is also known, the remaining sides are known because their ratios to the known side are known.

[189]   [Theorem]: If two sides of any triangle are known, and the included angle is also known, the remaining angles and sides are known. [190]   [Proof]: Consider triangle *ABG* whose sides *AB* and *BG* are known and angle *ABG* is also known; I say that line *AG* is known and that the remaining angles are known. [191]   If angle *ABG* is a right angle, this is clear from the preceding. [192]   Moreover, if it is acute as in the first figure [Fig. 4.11] or obtuse as in the second figure [Fig. 4.12], line *AG* is known. [193]   Let us draw perpendicular *AD* from point *A* to line *BG*, extended if necessary. [194]   In either case it is clear that angle *ABD* is known because either angle *ABG* or its supplement in two right angles is known. [195]   There remains angle *DAB* which is known because it is the complement in a right angle. [196]   Therefore all angles and one side of triangle *ABD* are known, and the rest may be found. [197]   Moreover, lines *AD* and *DG* are known in both figures. [198]   The amount of line *AG* in right triangle *ADG* is known, and thus the angles of triangle *ADG* are known including angle *AGD*. [199]   It was already assumed that angle *ABG* is known, there remains angle *BAG* which is known because it is the supplement in two right angles, and this is what we sought to demonstrate.

CHAPTER 5

[1] If a ray of the Sun, the Moon, or any other luminary among the stars enters the window, its [image] must be wider than the size of the window in all directions by the amount of the angular radius of the luminary at the place of the window. [2] We state it here as a principle but then we shall explain it, for if the luminary were only a single point, the ray that emerges from it would not expand perceptibly in the distance from the window to the wall that receives the ray. [3] Because of the greatness of the distance from the celestial bodies to the earth, it [the expansion] is not perceived at all in this small distance. [4] Indeed, even the radius of the earth is not perceptible for the stars on account of the great distance from them to us. [5] All the more so for this small distance, for it is impossible for it [the expansion] to be perceived even for the Moon, the nearest to us. [6] All this will be explained fully by the power of our words, God willing, in what follows.

[7] The proof for what we have mentioned [see Fig. 5.1]: first we draw AB for the wall with the window, and point E on it represents the top of the window. [8] We draw line GD for the opposite wall on which the ray falls. [9] The center of the circle of the luminary is point H, and the diameter of the luminary facing AB is line ZHT. [10] The line that extends from the center of the luminary through point E intersects line GD at point L, and we draw straight line HEL. It is clear that this ray falls on point L, and this is all that would reach the wall if the luminary were truly a point. [12] But since every part of the luminary gives off light, there emerges from point T on the diameter of the luminary a straight line through point E reaching line GD at point M. [13] Since lines HEL and TEM intersect, angle MEL is equal to angle HET which is the angular radius of the luminary at the place of the window. [14] Similarly, this matter is clear for all directions of the window, because the luminary gives off light from the ends of all the diameters marked on it on the side that faces us, and this is what we sought to prove.

[15] From this figure (*temuna*: Lat. *demonstratione*) it is clear that when a ray passes through a polygonal window, the image [*lit.*: ray] will not be polygonal because it expands to all sides of the angles in the amount of the angular

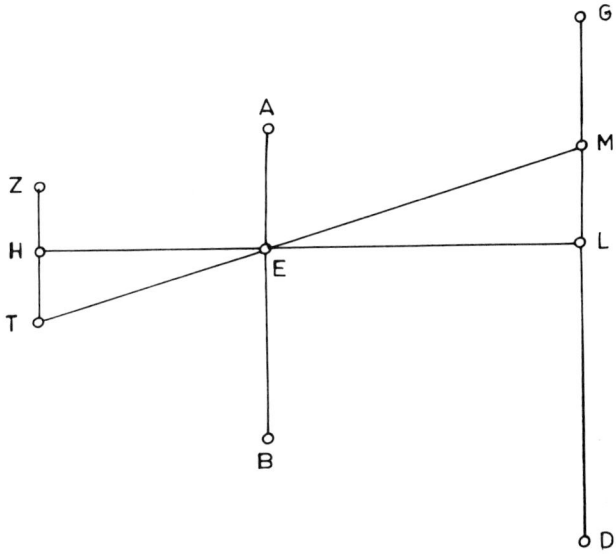

FIGURE 5.1. N 26b: *Z* and *T* are interchanged, a circle is drawn about *H* with radius *HT*, lines *TZ* and *GD* are interchanged, line *GD* is labelled *GLMD*; P 13b; Q 7b.

radius of the luminary. [16] A corner [*lit.*: angle] comes to be like a quadrant whose center is the point of the corner, and this matter is visually perceived for the rays that come from the Sun and the Moon through polygonal windows. [17] There is no difference perceived between the apparent size of the Sun from the point of the window and the size that would be seen from the center of the earth, on account of the smallness of the radius of the earth as compared with the great distance between us and the Sun, as we shall explain later, God willing.

[18] From this we shall determine the amount of the diameter of the Sun on the great circle on which it rotates at the time of observation. [19] This procedure (*sha‘ar*) is very useful for our research, because from it it will be clear whether or not the solar sphere is eccentric to the center of the world. [20] We shall also determine the amount of this eccentricity if there is any. [21] From this figure it should be clear to you that you may use it at the time of eclipses to find the amount of the digits of eclipse. [22] It is best for the hole of the window to be very small, for then the rays that arrive at the wall that receives the light take on the shape of the Moon according to the amount of the eclipse. [23] If you subtract the amount of the window from the diameters of the greatest and the smallest ray, you will find in the remainder the ratio of the eclipsed part to the body of the luminary, for it is equal to

the ratio of the surplus of the greatest diameter over the least diameter to the greatest diameter, or to the ratio of the least diameter to the greatest diameter. [24] From this it is also clear that when the eclipsed part is to one side of the luminary, the missing part of the image [*lit.*: ray] is on the opposite side. [25] This also follows from the figure, for the higher side of the luminary passes through the window and arrives at the lower side of the image [*lit.*: ray] on the wall. [26] Therefore, it is clear that every part of it requires an increment on the side opposite it, and from this what we said follows.

CHAPTER 6

[1] This will concern the way to use the instrument that we invented for taking observations with extreme accuracy. [2] First we must determine what point of the eye is the center of vision, i.e., where the angle [subtended by] the observed [object] terminates. [3] It is proper for us to investigate this now: we say that nothing prevents that it be at the center of the eye or on the periphery of the eye, and if so on the external periphery or the internal periphery, or in between. [4] This is necessarily the set of alternatives, for it is clear that the eye is the special instrument for vision.

[5] We say that it is clear with a little thought that it cannot lie on the external periphery, and it has been explained in the natural [sciences] that the observed thing comes to the eye (*ha-re'ut*) through the medium of the intervening transparent (*sappiri*) air that receives its impression (*rashum*), and it brings it to the eye. [6] If the center of vision were on the external surface of the eye, there would be no way for the power of vision to grasp observed things, for the observed things would reach there only at a point, and it is impossible for an image (*mar'e*) to be produced from a point without size or shape (*temuna*). [7] But the eye does grasp the image, and [its] size and shape, so it follows that the center of vision does not lie on the external periphery of the eye. [8] Further, the impression of the shape of the observed objects is perceived through the pupil (*bava*) of the eye in such a way that the impression of their shape takes place with a sharpened (*melutteshet*) image. [9] Therefore it is clear that when the eye is not suitable (*na'ut*) to be impressed by shapes, it will not see anything. [10] Thus it is clear without doubt that the center of vision is inside the eye, not on the external surface.

[11] Another proof that the center of vision is not on the external periphery of the eye comes from the science of medicine: the loss of vision may be due to a cataract [*lit.*: fluid (*mayyim*)] in the eye around which a case (*kis*) is generated on the crystalline lens [*lit.*: frosty (*kefurit*) moisture (*laḥut*)]. [12] When they move that case to one side in such a way that it does not cover the crystalline lens, vision returns, as is recorded in the science of medi-

cine. [13] Even though shapes are impressed on that eye, it does not grasp observed objects on account of there being a screen (*masakh*) that prevents this impression from reaching the place of vision. [14] Thus it is clear that the center of vision is inside the periphery of the crystalline lens. [15] If it were on the external periphery, vision would not be lost on this account, for that place would have no effect on grasping the observed object according to this hypothesis, as we have explained. [16] Thus it seems that vision takes place in the crystalline lens, and that its center is the center of vision. [17] In this way this humor [i.e., the crystalline lens?] receives (?: *yavi'*) the impression that reaches it from the observed object at the internal periphery of the crystalline lens, and the impression of the shape is transmitted to the front of the brain [*lit.*: the head], for it is there that the common sense (*ha-hush ha-meshuttaf*) is located, as is recorded in the natural [sciences].

[18] It is also possible to find the center of vision by a mathematical demonstration by means of the instrument we invented for taking observations at any time, and this shall be explained in what follows, God willing. [19] But for now, we will present a sufficient explanation for what is sought here. [20] We take a staff (*maqel*) with flat surfaces, and at one end we place the eye in the middle of it. [21] We also have plates (*luhot*) with holes (*nequvot*) in the middle of them whose surfaces are flat, and we draw the staff through them in such a way that their height on the staff is slightly less than the height of the eye on it. [22] We construct the two plates such that one is twice the size of the other, or in any other ratio. [23] We put the smaller plate closer to the eye and we draw it near to the eye until it hides the larger plate exactly. [24] If we do this carefully, we may ascertain with ease the place of the point at which there is the visual angle. [25] These two plates were parallel and at right angles to the staff, and parallel lines that intersect the sides of a triangle divide them proportionately. [26] Since the distance between the two plates is known and the ratio of one to the other is known, the place of the point of vision is known. [27] This is because the ratio of the line that reaches from it to the small plate, to the line that reaches from it to the large plate, is equal to the ratio of the small plate to the large plate. [28] When we alternate, separate, and invert the ratios, we find that the ratio of the small plate to the line from the center of vision to it is equal to the ratio of the excess of the large plate over the small plate to the excess of its distance from the center of vision over the distance of the small plate from the center of vision. [29] Since that second ratio is known by observation, the ratio of the small plate to its distance from the center of vision is determined. [30] But the amount of the small plate is known, and therefore its distance from the center of vision is determined.

[31] To illustrate this, let a straight line along the length of one of the surfaces of the staff on which vision takes place be line *AB* [Fig. 6.1].

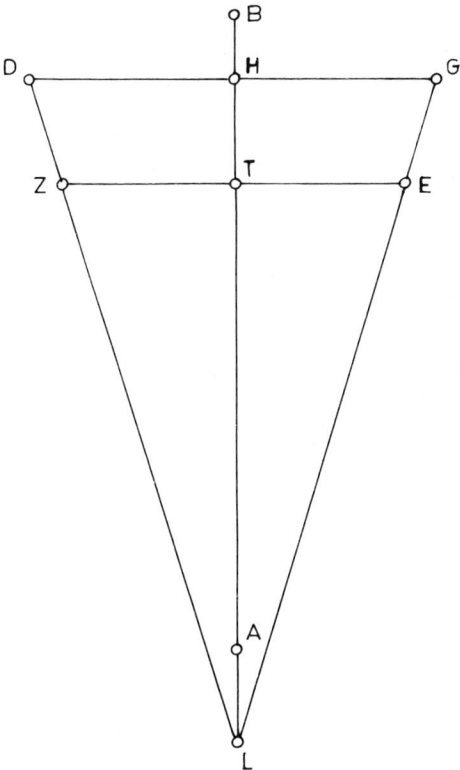

FIGURE 6.1. N 8b; P 14b; Q 28b.

[32] The parallel plates that intersect *AB* are represented by *GD* and *EZ* such that *GD* is greater than *EZ*, and *EZ* is closer to the eye. [33] Consider the situation where the larger plate is exactly hidden, as explained before; line *AB* intersects [line *EZ*] at *T* and line *GD* intersects line *AB* at *H*. [34] Since both lines, *GD* and *EZ*, are observed subtending the same angle by straight rays, it follows that line *GE* extended and line *DZ* extended intersect at the vertex of the angle of vision, say point *L*. [35] We now join points *A* and *L* with a straight line; it is clear that line *BAL* is a straight line, for the center of vision was placed on line *BA*. [36] Moreover, line *GH* is equal to line *HD* and line *ET* is equal to line *TZ*, line *HT* is common (*meshuttaf*), angle *DHT* is equal to angle *GHT*, and angle *HTZ* is equal to angle *HTE*. [37] It follows that if we superimpose figure *DT* on figure *GT* joining together corresponding parts, point *D* will fall on point *G* and point *Z* on point *E*. [38] Therefore, it is clear that angle *HDZ* is equal to

angle *HGE*. [39]   Since the two angles at the base (*toshevet*) of triangle *LGD* are equal, it is clear that triangle *LGD* is isosceles (*shaveh ha-shoqayim*). [40]   Therefore, the line from point *L* to the midpoint of the base, point *H*, meets line *GD* at right angles. [41]   Since line *HA* also intersects line *GD* at right angles, it follows that line *HA* coincides with *HL*. [42]   Since triangle *GDL* contains within it line *EZ* parallel to the base *GD*, the ratio of *EL* to *GL* is equal to the ratio of *EZ* to *GD*. [43]   Thus it is clear that the ratio of *ET* to *GH* is equal to the ratio of *LT* to *LH*. When we alternate and separate, it follows that the ratio of *ET* to *LT* is equal to the ratio of the excess of *GH* over *ET* to the excess of *LH* over *LT*, i.e., *HT*. [45]   But the excess of *GH* over *ET* is known, and the amount of *HT* is also known, as is the amount of line *ET*. [46]   There remains the amount of line *LT*, and it is determined, for the ratio of the amount of *ET*, which is known, to it, is also known.

[47]   We carried out this procedure many times and in as many ways as possible, and we found that the position of point *L* in this figure is at the center of the eye, i.e., at the middle of the crystalline lens. [48]   But we needed this investigation, for without it we could not truly find the amount of the angle of vision without error when we observe two stars with this instrument in order to determine the distance in arc between them. [49]   If we put this point somewhere between *L* and *T*, we would imagine that the angle was greater than it is at the angle of vision. [50]   The opposite would be the case if we put that point beyond *TL*, for then we would imagine that the angle was smaller than it is.

CHAPTER 7

[1] We have presented what was required to explain this instrument that we invented for taking observations at any time with greatest possible accuracy. [2] Now we will begin to describe the way to construct this instrument. [3] This is the way to make it: we take a straight (*yashar*) staff 6 spans (*zeratot*) in length making sure that its surface is plane (*yashar*), and it should be one digit wide for its entire length. [4] At one end we put a small plate (*luah*) carved (*haquq*) in such a way that there protrude from it two pegs distant from one another by a little more than one digit, such that one peg is placed in one corner (*ma'aq*) of the eye that is observing, and the other peg fall on the other corner of that eye without pressing on the eye. [5] When you follow these instructions, it will happen that the distance from the center of vision inside the head of the observer, to the surface of the plate adjacent to the eye is $\frac{1}{20}$ of a span for most people, as we determined by experiment (*bahannu*) with much diligence and effort. [6] We divide the staff into large units [*lit.*: degrees] such that there are 8 units in a span, and we mark them on the length [*lit.*: breadth] of the staff from one end to the other. [7] Begin the units on the staff from the place of the center of vision which is beyond the staff by about $\frac{1}{20}$ of a span. [8] Therefore, the first unit that is adjacent to the eye will be such that when it is combined with $\frac{1}{20}$ of a span it will be equal to the other units, and we write them down there in this way. [9] Then we divide each unit on one side into 6 equal parts, but on the other side we divide each of them into 12 equal parts. [10] We draw a diagonal line from the beginning of the unit line to the end of the first part of the line that is divided into 12 equal parts. [11] Then we draw a diagonal from the end of that part to the end of the first part of the line on which the unit was divided into 6 equal parts, and from the end of this part to the end of the third part on the line divided into 12 parts, and from the end of the third part to the end of the second part of those equal to a sixth of a unit and from the end of the second part to the end of the fifth part of those divided into twelfths of a unit, and so on for each and every unit on the staff. [12] All these diagonal lines bound $\frac{1}{12}$ of a unit which is 5 minutes. [13] To estimate [smaller parts of a unit], deter-

mine the part cut off on [these diagonals]; for if it is $\frac{1}{5}$, it is 1 minute, and if it is $\frac{1}{4}$ it is 1ᵐ15ˢ, and if it is $\frac{1}{3}$ it is 1ᵐ40ˢ, and if it is $\frac{1}{2}$ it is 2ᵐ30ˢ, and thus you can find minutes and seconds very precisely. [14] If you divide into five parts the breadth of the staff that is divided into units for its entire length, the diagonal lines will be divided into 5 equal parts that each bound 1 minute of a unit on the staff.

[15] We then make many plates such that each one has a round hole in the middle of it; the staff should go through it with pressure (*dohaq*), but we should be able to move the plate about it in any desired direction. [16] There should be then a plate whose size is 24 units of the units on the staff such that its upper surface is above the staff by the same amount that the center of vision is above (*govah*) the staff. [17] Similarly, we should have plates of 16 units, 8 units, 4 units, and 2 units, whose breadth on one side is 1 unit and on the other side $\frac{1}{2}$, or $\frac{1}{4}$ of a unit or even less, in order that we may observe stars with it that are very close to one another in longitude or latitude. [18] These plates should all be plane and the surfaces should be perpendicular to one another.

[19] When we wish to observe two stars with this instrument to determine the distance between them, we take a plate appropriate for this distance, e.g., if the distance is 25° or more, we take the largest plate. [20] Moreover, we try to put the plate as close as possible to the [far] end of the staff. [21] We introduce the staff through the plate in such a way that it is perpendicular [to the staff] and we put the pegs at the end of the staff near the [center of] vision at the corner of the eye of the observer, and we ought to shut the other eye so that it will not confuse the observation. [22] We bring the pegs as near to the eye as possible and then we bring the plate nearer to, or farther from, the eye until we see from the top [*lit.*: head of the upper surface] of the plate on one side one star and on the other side the second star such that the two stars touch this plate at its two ends according to the observation. [23] Since it is not possible to accomplish this unless the ends of the plate can be seen clearly, we put a candle behind us in such a way that it illuminates the surface of the plate but its light should not prevent our vision from seeing the stars. [24] When we have accomplished this, we should write down which plate was used in the observation and note the units, minutes, and parts of a minute, on the staff where the plate was located at the time of the observation. [25] We call this the distance—say these stars were observed with a plate of 10 units at a distance of 40 units at some time. [26] When you wish to find the distance between these two stars in degrees on the great circle that passes through them from the units recorded in the observation, add (*habber*) the square of the distance to the square of half the plate, and take the square root of the sum, and it is the corrected radius. [27] Multiply the units of the plate by 60 and divide the result by the corrected radius, and the result is the cor-

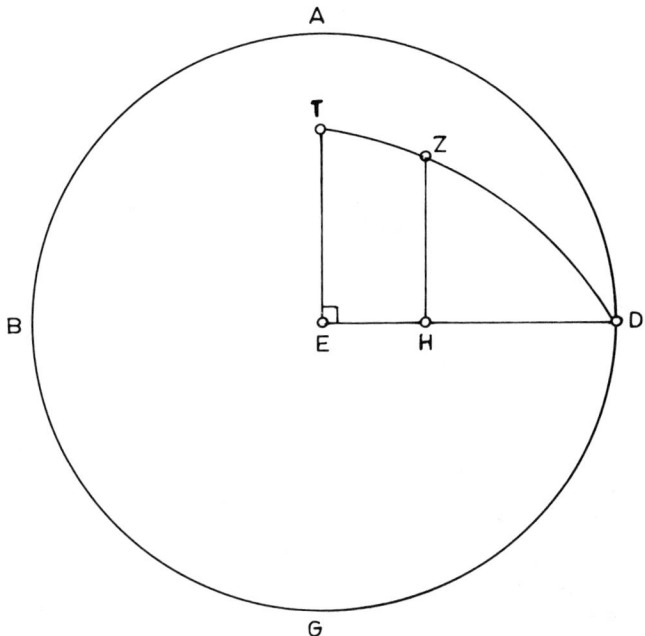

FIGURE 7.1. N 31b: angle *TED* is drawn as an acute angle; P 15b: angle *TED* is drawn as an acute angle; Q 9b.

rected chord (*metar*), for that is what we call it. [28]   Find the corresponding arc in the table of arcs and chords, and it is the distance between the two stars on the great circle passing through them.

[29]   If both stars were on the ecliptic, this arc is the distance between them in longitude, whereas if they were on the same place in longitude, this is their distance in latitude. [30]   If one of the stars was on the ecliptic, the aforementioned arc is the latitude of the other from the ecliptic in its proper direction, i.e., if the star is north of the ecliptic, its latitude is to the north, and if the star is to the south of it, this latitude is to the south. [31]   If one of the stars has a known latitude and they are both at the same longitude, the aforementioned arc is the excess of the latitude of the other star over its latitude. [32]   If both stars are on the same side [of the ecliptic] and one star's latitude is known, add the aforementioned arc to it and the result is the unknown latitude of the second star; but if they are on opposite sides [of the ecliptic], subtract the smaller latitude from the greater [sic][1] and the remain-

---

[1] Read: subtract the smaller latitude from *the aforementioned arc*....

der is the latitude of the star whose latitude was unknown in the direction of the greater latitude. [33] It may happen that they do not have the same longitude, and one or both of them have some latitude—we will prove to you that if the latitudes are known, the distance between them in longitude may be found.

[34] To prove this, we first present a proposition: the sine of the arc of latitude of a star is perpendicular to the plane of the ecliptic—if we draw the sine beginning at the position of the star. [35] Proof: we consider the plane of circle $ABGD$ to be the plane of the ecliptic, and its center is at E [Fig. 7.1]; it is clear that the plane of the circle of latitude passes through its poles. [36] We consider arc DZ to be part of the circle of latitude; its sine which comes from point Z is line ZH. [37] We draw line EHD, and it is clear that line EHD is a single straight line, for the line from the center to the versine [or: half] of a chord bisects its arc. [38] We let arc DZT be a quadrant; therefore point T is the pole of circle $ABGD$. [39] We draw line TE; it is clear that line TE is perpendicular to plane $ABGD$ since circle DZT passes through the poles of $ABGD$. [40] We say that line ZH is parallel to line TE because angle ZHE is a right angle inasmuch as line EH goes out from the center of the circle and bisects the chord of twice arc ZD, and angle TEH is also a right angle inasmuch as line TE is perpendicular to the plane of circle $ABGD$. [41] Therefore, it follows that it is perpendicular to every line passing through point E in plane $ABGD$ including the line ED; since line ZH is also perpendicular to line ED, lines ZH and ET lie in the same plane, the plane of circle DZT, and lines ZH and ET are parallel. [42] Since line ET is perpendicular to the plane of circle $ABGD$, line ZH which is parallel to it is also perpendicular to the plane of circle $ABGD$; similarly, it is clear that the sines of all the arcs of latitude are perpendicular to the plane of circle $ABGD$, and they are all parallel because they are parallel to line ET.

[43] We will now explain how one may find the distance between two stars in longitude from such an observation when the latitudes of the stars are known. [44] If only one star has latitude, subtract the square of the sine of the arc of latitude from the square of the corrected chord, for so we will call it in this chapter. [45] When this is done, you have a triangle whose three sides are known: one is the radius, the second is the second chord, and the base is the remainder of the versine of the arc of latitude of the star. [46] When we subtract it from 60, the perpendicular that falls from the apex of the triangle to the base is known and it is the sine of the arc of the distance; find its arc, and it is what was sought.

[47] To illustrate this, let the plane of the circle of the ecliptic pass through center E, and circle $ABGD$ be drawn about center E [Fig. 7.2]. [48] Let point A be the position of the star that has no latitude, and let point B be the position on the ecliptic for the second star; but it is not on the

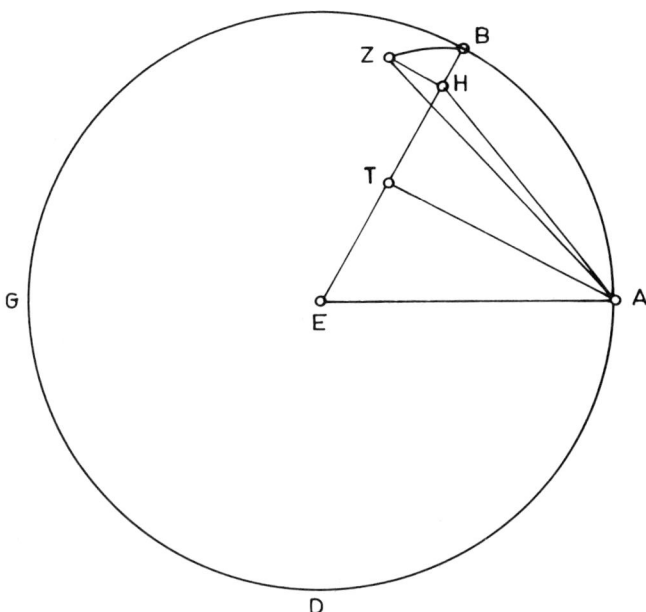

FIGURE 7.2. N 32a: angle *ATB* is drawn as an acute angle, *G* and *D* are missing; P 16a: *G* and *D* are missing; Q 10a.

ecliptic: its latitude is arc *BZ* and the star is at point *Z*. [49] We draw lines *EA* and *EB*, and we draw perpendicular *ZH* from point *Z* to line *EB* and it is the sine of arc *BZ*; line *HB* is its versine, and it is also known; the amount of line *HE* is therefore determined, for it is the remainder of the semidiameter. [50] We draw line *AZ* and it is the corrected chord for the distance between these stars, and we draw line *AH*. [51] It is clear from before that angle *AHZ* is a right angle for line *ZH* is perpendicular to the plane of circle *ABGD*; therefore, line *AZ* is the square root of the sum of the squares of *AH* and *ZH*. [52] When we subtract from the square of line *AZ*, which is known, the square of line *ZH*, which is known, the remainder is equal to the square of line *AH*, and it is the second chord. [53] Since all the lines in triangle *AHE* are known, the perpendicular *AT* from point *A* to line *EH* is also determined. [54] Thus, the amount of line *AT* is known and it is the sine of the arc corresponding to the distance between the two stars, and this is what was sought. [55] This may be clarified in another way—the angles of triangle *AHE* are determined because all the sides are known. [56] Therefore, angle *AEH* which bounds the arc corresponding to the distance between these stars is known.

[57] If both stars have latitude, there are two cases: if they are in different

directions, add the sines of their arcs of latitude, and keep the sum; if they are in the same direction, take the difference between their sines and keep it. [58] Subtract the square of the sum or difference from the square of the corrected chord and take the square root of the remainder, and it is the second chord. [59] When this is done, you have a triangle all of whose sides are known: one side is the second chord, and the two lines that remain are the remainders of half the arcs of latitude of the stars after we subtract them from 60. [60] From this, the angle that bounds the arc corresponding to the distance between the two stars in longitude is known.

[61] You may do this in another way by finding the amount of the perpendicular from the apex of the triangle to the base. [62] We consider the base of the triangle to be a line that joins the center of the ecliptic and the place where the sine of the arc of latitude for one of the stars falls. [63] When you find the amount of that perpendicular multiply (*ta^crokh*) it by 60 and divide it by the line that joins the center of circle of the zodiac and the place where the sine of the arc of latitude of the second star—that was not used as the base of the triangle—falls, and the result is the sine of the arc corresponding to the distance between the stars, from which one may find the arc corresponding to the distance. [64] If the two latitudes are in the same direction and they are equal, the corrected chord is also the second chord and you may continue as before to find the sine of the arc corresponding to the distance between the two stars. [65] If you multiply the second chord by 60 and divide the result by the remainder in 60 of the versine of the arc of latitude, the result is the chord of the arc of distance: find its arc and it is what was sought.

[66] We now present examples to illustrate each of these procedures. [67] Let the plane of the ecliptic be circle *AB* about center *G*, and the arc corresponding to the distance between the two stars in longitude be arc *AB* [Fig. 7.3]. [68] Let the arc of the latitude of one star be arc *AD* and of the second star be *BE* such that arc *AD* is to the north and arc *BE* is to the south; line *DE* is the corrected chord. [69] We draw lines *AG* and *GB*; line *DZ* is the sine of arc *AD*, and line *EH* is the sine of arc *BE*, and both are known because the latitudes are known. [70] It is clear that the versines of these arcs are also known, and they are lines *AZ* and *BH*; their remainders, lines *GZ* and *GH*, are known because they are the complements of known quantities. [71] We draw line *ZH* which is the second chord and we extend line *DZ* to the south[2] by the amount of line *HE*, and it is line *ZT*. [72] Line *DZT* is parallel to line *HE*, and therefore line *ZT* is parallel and equal to line *HE*. [73] We draw line *TE*, and it is clear that lines *ZH* and *TE* are parallel and equal, for they lie between parallel and equal lines *TZ* and *HE*. [74] Angle

---

[2] With P supra, and N mg; P, N, Q: north.

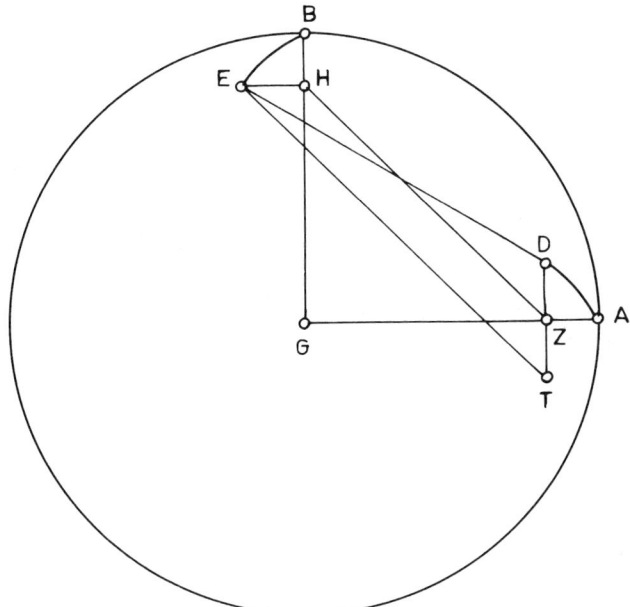

FIGURE 7.3. N 33a; P 16b; Q 10a. In all MSS angle *HGA* is a right angle, whereas the text refers to an arbitrary angle.

DZH is a right angle, for line *DZ* is perpendicular to the plane of circle *ABGD*, and angle *DTE* is a right angle because line *TE* is parallel to line *ZH*. [75] Therefore, it is clear that when the square of line *DT* is subtracted from the square of line *DE* which is the corrected chord, the square root (*yesod*) of the remainder is equal to line *TE*, and thus the amount of *TE* is determined. [76] Therefore, the amount of line *ZH* is known and it is the second chord, for it is equal to line *TE*.

[77] If the latitudes are both in the same direction, and unequal, it is clear that the same method may be applied to find the second chord. [78] Let us consider a similar figure [Fig. 7.4], wherein we put lines *ZD* and *HE* on the same side; we draw line *ZH*, the second chord, and we assume that line *ZD* is longer than line *HE*. [79] We mark off line *ZT* equal to line *HE* on it, and we draw line *TE*. [80] Since both lines *ZT* and *HE* are parallel and equal, both lines *ZH* and *TE* are parallel and equal. [81] It is clear, as before, that angle *DTE* is a right angle, and it follows that when the square of line *TD* is subtracted from the square of line *DE*, the corrected chord, the square root of the remainder is equal to line *TE* which in turn is equal to line *ZH*, the second chord. [82] It is also clear that if the latitudes are equal and in the same

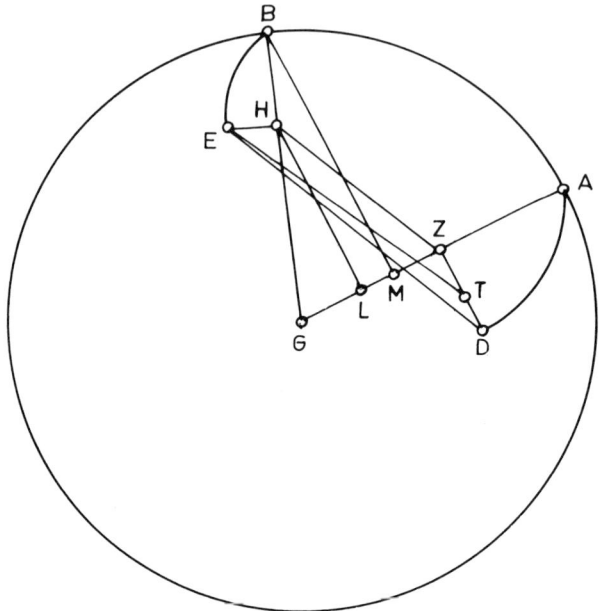

FIGURE 7.4. N 33b: *M* lies at the intersection of *AG* and *ET*; P 16b: *M* lies at the intersection of *AG* and *ET*; Q 10b.

direction, the corrected chord will be equal to the second chord. [83]   Once it is established (*hityashev*) that the second chord, line *ZH*, is known, and that lines *HG* and *ZG* are known, you may find all the angles of triangle *HGZ*, and in particular angle *HGZ* which bounds the arc corresponding to the distance [in longitude] between the two stars.

[84]   This may be found in another way—since all the lines of triangle *HGZ* are known, the amount of the perpendicular *HL* from point *H* to line *AG* may be found. [85]   We draw [perpendicular] *BM* from point *B* to line *AG* extended if necessary, and it is parallel to line *HL*; it is clear that line *BM* is the sine of the arc corresponding to the distance between the two stars. [86]   It follows that the ratio of line *GH* to line *HL* is equal to the ratio of line *GB* to line *MB*, for the two triangles *HGL* and *BMG* are similar (*mitdamim*). [87]   It is clear that the product [*lit.*: area] of the second by the third is equal to the product of the first by the fourth. [88]   It follows that if we multiply line *HL* by line *BG* and divide the product by line *HG*, the result is equal to line *BM*, the sine of the arc corresponding to the distance between the two stars.

[89]   When the latitudes are equal and in the same direction, it is clear

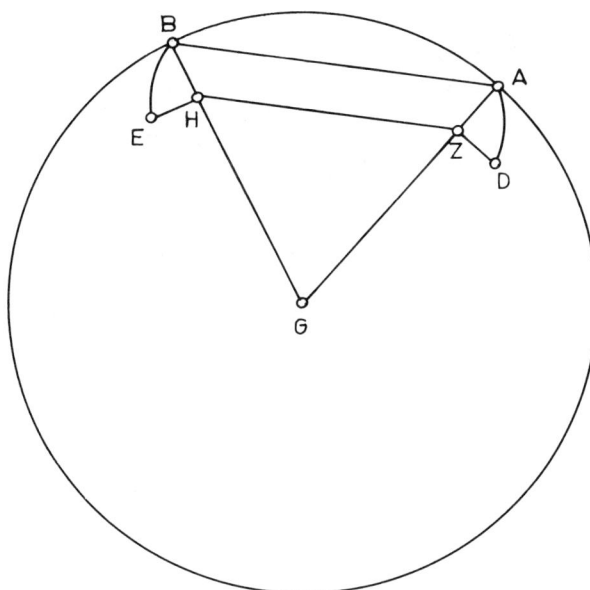

FIGURE 7.5. N 34a; P 17a; Q 10b.

from this proof that the corrected chord is equal to the second chord, and therefore line *HZ*, the second chord, is known. [90] We draw line *AB* [Fig. 7.5]—since line *AZ* is equal to line *HB*, and line *AG* is equal to line *GB*, it is clear that lines *AG* and *GB* are divided in the same ratio by line *ZH*. [91] Therefore, lines *AB* and *ZH* are parallel, and triangles *ZGH* and *AGB* are similar, from which it follows that the ratio of *ZG* to *ZH* is equal to the ratio of *AG* to *AB*. [92] By the previous procedure, it is clear that if we multiply line *ZH* by line *AG* and divide the product by line *ZG*, the result is equal to line *AB*, the chord of the arc corresponding to the distance in longitude between the two stars.

[93] You may determine the chord of the arc corresponding to the distance [in longitude] more conveniently by the [following] approximation without harming the observation when one or both have latitudes, and they are unequal in one or both directions. [94] Find the second chord by the aforementioned procedure, square it, and then subtract the square of the excess of the versine of the greater over the lesser. [95] Take the square root of the remainder, and it is the second corrected chord, as if the latitudes were equal. [96] Add half the excess of the greater versine over the lesser versine to the remainder of the greater versine in 60 if both of them had latitude, and this remainder is the remainder of the corrected versine. [97] But if only

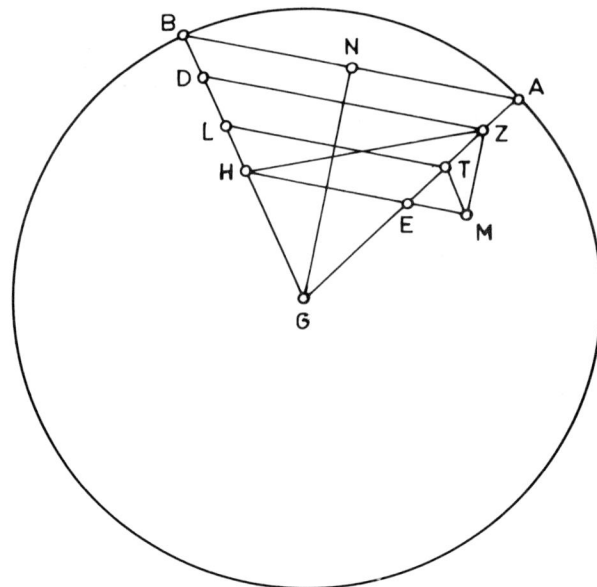

FIGURE 7.6. N 34b: only arc *AB* of the circle has been drawn; P 17a; Q 10b.

one has no latitude, add half the versine of the latitude of the arc that has latitude to the remainder of the versine in 60, and the result is the remainder of the corrected versine. [98] Multiply the second corrected chord by 60, and divide [the product] by the remainder of the corrected versine: the result is approximately the chord of the arc corresponding to the distance, and from it you may find the arc corresponding to the distance.

[99] To prove this we consider the arc corresponding to the distance in this figure to be arc *AB* about center *G*, and we draw lines *AG* and *GB* [Fig. 7.6]. [100] We can find the second chord as before, and this method brings us close to the truth without harming the observations. [101] This will become clear from the following explanation: we draw lines parallel to the chord of the arc corresponding to the distance, which is line *AB*, from the two ends of the second chord in triangle *AGB*, and we draw a perpendicular from one to the other. [102] When we subtract the square of the perpendicular from the square of the second chord and take the square root of the remainder, the result is equal to the parallel line that goes through the point [that marks] half the excess of the greater versine over the lesser. [103] Proof: we consider the second chord in this figure to be line *ZH*, and line *AZ* is shorter than line *BH*. [104] We draw lines parallel to *AB* from the two ends of line *ZH* in triangle *AGB*, namely lines *ZD* and *HE*; line *ZE* is the excess of *BH* over

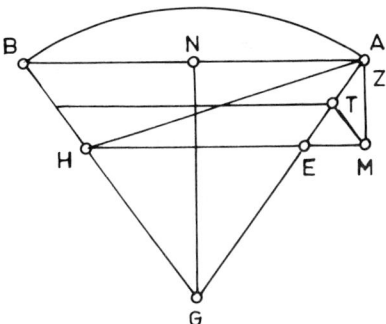

FIGURE 7.7. N 35b; missing in P and Q.

AZ. [105]   We bisect line *ZE* at point *T*, and draw line *TL* from point *T* in triangle *ABG* parallel to line *AB*. [106]   We draw a perpendicular *ZM* from point *Z* to line *HE* extended, and we join *HEM* by a straight line. [107]   We draw line *TM*, and it is clear that line *TM* is equal to line *TE* because triangle *ZME* has a right angle and line *ZE* is the hypotenuse. [108]   It was bisected at point *T* which is the center of the circumscribed circle about triangle *ZME*. [109]   Therefore, line *TM* is equal to line *TE*, and angle *TME* is equal to angle *TEM*. [110]   Since triangle *EGH* is isosceles, angles *HET* and *EHL* are equal. [111]   Since angles *TEH* and *TEM* are supplementary, angles *TME* and *EHL* are also supplementary. [112]   Since the two lines *MT* and *HL* are joined to line *MH*, and the interior angles are supplementary, the two lines *MT* and *HL* are parallel. [113]   Since lines *MH* and *TL* are parallel, they are also equal, from which it follows that when we subtract the square of line *ZM* from the square of line *ZH*, the square root of the remainder is equal to line *MH*. [114]   The square root of the remainder is also equal to line *TL* which is equal to line *MH*, as we explained. [115]   It follows that if we subtract the square of line *ZM* from the square of line *ZH*, the square root of the remainder is equal to line *TL*. [116]   If we take line *ZE* instead of line *ZM*, there will be some approximation. [117]   But we will prove that this approximation does not harm the observations made with this instrument with the following conditions: the stars should not be more than 7° from the ecliptic, and the distance between them should not be greater than 30° nor less than 10°.

[118]   It is clear that even if the latitude of one of the stars is 7° and the other has no latitude, this approximation will not harm the observations, because the versine of 7° is less than 0;27. [119]   Let us assume that the corrected chord [corresponds to an arc of] only 10°, i.e., line *AB* in this example, and otherwise we leave the figure alone; we draw line *GN* from point *G*

perpendicular to *AB* [Fig. 7.7]. [120]   Since the versine of half arc *AB* is less than 0;14, line *GN* is greater than 59;46 where line *AG* is 60. [121]   Since lines *HM* and *AB* are parallel and line *AE* is the [transversal] between them, the alternate [interior] angles are equal, i.e., angle *ZEM* equals angle *GAN*. [122]   Angle *ZME* is a right angle and it is equal to angle *GNA*; there remains angle *MZE* in triangle *ZEM* which is equal to *AGN* in triangle *GAN*. [123]   Therefore, triangles *ZEM* and *GAN* are similar, and the ratio of *ZM* to *ZE* is equal to the ratio of *GN* to *GA*. [124]   It is clear that line *ZE* is longer than line *ZM* by less than 7 seconds,[3] and this adds to the square of line *ZM* less than 6 seconds which introduces an error in line *MH* which is less than 1 second.

[125]   If the arc corresponding to the distance is 30°, there will also be no effect due to this approximation. [126]   This may be determined according to the preceding procedure, i.e., line *ZE* is greater than line *ZM* by less than 1 minute, and this adds to the square of *ZM* only about 53 seconds, which introduces an error of less than 1 second in line *MH* and this does not harm the observations. [127]   It is thus proper to use the latter more convenient procedure for the observations. [128]   Now we will complete our explanation of the way to use this instrument.

---

[3] N mg. adds: 6 seconds 18 thirds.

CHAPTER 8

[1] You should know that we can use this instrument for accurate observations of the altitude of the Sun, Moon, or any star that is seen on the meridian, and this is very useful for our subsequent investigations. [2] For this we must make fixed legs for the staff to stand on the ground, two legs in the middle and two at the end near the eye. [3] At the end near the eye we put a vane (*daf*) pierced with a small hole adjacent to the surface of the staff, and from the place of the hole we begin to write the number of degrees on the staff, rather than from the center of vision. [4] This staff is placed parallel to the horizon in the meridian; we shall explain later how to find the meridian.

[5] We should have many plates (*luḥot*) with holes at one end of them in such a way that when we introduce the staff through them, they are as close to perpendicular as possible. [6] We make the length of one plate 60 units [*lit.*: degrees] or greater, and the length of the second plate 40 units, and the length of the third plate 30 units, and the length of the fourth plate 20 units, and the length of the fifth plate 15 units, and the length of the sixth plate 10 units. [7] The staff should be parallel to the plane of the horizon when we introduce the staff through the plate, and the plumb-line (*anakh*) is placed at the middle of the top of the plate such that the plumb-bob falls on the middle of the plate. [8] When this is perfected, we draw the plate nearer to, or farther from, the eye until we see the star through the hole as if it touches the top of the plate. [9] We should always make sure that the plate is perpendicular to the staff, and note how many units and minutes of the staff are reached, and this is [what we shall call] the distance. [10] We know the height of the top of this plate above the plane of the staff, say 50 units more or less according to the size of the plate used in that observation, and it is what we shall call the altitude in this chapter. [11] When we know the altitude and the distance, we add the square of the altitude to the square of the distance, and we take the square root of the sum and it is the corrected semidiameter. [12] We divide it into the units of altitude multiplied by 60, and the result is the sine of the arc of altitude of the star over the earth, and from it we determine the arc of its altitude over the earth, and that is what we sought. [13] We use the

plate that is proper for a particular observation, according to the altitude of the star over the earth, and in general the plate that yields the furthest distance on the staff is the most accurate. [14] In this way we may use this instrument to find the distance of a star from the equator by taking its altitude on the meridian. [15] In this way we may also use this instrument to find the time of day or night: from the altitude of the Sun by day, and from the altitude of some star at night.

CHAPTER 9

[1]   This instrument may also be used to find the measure of the diameter of a star in relation to the circle on which it travels, and this may be done in three ways. [2]   The first way is to observe the distance of the star from some other star without introducing into this distance any of the diameter of the star whose size we wish to find. [3]   Rather, we should take the distance between the [other] star and the periphery of this star closest to it. [4]   When this is done, we should observe the distance from this star to the other star, and include in this distance the entire diameter of the star whose size we wish to find. [5]   To do this we take the distance between the [other] star and the opposite periphery of this star. [6]   The excess of this distance over the first distance is the measure of the diameter of the star in relation to the circle on which it travels.

[7]   The second way is for there to be a hole [whose sides are at] right angles of known size in relation to the degrees of the staff on the plate through which the staff passes. [8]   The height of the hole above the plane of the staff should agree with the height of the center of vision above it; draw the plate near to the eye until the diameter of the star fills the breadth of the hole. [9]   Since you know the distance to the plate at the time of the observation and the size of the breadth of the hole, you can find the size of the diameter of the star in relation to the circle on which it travels. [10]   The size of the breadth of the hole here corresponds to the amount of the plate used for the observation.

[11]   The third way may only be used for the luminaries, like the Sun and the Moon. [12]   For Venus and Jupiter, despite the difficulty due to the weakness of their light, we will describe [an experiment to be undertaken] in utter darkness when the light of the remaining stars is dim [*lit.*: absent], i.e., in the dark of night, in which their light enters through a window without mixing with the light of another luminary. [13]   For this we should make the staff 16 spans or more, and at one end we should put a plate at right angles to it, i.e., its surface at right angles to the surface of the length of the staff. [14]   The plate should have a round hole of known diameter in relation to

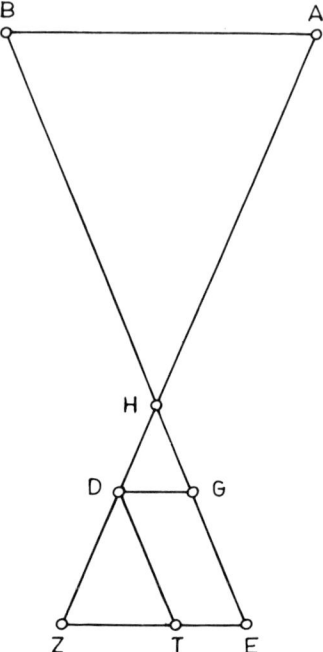

FIGURE 9.1. N 37b; P 18b; Q 12a.

the measuring units [*lit.*: degrees] on the staff, say it is 1 or 2. [15] At the other end of the staff there should be another plate parallel to the first on which is seen the image [*lit.*: ray] of the luminary that enters the hole in the first plate. [16] If the amount of that image is known, the ratio of the diameter of the luminary to the circle whose center lies on the surface of the earth, where the observation is made, is known; its semidiameter is the distance of the luminary from that surface. [17] For, if we know the excess of the breadth of the image over the breadth of the hole, we can find the arc that bounds that chord in the measure where the length of the staff is 60, and it will be the diameter of the luminary in relation to the aforementioned circle.

[18] To illustrate this, let the length of the staff between the two plates be 100, and the breadth of the hole 2, and the breadth of the image 3. [19] The excess of the breadth is 1 in relation to the length of the staff of 100, but in the measure where the length of the staff is 60, this excess becomes 0;36. [20] The corresponding arc is 0;34,22°; therefore, the diameter of the luminary in the aforementioned unit is 0;34,22°. [21] To prove this, let line *AB* be the diameter of the luminary, line *GD* the diameter of the hole parallel to it, and line *EZ* the diameter of the image parallel to it on the second plate; draw

lines *ADZ* and *BGE* and they intersect at point *H* [Fig. 9.1]. [22] Angle *AHB* is the angular diameter of the planet in relation to the circle whose center is at point *H* on the surface of the earth, and angle *EHZ* is equal to it. [23] We mark off line *ET* equal to line *GD* on line *EZ* and we draw line *TD*; lines *ET* and *GD* are parallel and equal. [24] Hence, line *TD* is parallel to line *EH* and angle *TDZ* is equal to angle *EHZ* which is equal to angle *AHB*, the angular diameter of the star in relation to the aforementioned circle.

[25] It is clear to anyone who is learned in this science that for the Sun, Jupiter, and Venus, there is no perceptible difference between point *H* and the center of the earth. [26] But for the Moon it is clear that there is a difference because of its closeness to the earth; we must find the ratio of the diameter of the Moon to this aforementioned circle and then find the ratio of its diameter in relation to the circle whose center is the center of the earth. [27] You ought to know that the longer the staff the more accurate will be the measurement of the diameter of the luminary. [28] This is all we shall explain concerning the instrument that we invented for taking observations and it is most wondrous, as you will see. [29] We call this instrument *The Revealer of Profundities* for with it many profundities in this science can be verified, with the help of God. [30] In order to bring to the attention of scholars the great value of this instrument and to draw their hearts to it, I have written two poems about this instrument, and here they are.

[A]

[31] So that man might acquire benefit God granted him intelligence, with it man beholds His pleasantness and may visit His temple.[1]

[32] Every instrument is provided to him to grant him understanding, to know every obscurity in the secrets of man and his creator.

[33] By means of all the stars in the firmament He teaches man His secret, He teaches man His way, His distance, and His greatness.

[34] Man may consult his rod concerning the structure of the heavens and the motions in their paths; his staff will inform him.

[B]

[35] Why do you call me "destroyer?"[2] the staff cried out bitterly,

[36] When my name is "pleasantness"; the beasts of the earth and flying birds will vouch for me.

[37] A shoot has come forth from the house of David;[3] to [*lit*.: at] the ends of the earth it will give forth its fragrance.

[38] Of my glories I will tell you a little, and I will boast of them.

---

[1] Cf. Ps. 27:4.
[2] Cf. Zach. 11:7. I took two staffs, one I called "pleasantness", and the other "destroyer" ....
[3] Cf. Is. 11:1.

[39] With me is education [*lit.*: rebuke, reproof]; through me a student who has forgotten his learning may become wise.

[40] In me there is support for a weak man, thanks to me light shines in the darkness.

[41] With me one may walk through water and mud; with me the wanderer may find his way.

[42] I will be eyes for the blind; feet for the lame.

[43] Through me vessels (*kelim*) become useful [*lit.*: have hands], with me the dumb will shout again.

[44] A man may lift me in battle against a multitude and be victorious.

[45] Even if [part of me] is cut off, a new branch will grow from my trunk.

[46] Because of my great power a man may err regarding me, he who worships me and trusts in me.[4]

[47] He does not understand that I am not a god, may God [*lit.*: the Forgiver] forgive his sin.

[48] I was present at the carrying of the holy ark,[5] and I was attached to its inheritance.[6]

[49] I was there with the daily sacrifices when they were offered on the altar.

[50] Miracles were accomplished with me before Pharaoh,[7] they made the Man of Rahab [i.e., Pharaoh] lie low.

[51] Were it not for me, Laban would have sent Jacob away with nothing.[8]

[52] A man who observes the stars of the sky with me can open the gates of heaven.

[53] He will know the pattern of their orbs, and the paths of the Sun and the Moon.

[54] With me he can measure that which is measurable, [with] my right hand span the firmament.

★

[55] You should know that we first invented this instrument to determine whether there is an eccentric sphere, for with it we can determine accurately the apparent size of the diameter of the Moon at all four of its distances according to Ptolemy's model. [56] The arguments of the physicists (*tiv^{c}iim*) against Ptolemy's model brought this question to our attention. [57] When we verified that this matter is not in accord with what Ptolemy assumed, we had to investigate models for the motions of the celestial bodies that would be

---

[4] I.e., a Christian (he who worships the crucifix).
[5] According to Carlebach, an allusion to Nu. 4.
[6] Cf. I Sam. 26:19.
[7] Cf. Ex. 7:9 ff.
[8] Cf. Gn. 30:37.

in agreement with observation and in particular with the variations in the observed distances. [58] We shall use this instrument, to which God directed us, to find the truth in this science, as you will see.

[59] It is clear from before that if we know the longitude and latitude of one star and the latitude of a second star seen with it, this instrument allows us to find the longitude of the star seen with it. [60] Similarly, if we know the longitude of some star and we use this instrument to find the amount of its altitude at mid-heaven we can determine the amount of inclination of that star from the ecliptic once we find the altitude of the degree in which it is located at mid-heaven. [61] We shall explain later, God willing, how to find this for any horizon [i.e., terrestrial latitude], for we have not been correctly informed of the longitude or latitude of any star. [62] We determined this from many observations of the positions of the stars in longitude and latitude, for it follows from those observations in which there is no doubt that their positions in longitude and latitude do not agree with what one finds in the tables. [63] This adds to the difficulty, for it is thus necessary for us first to investigate the positions of the fixed stars in longitude and latitude. [64] After that we can use them to find the positions of the planets by observations with this instrument.

Chapter 10

[1] We explained that we must first investigate the positions of the fixed stars in longitude and latitude at the time of our observations before we go on to find the positions of the planets which are seen with them. [2] The procedure for this, as will be explained later, is first to determine the position of the Sun because we can determine its true position precisely at any time. [3] It is clear that celestial bodies [*lit.*: a star] are only observed together with it with great difficulty, except for the Moon. [4] We will proceed to find the position of the Moon in longitude with this instrument by observing the Moon together with the Sun, as we shall explain later. [5] We take such an observation a little before sunset, and then observe the Moon at the beginning of the night, or shortly thereafter, with some fixed star determining the distance between them in longitude on the ecliptic. [6] From the mean time between the two observations we can determine the true motion of the Moon in that time, taking account of its position in anomaly and with respect to the apogee according to Ptolemy's model; this involves an imperceptible approximation [i.e., error] in a short interval, as we shall explain later at length. [7] In this way we can determine the place of that fixed star from the place of the Sun at that time.

[8] This is the description of the plate (*luah*) we made: its surfaces were plane with a hole in the middle, as were the aforementioned plates. [9] Some distance from the hole we fixed another plate at right angles to the first plate. [10] We set the length of the [second] plate equal to the length of the remainder of the first plate in the longer direction. [11] Then we fixed a third plate to these two plates, attached to their ends, to form an isosceles triangle. [12] Then we made a hole in the third plate such that when the staff is inserted into the two holes [i.e., made in the first and third plates], the first plate is perpendicular to the staff. [13] We mark a right triangle in the middle of the thickness of the plates of the triangle and on each leg of this triangle, we mark off a line whose ratio to that entire line is equal to the sine of 15° to the sine of 75°. [14] We mark these two points on both sides of the right angle dividing those lines according to the aforementioned ratio. [15] It fol-

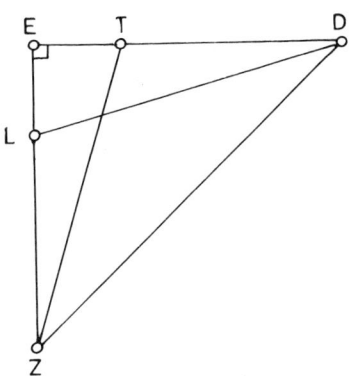

FIGURE 10.1. N 39b; P 20a; Q 12b.

lows that when this is done and lines are drawn from these points to the ends of the sides of the right triangle that was marked on these plates, two right triangles are generated each of which has one acute angle of 15° and the other acute angle of 75°.

[16] To illustrate this, let triangle *DEZ* be the triangle formed by these plates, and the angle at *E* be the right angle [Fig. 10.1]. [17] We mark line *ET* on line *DE* such that its ratio to line *EZ* is equal to the ratio of the sine of 15° to the sine of 75°; therefore angle *ETZ* is 75° and angle *EZT* is 15°, and we take line *EL* on line *EZ* equal to line *ET*. [18] Similarly we mark off lines on both sides of the right angle in triangle *DEZ* whose ratio to that entire line on which it is marked is equal to the ratio of the sine of 30° to the sine of 60°. [19] It follows that if we draw lines from these points to the two points *D* and *Z*, two right triangles will be generated such that one of their angles is 30° and another 60°. [20] It is clear that the acute angles of triangle *DEZ* are both 45° because they are both half of a right angle. [21] In this way we can draw angles up to 90° at 15° intervals in triangle *DEZ*, i.e., angles of 15°, 30°, 45°, 60°, 75°, and 90°.

[22] After we have found these points, we draw straight lines from the acute angle of this triangle at the end of the first plate that pass through these points that were marked on the second plate. [23] We do the same from the acute angle at the end of the second plate to the points marked on the first plate. [24] We make two holes at the vertices of the acute angles of this triangle with these points at their centers, and they should be perpendicular to the plane of the triangle. [25] We place two pegs in them perpendicularly on top of which we put a vane (*daf*) with a small hole in it such that it can move in any direction, from right to left or from left to right; and on the places marked by the aforementioned points on these plates, we also place pegs

perpendicular to the plane of the triangle bearing vanes with holes. [26] The roots of these pegs should be perpendicular to the lines marked through these points. [27] We draw a line on each peg perpendicular to the plane of the triangle from the place of intersection of each peg with the lines at their root; the center of the hole on the vane belonging to that peg should lie on that line, and the holes [in the plates] belonging to these pegs should all be [bored] perpendicular to the surface of the peg. [28] We put the holes [in the vanes] at equal heights above the plane of the triangle.

[29] We do the same on the other side such that these plates form another triangle, and in this way with this instrument we can observe the Sun and the Moon together, whether the Sun is to the east of the Moon or to the west of the Moon. [30] It is clear that this plate [can be put] in two positions (*maṣavim*): in the first position the diagonal (*ha-alaksoni*) plate lies between the eye and the straight plate that is perpendicular to the staff; and in the second position the straight plate lies between the eye and the diagonal plate. [31] We observe the Moon at the place of vision with the staff such that the Moon touches the other end of the [straight] plate, [the end] that has no triangle on it, and [the rays of] the Sun at that time enter a hole at one of the acute angles of the triangle and continue to another hole on the plate that marks the sine [*lit.*: chord] of this angle. [32] When the Sun is east of the Moon we put the end with the triangle to the right, and vice versa when the Sun is to the west. [33] When the distance of the Sun from the Moon is up to about 90°, we put the diagonal plate between the straight plate and the eye. [34] When the distance from the Sun to the Moon is between 90° and 180°, we put the straight plate between the diagonal plate and the eye. [35] We claim that when the ray of the Sun passes from a hole at one of the acute angles to one of the holes on the plate that marks the sine for it, and the Moon is observed from the place of vision with this staff, then the apparent distance from the Moon to the Sun has been determined. [36] From this we may find its true distance from the Sun provided that we know the amount of the correction for parallax. [37] We shall determine the true amount for parallax later, God willing.

[38] To illustrate this, let the staff be represented by line ABG, and let point A be the place of the center of vision, as before, and point B the place of intersection of the plate and the staff at the time of the observation [Fig. 10.2]. [39] The right triangle formed with this plate is triangle DEZ; angle E is the right angle whereas angles D and Z are acute. [40] Point D is the place of the acute angle at the end of the first plate; at the points on the first plate we write DHT, and at the points on the second plate we write ZKL. [41] We first assume that the ray passes from point D to point L such that angle EDL is 15°. [42] In this case the hypotenuse of the triangle lies between the eye and the straight plate and we put point M at the end of the first plate—

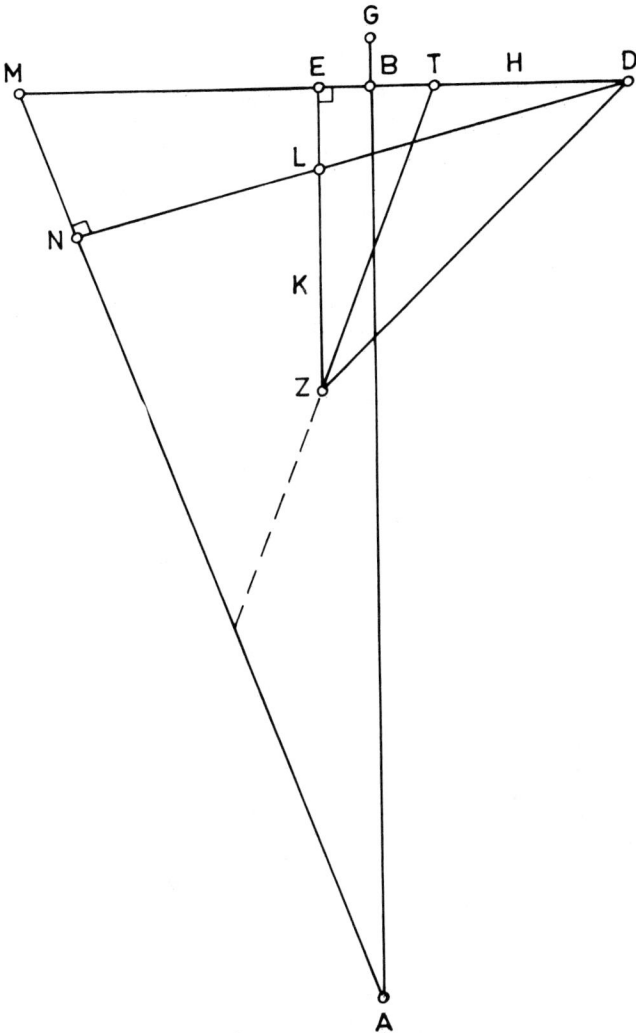

FIGURE 10.2. N 41a; P 13a; Q 20b.

opposite the end where the plates [that form] the triangle were put—and this is where the center of the apparent Moon touches it. [43] Angle *ABM* is a right angle where the amounts of lines *AB* and *BM* are known. [44] Therefore, by the preceding method, you can find angle *BAM*, and from it angle *BMA*, for it is its complement. [45] Let us assume that angle

BMA is 75°, and angle EDL is 15°: I claim that the distance from the Moon to the Sun at the time of observation is their supplement in 180°, i.e., 90°.

[46] Proof: lines DL and MA in the same plane lie at the ends of line DM, and the two angles formed with line DM are less than two right angles. [47] These two lines then meet, and at the place of their intersection the angle will be their supplement in two right angles. [48] Their place of intersection may be likened to the center of the angle of vision, because of the small distance from that point to the center of vision; thus, there will be no perceptible difference in this observation whether the distance observed between the Moon and the Sun is observed from that point or from the place of the center of vision. [49] This angle will be the apparent angular distance from the Moon to the Sun and, when the position of the Moon is corrected for the parallax appropriate at that time of observation, we will then know the true place of the Moon at the time of observation. [50] It follows that if the periphery of the Moon in the observation was put touching point M, it would be necessary to add to the distance from the Moon to the Sun the amount of the semidiameter, i.e., about 0;15° as we shall explain later, God willing.

[51] Next we assume that the ray passes from point T to point Z; in this illustration angle EZT is 15°, and there remains angle ETZ equal to 75°. [52] We assume, as in the preceding example, that angle BMA is 75°; thus, the apparent angular distance from the Moon to the Sun at the time of observation is their supplement in 180°, namely 30°, the analogy (*heqqesh*) is based on this.

[53] Now let the straight plate in the triangle lie between the eye and the diagonal plate. [54] We leave the figure alone, and I claim that the apparent distance from the Moon to the Sun can be found with this instrument. [55] We assume that the ray [of sunlight] passes from point D to point L, as in the previous example, and at that time the apparent Moon is [seen] at point M. [56] I say that the apparent distance from the Moon to the Sun is determined.

[57] To prove this, we draw lines DL and AM [Fig. 10.3]; from point A we draw line AN parallel to line DL, and line AS parallel to line DM. [58] We claim that the apparent distance from the Moon to the Sun is equal to angle NAM. [59] Since lines DL and NA are parallel, at the time when the ray of the Sun comes from point D to point L, it comes from point N to point A. [60] This is because the semidiameter of the terrestrial globe is like a point in relation to the distance from the Sun to the earth, and all the more so for the distance between these two lines; therefore, the Sun and the Moon are seen at angle NAM. [61] I say that angle NAM is equal to angle LDM plus the supplement in 180° of angle BMA. [62] This is because [the sum of] angles BMA and MAS is equal to two right angles, and angle SAN is equal to angle LDM, for the lines surrounding these two angles are parallel to each

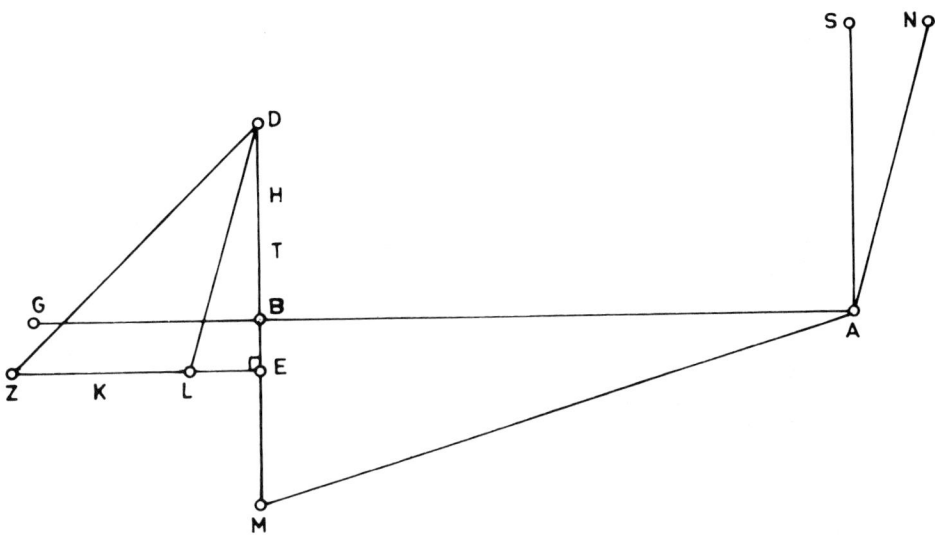

FIGURE 10.3. N 42a; P 20b: line segments *DL* and *ZL* are omitted, *Z* and *K* are omitted; Q 13b.

other. [63] Thus angle *NAM* is equal to angle *LDM* plus the supplement of angle *BMA* in 180° which is equal to angle *MAS*.

[64] Further, let the ray of the Sun pass from point *D* to point *E* at the time that the Moon appears to be on line *AM*. [65] I claim that the angular distance from the Moon to the Sun is equal to the supplement of angle *BMA* in 180°. [66] Let us reconsider the same figure where line *AS* is parallel to line *DE*. [67] It is clear by the previous procedure that angle *SAM* is the angular distance from the Moon to the Sun, and it is the supplement of angle *BMA* in 180°. [68] Hence we have proved what we wished to prove about this instrument.

Chapter 11

[1] You should know how to use this instrument without introducing errors. [2] First put a candle-light behind you when you are observing at night such that the light illuminates the surface of the plate visible to you, with which the observation is performed, in order for it to be clear to you whether the stars under observation touch the two ends of the plate, i.e., such that one star touches one end of the plate and the other star touches the other end of the plate. [3] The line reaching from the center of the body of one star to the center of the body of the other star must be parallel to the lines of the plate used in this observation without any perceptible inclination; the reason will be clear to anyone who reads this book carefully. [4] You should know that the observation will not yield the truth when clouds are near the star, for it happens that the cloud makes them appear to be removed from their proper place. [5] This remark holds for any observational instrument; even when you observe the altitude of the Sun on a cloudy day with the astrolabe it is not the same, for when the clouds move away the altitude of the Sun also changes. [6] For the same reason, you should not observe a star very near to the horizon, on account of the thickness of the vapors there. [7] Moreover, you should know that the cloud will also distort the size of the star, for when the cloud is very thick, the star will appear to be smaller than its proper size, and when the cloud is thin, it will appear to be greater than its proper size. [8] Similarly, in sunlight the stars close to it [the Sun] appear smaller than their proper size until they approach so closely that they are completely invisible.

[9] You should know that when you wish to use this instrument to find the position of some planet with respect to a fixed star, the distance between them should not be more than 40°, for at such great distances the eye is not able to see both stars together. [10] If their latitudes are not exactly known, the distance between them should not be less than 20°, for at a smaller distance the error introduced by a small error in the latitude produces a large error in the longitude of the star; all the more so when [the error in] the latitude [*lit.*: the distance] is a great amount. [11] Therefore, the distance between the

stars under observation should be such that even a large error in the latitudes would produce a small error [in the longitude]. [12]   This will not be obscure to those who have seen our remarks on the procedure for observing the distance between stars with this instrument.

CHAPTER 12

[1] There is great benefit to be gained from the use of the astrolabe because of the ease of using it, particularly for taking the altitude of a star at any place. [2] The errors in constructing it arise from many causes [*lit.*: sides]: first, due to the inclination of the diameter of the instrument; second, due to faulty construction of the alidade (*luaḥ*) with the two pierced vanes (*dappin*); and third due to the difficulty in determining the direction of the alidade to [an accuracy of] parts of a degree at the moment of the observation.

[3] We decided to present instructions for you to make this instrument in such a way that these errors may be avoided. [4] To do so, make the alidade that has the two pierced vanes before you draw the diameter of the instrument and before you divide it into degrees, and mark one end of it. [5] Then suspend the instrument when the Sun is on the meridian, approximately; there [should] be at the lower part of the instrument a ring whose width is equal to the thickness of the instrument. [6] Suspend a plumb-bob (*even ha-oferet*) from it such that the place of the lower part of the instrument stays at one point and is not disturbed for any reason. [7] Set the alidade in such a way that the Sun's ray enters one hole and continues through the second hole, and make a mark at the place where the alidade falls on that side. [8] Then turn the alidade to the other side, and let the hole that was first set underneath remain underneath, and make a mark at the place where the alidade falls when the Sun's ray enters one hole and continues through the second hole. [9] Take this observation immediately after taking the first observation so that the altitude of the Sun will not have changed sensibly. [10] Then bisect [the interval] between the two marks on the lower part of the instrument, and make a mark at the place of the bisection. [11] It will correspond to the maximum altitude from both sides, i.e., the alidade will fall there when that which is observed (*ha-mubaṭ*) through the two holes is at the zenith. [12] Draw a line through the place of the division[-mark] and through the center of the instrument; this line is the diameter of the instrument for this alidade, and for the condition (*maṣavo*) according to which these marks were marked [?]. [13] Then divide the circle from the place of the division[-mark]

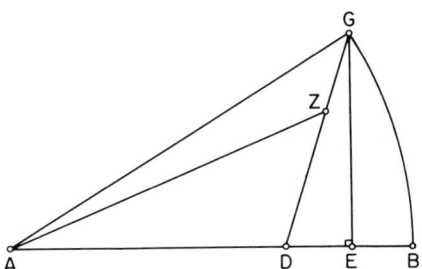

FIGURE 12.1. N 44a; P 21b; Q 14a.

into 4 equal quadrants with the greatest possible accuracy, and divide each quadrant into 90 equal degrees with the greatest possible accuracy, and this will allow you to determine the degrees of altitude without error.

[14] But the determination of subdivisions of a degree requires that we divided the semidiameter of the circle [*lit.*: ecliptic] into 6 equal parts. [15] We draw a circle about the center of the instrument with a radius of $\frac{5}{6}$ of the radius of the circle. [16] We divide it in the same way that we divided the outer circle, i.e., we put the alidade which is attached to the center of the instrument at the division[-marks] for the degrees on the outer circle, and we mark the place that the alidade intersects the inner circle until the inner circle is divided in this way into 360 degrees, and its lower half is cut into 180 degrees. [17] Then we put a straight edge (*qav ha-mishor*) on the mark for the degree of the inner circle and on the mark of the degree before and after it on the outer circle, and we draw these lines in both directions. [18] Then we skip one mark on the inner circle in both directions, and we treat the third from it in both directions in the preceding manner, and this is what we do until this operation is complete for the entire lower half.

[19] After this, you can find for each part of the degrees on the instrument the distance [from the center] on the 10 parts [of the radius]. [20] We will explain this to you by dividing each degree into 15 equal angles in such a way that the amount of each of them is 4 minutes. [21] Let point *A* be the center of the instrument and let one degree on the outer circle be arc *BG* [see Fig. 12.1]. [22] We draw lines *AB* and *AG* and drop perpendicular *GE* from point *G* onto line *AB*; let the amount of line *BD* be $\frac{1}{6}$ of the line *AB* and draw line *DG*. [23] It is clear from the table of arcs and sines that in the measure where *AG* is 60, *EG* is 1;2,50, very nearly, and *EB* is 33 seconds, very nearly; then remains *ED* which is 9;59,27, very nearly. [24] Since the square of *GD* is greater than its square by the amount of the square of *GE*, *GD* is 10;2,47,[1]

---

[1] For 47, N: 46, N mg. 44.

very nearly. [25] We mark point Z on line GD, and we draw AZ first assuming that angle GAZ is 4 minutes. [26] We wish to determine from this the amount of line AZ; it is clear that because angle GAE is 1°, angle EGA is its complement in a right angle, i.e., 89°. [27] When we subtract from it angle EGD, there remains angle DGA which is known, and angle EGD was known because the sides of triangle GED are known. [28] In the measure where line GD is 10;2,47[2] line GE is 1;2,50, and in the measure where line GD is 60, line GE is 6;15,16.[3] [29] Therefore angle GDE is 5;59°; angle EGD is its complement in a right angle, i.e., 84;1°, and angle DGA is 4;59°. [30] Angle GAZ was given as 4 minutes, the remaining angle in the triangle, angle GZA, is the supplement in 180°, i.e., 174;57°. [31] Therefore, it is clear that in the measure where line GA is 5;16,53, line AZ is 5;12,43, and in the measure where line GA is 60, line AZ is 59;12,31. [32] Thus, if we draw a circle about the center of the instrument, its radius should be 59;12,31 in the measure where the radius of the circle marking the degrees of the astrolabe is 60.

[33] We divide all the degrees in this way, i.e., the angle of the smallest part of them is 4 minutes. [34] Now let angle GAZ be 8 minutes, and angle ZGA is still 4;59°; there remains angle GZA in the triangle and it is the supplement in 180°, i.e., 174;53°. [35] Thus, it is clear that in the measure where line GA is 5;21,3,[4] line ZA is 5;12,43. [36] But in the measure where line GA is 60, line ZA is 58;26,45.[5] [37] We set the radius of the third circle at this amount. [38] By this procedure it follows that the radius of the fourth circle is 57;41,41. [39] The radius of the fifth circle is 56;57,50.[6] [40] The radius of the sixth circle is 56;14,59.[7] [41] The radius of the seventh circle is 55;33,20. [42] The radius of the eighth circle is 54;52,53. [43] The radius of the ninth circle is 54;13,14.[8] [44] The radius of the tenth circle is 53;34,22.[9] [45] The radius of the eleventh circle is 52;56,26. [46] The radius of the twelfth circle is 52;19,31.[10] [47] The radius of the thirteenth circle is 51;43,28.[11] [48] The radius of the fourteenth circle is 51;8,13.[12] [49] The radius of the fifteenth circle is 50;33,45.[13] [50] The radius of the sixteenth

---

[2] For 47, N: 47, N mg. 44.
[3] For 16, P: 10.
[4] For 3, N: 53, N mg. 3.
[5] For 26, N: 9; N mg. 26, 33.
[6] For 56, N: 57.
[7] For 59, N: 15.
[8] N: 54;53,54.
[9] N: 53;32,52.
[10] N: 59,19.
[11] For 28, N: 58.
[12] N: 51;43,28.
[13] For 33, N: 37.

circle according to this calculation is 50 and that is its amount; similarly, the division of all the degrees into 15 equal angles may be completed.

[51] The division of the degrees in this way is not difficult according to the method Euclid[14] expounded for dividing any line in proportion to the division of a given line. [52] To do this we divide a long line according to this division, and thus this line which is the radius of the circle of the instrument, is divided according to its proportion in the manner that was described there. [53] You should understand that the larger the astrolabe, the greater the precision found in it; its radius should not be less than a span (*zeret*) so that you can construct it to the desired precision. [54] As for the rest of the [instructions] needed to make an astrolabe, others have said enough about it.

---

[14] N mg. adds: in VI.13 [of Euclid: trans. Heath, VI.12].

CHAPTER 13

[1] We have already instructed you concerning the perfection of the construction of the astrolabe in such a way that an error will not occur in finding the altitude with it and our precision can reach minutes and even part of minutes; thus it is clear that with this instrument we can easily determine the meridian, very nearly. [2] Observe with it the Sun's altitude close to noon and note that its altitude continues to increase a little until it no longer is increasing; this is noon approximately. [3] If you draw a straight line on the floor of a house along a ray of the Sun that enters through a window at that time, this line is the meridian approximately. [4] We also can determine the meridian in another way with extreme precision. [5] We first investigate the altitude of the pole for our location with the aforementioned astrolabe, observing the Sun at Cancer 0° and Capricorn 0°, or close to them. [6] Under these conditions the error due to approximation in determining the position of the Sun in longitude will have no perceptible effect on the altitude of the Sun because of the smallness in the change of declination corresponding to each degree there. [7] If we add to the altitude of the Sun when it is at Capricorn 0° half the excess of its altitude at Cancer 0° over its altitude at Capricorn 0°, the result is equal to the altitude of Aries 0° for that horizon [i.e., geographical latitude]. [8] If the altitude of Aries 0° is subtracted from 90°, the remainder is equal to the altitude of the pole at that horizon.

[9] Half the excess of the altitude of Cancer 0° over the altitude of Capricorn 0° is equal to the declination of Cancer 0° from the equator. [10] In this way we found the value for the obliquity to be 23;33°, very nearly.

[11] After verifying this, we will observe the rays of the Sun in a house whose [opposite] walls are parallel and whose floor is parallel to the plane of the horizon, when the Sun is at Cancer 0° and Capricorn 0°, very nearly, so that a perceptible error will not occur in our observations due to the true position of the Sun. [12] There should be a window facing east or west, very nearly, whose [edges] are straight lines at right angles such that the upper edge of the window is parallel to the floor of the house, and whose left and

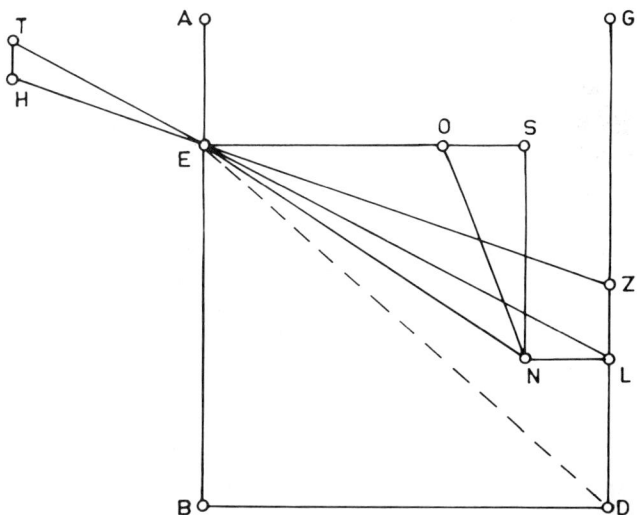

FIGURE 13.1. N 46a: *ED* and *NL* are omitted, angle *OSN* is acute; P 22b: *ED* and *NL* are omitted, angle *OSN* is acute; Q 15a: *ED* is unbroken.

right sides are perpendicular to the plane of the floor of the house. [13] If the window faces east, more or less, we mark a line on the western wall opposite it parallel to the horizon a certain amount below the height of the top of the window, say 5 spans or more. [14] A span or more below that line or that wall we mark a line parallel to the first line. [15] In the morning when the upper ray reaches the upper line, we mark the lower line at the place that the ray crosses the lower line such that the mark reaches straight to the upper line, and wherever it crosses it we consider that to mark the top of the image [*lit.*: ray (*niṣoṣ*)]. [16] From this we can determine the position of the image of the Sun on that wall at the true time of sunrise.

[17] To illustrate this, let us consider line *AB* to lie on the eastern wall perpendicular to the floor of the house such that this line touches the northern edge of the window, and point *E* on it to represent the topmost point on the northern edge of the window [Fig. 13.1]. [18] We mark line *GD* on the western wall parallel to line *AB* such that this line touches the ray that enters from the northern edge of the eastern window. [19] Let point *Z* on it represent the top of the image [*lit.*: ray] on this side; and let point *T* be the center of the Sun whose lower semidiameter, parallel to line *GD*, is line *TH*. [20] It follows from the preceding that the true ray from the center of the Sun reaches a point below point *Z* by the amount of the angular semidiameter of the Sun and to the south of point *Z* by the amount of the angular semidiameter of the Sun. [21] We will explain how we can find the point that lies

below point Z by the amount of the angular semidiameter of the Sun and from this we will also be able to find the point to the south of it by the amount of the angular diameter of the Sun. [22] We extend line ZE to point H and we consider angle ZEL to be the angular semidiameter of the Sun that was determined with the instrument we invented for this [purpose] described above. [23] At Capricorn 0° it is 0;15°, very nearly, according to what we determined by observation. [24] Point L lies on line GD between points Z and D; we draw line ED and measure it with a ruler (*qeneh ha-midda*), and similarly we find the amount of line ZD. [25] Triangle ZED has known sides and therefore its angles are also known including angle EZL. [26] Since angle ZEL was also determined, there remains angle ZLE in the triangle and it too is known. [27] Since the amount of line ZE is known, the amount of line ZL can be found, as was explained previously. [28] After we have found the amount of line ZL, we draw line LN on the wall parallel to the floor of the house such that point N is to the south of point L. [29] We set angle LEN equal to the angular semidiameter of the Sun, and the amount of line LN can be found by the same procedure as before. [30] Point N is the position of [i.e., corresponding to] the true rising of the center of the Sun in this example, and we set line NS on the western wall perpendicular to line LN such that the height of point S is equal to the height of point E. [31] We measure the amount of line NS with a ruler and it is clear that the altitude of the image [*lit.*: ray] of the true Sun at the time of its true rising is equal to the altitude of point S, but it inclines from point S to the south on the western wall.

[32] Now we have to determine the position on the western wall [corresponding to] the true rising. [33] We will present a clear argument (*haṣaʿa*), namely that the amount of the inclination of the plane of the equator to the horizon is equal to the amount of inclination of all the circles parallel to it. [34] The Sun rises on such an inclined plane. [35] Further, we argue, and it will be clear to the reader of this book, that the ray of the Sun that reaches us when it lies on a circle parallel to the equator is not in its parallel circle because it passes through the center of the equator, not the center of its parallel circle. [36] This is because the earth lies at the center of the equator, as the ancients proved, and it follows from our forthcoming remarks that all of it is like a point, approximately, in relation to the orb of the Sun. [37] Since this ray passes straight to the center of the equator, it is clear that it reaches the parallel circle on the other side of the circle on which the Sun lies. [38] Let us consider the window in this example to lie at the center of the earth because this does not harm the argument and, for the reason already mentioned, you will find that however much the Sun inclines to the north, its image (*niṣoṣ*) on the western wall lies to the south. [39] Since the Sun rises on an inclined circle at the horizons [i.e., at *sphaera obliqua*], not on a circle perpendicular to the horizon, we must explain [how to find] the amount of the

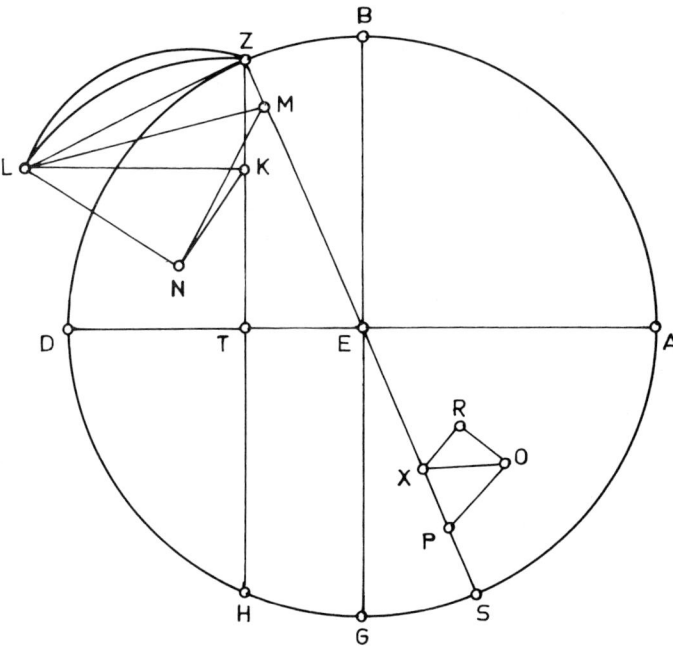

FIGURE 13.2. N 47b: the inner arc from point Z meets line LM at a point between L and M, and from that point a line is drawn to point K, line KL is omitted, quadrilateral XROP is omitted here and is drawn elsewhere (N 48a) with circle ABGD and lines AD, BG, ZH, ZS; P 22b agrees with N, the second figure appears on 23b; Q 15b.

angle at which the Sun rises, i.e., from its true rising until it reaches point Z in our example which corresponds to what is observed, because it is not the angle that its parallel circle makes with the horizon.

[40]   Let us consider the plane of the circle of the horizon to be circle ABDG whose center is point E; we draw straight line BEG such that line BEG represents the diameter of the equator and line AD intersects it at right angles at point E [Fig. 13.2]. [41]   Line ZH represents the intersection of the horizon with the Sun's day-circle when it is at Capricorn 0°, and it intersects line AD at right angles at point T. [42]   We assume that the Sun rises to the east of the parallel circle by arc ZL where the Sun is at point L and line LK is perpendicular to line HZ intersecting it at point K. [43]   We draw a perpendicular LN from point L to the plane of circle ABGD and draw line KN in the plane of circle ABGD. [44]   Angle LKN is equal to the inclination to the horizon of the rising of all the parallel circles and it is the complement in 90° of the altitude of the pole: there remains angle KLN and it is equal to the angle of the arc of the altitude of the pole. [45]   You can find the amount of

line *LN* if you determine the altitude of the Sun at that time with an astrolabe, mentioned above, because the sine (*metar mehuṣeh*) of that altitude is equal to line *LN*. [46] It is clear that if the window were at point *T*, the inclination of this parallel circle to the horizon would be the apparent inclination at which the Sun rises. [47] But the position of the window is at point *E*, very nearly. [48] Therefore, when we wish to find the amount of the apparent inclination at which the Sun rises from point *Z* to point *L*, we must draw straight line *ZES*. [49] Arc *ZL* is an arc of a great circle from point *Z* to point *L*, and so we draw its sine, line *LM*, from point *L* falling on diameter *ZES* at point *M*. [50] We join line *MN*, and it is clear that angle *LMN* is the apparent inclination of the Sun at this rising and it is not equal to angle *LKN*.

[51] From the solar altitude you can find the arc that rises on the parallel circle, namely arc *ZL*, by the method for the calculation of paths (*mahalakhot*) which will be explained later on. [52] You can find the amount of its entire chord *ZL* in this figure, in the table of arcs and chords, in the measure where the semidiameter of this parallel circle is 60. [55] Therefore, you can find the amount of line *ZL* in the measure where the semidiameter of the equator is 60 because the ratio of these diameters to one another is known. [54] If you draw the entire arc for this chord, you will have the arc that extends from *Z* to *L* on a great circle. [55] Its sine, namely line *LM* in this figure is known; since line *LN* has already been found, the amount of line *MN* is known because line *LM* squared is greater than line *LN* squared by the amount of line *MN* squared. [56] Since all three sides of triangle *LMN* are known, all its angles are known including angle *LMN* which is the angle of the apparent inclination at the rising of the Sun when its arc is assumed to be arc *ZL*.

[57] In the first figure we now draw at point *N*, on the western wall to the south, angle *SNO* equal to angle *MLN* in the second figure such that line *SO* is parallel to the floor of the house which is parallel to the horizon. [58] It follows that the true rising for the image of the center of the true Sun is at point *O* because angle *SON* is equal to angle *LMN* which is equal to the angle of the apparent inclination for the rising of the Sun, as was already explained. [59] You should know that this is correct if the western wall is perpendicular to line *OE* in this figure, and this is because line *MN* is perpendicular to line *ZS* in the second figure. [60] If the western wall is not perpendicular to line *OE*, this inclination will be a little greater to the south on the [western] wall, and this will be clear from what I say.

[61] Let us consider the line of the western wall parallel to the line of the floor of the house in which the position of the true rising is not perpendicular to line *ZS* in this second figure, namely line *OP* in the second figure [Fig. 13.2]. [62] Point *P* is the position of the true rising when the Sun is at point *Z*; point *E* represents (*mashal*) a point on top of the window and point

O represents point S in the figure of the window. [63] Line RO is perpendicular to line OP, and it is marked on the wall from above to below as was done for line SN in the preceding figure such that line OR is equal to line SN in the preceding figure. [64] We draw perpendicular OX from point O to line SE and then draw line RX. [65] It is clear from the preceding that angle OXR is equal to angle LMN which has been determined; since angle ROX is a right angle, the remaining angle ORX is known. [66] Since line OR is known, the remaining lines are known including line OX. [67] It is clear that line OP squared is greater than its square by the amount of line PX squared. [68] Therefore, the inclination for the true rising of the Sun is greater than what was assumed by the excess of line OP over line OX; but we do not require all that much accuracy in this place because the altitude of stars near the meridian will not differ perceptibly as a consequence of the approximation that may be found in the meridian. [69] If you wish, you can be more precise than this, reaching what is sought with great accuracy. [70] You should see whether or not the line reaching from the window to the place of the apparent rising (*zeriha*) on the wall is perpendicular to line SO in the first figure. [71] If it does not meet it at right angles, find its angle if it is acute, or its supplement in 180° if it is obtuse. [72] The ratio of the sine of that angle to 60 is equal to the ratio of line OX in this figure to line OP. [73] Therefore, add to line SO according to the ratio of the excess of line OP over line OX and you will reach what was sought very precisely. [74] If you determine the aforementioned ratio again very carefully with this procedure [to find] the increment to line SO, you will reach the truth in this, very precisely.

[75] When you find the point of true rising (*zeriha*) on the western wall, place a plumb line on that point and mark the point on the floor of the house where the plumb line falls. [76] Similarly, place a plumb line at the top of the window on the side used to define its image and mark the point on the floor of the house where the plumb line falls. [77] Then draw a straight line on the floor from one mark to the other; this line corresponds to line ES in the second preceding figure. [78] Then find in the tables the ortive amplitude (*merhav ha-zeriha*) of the degree in which the Sun lies for your horizon, namely arc BZ in the second figure. [79] Draw a line on the floor intersecting the first line on the floor such that they include an angle equal to angle BEZ in this figure which is the ortive amplitude of the degree in which the Sun lies. [80] Consider the inclination of this second line on the floor from the first line on the western side to the north when the Sun is in the northern signs, and to the south when the Sun is in the southern signs, and vice versa if the window is on the western wall; and this is entirely obvious to the reader of this book. [81] This second line represents line BEG in this figure; its eastern end (*rosho*) is the position of the rising of Aries 0° and Libra 0°, and its western end is the position of their setting (*ariva*). [82] If the south wall of

the house has a window, drop a plumb line from the top of the window at one of its edges to the floor and mark a point there. [83] From the place of the mark draw a line at right angles to the line that represented line *BEG*, and it will be the meridian, i.e., when the ray of the center of the Sun passes from this window to that line it will be noon. [84] If the length of the house is 40 or 50 spans, it will be of great benefit for finding most of what is sought concerning the Sun, as will be explained in what follows.

CHAPTER 14

[1] You should also be aware of the difficulty in finding the exact positions of the fixed stars in longitude and latitude; yet this is the gate through which we enter in order to know the planetary positions by observation. [2] It may be thought that we have already dealt with the difficulties in the observations, for the motion of the Moon is thought to be known without doubt, but our investigation has led us to understand that the position of the Moon is not sufficiently accurate to determine the positions of fixed stars seen together with it. [3] Even at conjunction and opposition the Moon is not found at the place that follows from the tables constructed according to Ptolemy's model. [4] This fact has been determined from the times of lunar eclipses which were sometimes before the time predicted from Ptolemy's model and sometimes after it, and the error is perceptible. [5] All the more so for the other days of the month, and the error in those cases is sometimes more than plus or minus $1\frac{1}{2}°$. [6] Moreover, the amount of lunar parallax is full of confusion and doubt, and this has become clear to us on the basis of many observations. [7] Indeed, the [observed] amounts [of parallax] do not agree with what follows from Ptolemy's computations as we shall explain later, God willing. [8] This introduces a large uncertainty in the positions of the stars seen together with it. [9] Moreover, even if the elongation of the Moon from the Sun verified by eclipses is accurate in Ptolemy's model, we cannot determine the true position of the Moon without knowing the true position of the Sun. [10] It is extremely difficult to find the true position of the Sun, and even Ptolemy wrote that he did not know the [exact] moment when the Sun entered the beginning of Aries, nor did his predecessors, or his successors. [11] The reason he gave for this statement was that the observational instruments made at that time for measuring altitude, like the astrolabe, had approximations due to many causes, as we have already mentioned, and a small error in the altitude of the Sun produces a large error in its longitude. [12] Moreover, there is uncertainty in the correction for solar mean motion; Ptolemy believed that the maximum equation was about 2;23°, but his successors believed that it only reaches 2°. [13] Furthermore, there is uncertainty

in the apogee of the solar orb; Ptolemy believed that it was fixed at Gemini $5\frac{1}{2}°$, and some of his successors believed that it was at Gemini $17\frac{1}{2}°$[1] about 12 centuries after the observations of Ptolemy. [14] According to al-Battānī it is to be found at about Gemini 29;15° at the aforementioned time. [15] These matters all contribute to the difficulty in finding the true place of the Sun.

---

[1] With N; P: $7\frac{1}{2}°$, Q: $18\frac{1}{2}°$.

CHAPTER 15

[1] We have explained that the only way to find the longitudes and latitudes of the fixed stars is first to find the true position of the Sun. [2] Finding the true position of the Sun is quite difficult because of the uncertainty in its mean position in longitude, in the inclination of its orb to the sphere of the equator, in the amount of the [maximum] equation, and in the position of the apogee. [3] We endeavored to find a procedure that would allow us to determine the true position of the Sun for every day, the maximum equation [for correcting] its mean motion, the inclination of the pole of its orb [i.e., the ecliptic] from the pole of the equator, and the apogee of the solar orb.

[4] To do this we made a high window, in the south [wall] of a house, [whose edges were] perpendicular straight lines. [5] The edge at the top [of the window] was parallel to the floor of the house, and the western edge was perpendicular to the floor, i.e., the plane of the horizon. [6] We dropped a plumb-bob to the floor at the western edge of the window, and made the floor of the house plane such that the wall with the window was at right angles to the floor. [7] We marked the place on the floor at the foot of the plumb line and from it we drew a line on the floor to represent the meridian, according to the previously described procedure. [8] To the west of this line we drew a line on the floor parallel to it, about half a span or a little less away from it. [9] When the western edge of the image reaches this second line, we mark the first line wherever the outer limb [*lit.*: head] of the image from the top of window intersects it, and we mark this spot for each day that we wish. [10] You will find that as the Sun approaches the beginning of Cancer, the outer limb of the image approaches the wall with the window, and you should notice on this line [the change in] solar declination due to its motion in one day even when it is close to the beginning of Cancer. [11] If you investigate this carefully, you will find that if the height of the top of the window is 24 spans, the [change in] solar declination of 1° at the beginning of Cancer produces a [difference] between the marks on the meridian [on the floor] of this house that is close to half a span for the horizon [i.e., geographical latitude] where the altitude of the pole is 44°. [12] The amounts [of these distances]

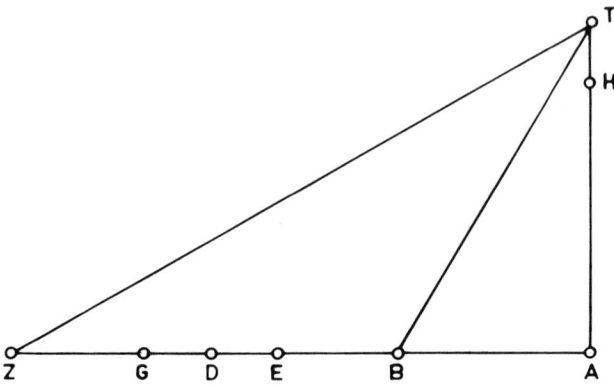

FIGURE 15.1. N 51a; P 25a; Q 16b.

can be divided into minutes and even possibly into seconds approximately, by means of the method used to divide the units [*lit.*: degrees] on the staff into minutes.

[13] For about ten days or a little less [before] the Sun reached the beginning of Cancer we observed, every day, the image of the Sun in the aforementioned manner, and we wrote the day of the observation on the place of the mark [on the floor]; we continued to do so until the mark [came] as close as possible to [the wall with] the window, and we marked that place. [14] Afterwards, when we observed this during the time when the Sun again descended to the south [and its image returned] towards the place that we marked about 10 days before the Sun reached the beginning of Cancer, we saw that it either returned to that very spot or close to it. [15] However it may be, we can use this to determine very accurately the moment when the Sun enters the beginning of Cancer.

[16] To illustrate this, let the top of the window be marked by point T, and the western edge of the window be line TH [Fig. 15.1]. [17] We extend that straight line to the floor of the house, and it is line THA, perpendicular to the plane of the floor; let the meridian marked on the floor be line ABG. [18] We first assume that the Sun returned exactly to the same place where it was about 10 days before it reached the beginning of Cancer. [19] Let point B represent the mark when it reached the beginning of Cancer, and point G represent the point about 10 days earlier. [20] We assume that it returns to that mark 19 days later. [21] We claim that it is clear that in 9 days and 12 equal hours after the first observation, the Sun was at the beginning of Cancer, for the true motion of the Sun in the first half of this time interval must equal very nearly its true motion in the second half of this time inter-

val inasmuch as the amount of the correction does not differ perceptibly in this small interval—even if the apogee were up to 10° before or after the beginning of Cancer.

[22] We now assume that it does not return exactly to point G, but close to it, say at D. [23] Let us assume that on the day following the day when it was at point G, the mark was at point E; we know the ratio of GD to GE and we divide the day according to this ratio. [24] For example, if GD were $\frac{1}{6}$ of GE, we would know that in 4 hours, $\frac{1}{6}$ day, the Sun's motion is represented by the distance between points G and D. [25] If point D lay between points B and G, we add these 4 hours to the time interval between mark G and mark D; if point D lay on the other side, we subtract these four hours from that time interval. [26] If we bisect the interval that results from the addition or subtraction, we obtain the moment when the Sun was at the beginning of Cancer, very accurately. [27] In the same way, we observe the Sun when it is near the beginning of Capricorn to determine the moment when the Sun entered the beginning of Capricorn. [28] If the interval from the moment that the Sun was at the beginning of Cancer to the moment that the Sun was at the beginning of Capricorn were equal to the interval from the moment that the Sun was at the beginning of Capricorn to the moment that it was at the beginning of Cancer, we would know that the apogee was at the beginning of Cancer. [29] We shall see whether the interval that we found from the moment that the Sun was at the beginning of Cancer to the moment that it was at the beginning of Capricorn is equal to half the length of the year. [30] If they are not equal, we know that the apogee lies in the part that the Sun took a longer time to traverse. [31] Since the matter is so, we shall determine the place of the apogee by a method that will be explained later.

[32] When we compare the mark for the Sun at noon at the beginning of Cancer with its mark at the beginning of Capricorn, we can find the obliquity of the sphere of the zodiac from the sphere of the equator and the amount of the altitude of the pole for that [geographical] place. [33] For this we use the previous figure, where point B marks the beginning of Cancer, and point Z marks the beginning of Capricorn. [34] We draw line $ABGZ$ and point T, the top of the window, is the point from which the ray comes, as before; we then draw lines TB and TZ. [35] We determine the amounts of lines TB, TZ, and BZ with [the aid of] a ruler (*qeneh midda*), and so we can find angle BTZ which is twice the obliquity, and angle TZB, the altitude of the Sun when it is at the beginning of Capricorn. [36] According to the preceding argument, you should add the solar radius to the altitude of the Sun at the beginning of Capricorn and then add half of angle BTZ to the sum; the result is the [solar] altitude at the beginning of Aries, and its complement in 90° is the altitude of the pole for that horizon. [37] Once you know all this, you can find by this method and with the use of this line, the altitude of the Sun on

any day you wish. [38] If you observe the altitude of the Sun and know its obliquity, you can easily determine the degree and fraction of a degree [of longitude] that corresponds to that declination with the help of a declination table; thus, the true position of the Sun may be determined for any day desired. [39] If you investigate in this way the true position of the Sun on the ecliptic about 40 or 50 days before the Sun is at the beginning of Cancer and then about another 40 or 50 days after the Sun is at the beginning of Cancer, the position of the Sun's apogee and its maximum equation can be found by a procedure similar to the one Ptolemy described for the planets [*lit.*: stars], and we shall give you this proof, God willing, when we come to describe the orb of the Sun.

[40] Whether or not the Sun has a sphere eccentric to the center of the world will become clear to you after you find the position of the apogee, for you can investigate the size of the solar diameter there by means of the ray[s] that enter the window of the staff. [41] Moreover, you should investigate its size at perigee: if you find that the sizes are equal, you can conclude that the solar sphere is not eccentric to the center of the world as Ptolemy assumed; but if you find that the sizes are different, you can conclude that its center is eccentric to the center of the world. [42] In general, if the apparent size of the solar diameter at various places on its sphere varies in amount, its sphere must be so arranged, i.e., we would know that its center is eccentric to the center of the world. [43] You can find the true position of the Sun on any day even if it is not on the meridian, determine the moment that the Sun is at the beginning of Cancer and of Capricorn, the place of the solar apogee, and the maximum solar equation, once you have found the altitude of the pole for your horizon and the obliquity of the ecliptic with the astrolabe, as described earlier.

[44] Consider a house with a window, whose edges are straight lines, on the east or west [wall]; let the length [of the north wall of the house] be about 40 spans or more, for the longer it is, the greater the resultant accuracy. [45] The walls of the house should be perpendicular [to each other]; we draw a line on the wall opposite the window parallel to the line of the floor of the house and lower than the top of the window by about 6 spans or more, and under it about a span or more we draw a second line parallel to the first. [46] When the top of the image reaches the upper line, seen at some place, it then crosses the lower line to one side of the image, say the north side; we mark the lower line and draw a line through that place [on the lower line] perpendicular to it reaching the upper line and intersecting it: the mark of this image corresponding to the time of the observation belongs there. [47] Note the declination for each day most carefully: each degree of declination corresponds to a very great amount on the line for the image, and [this amount] is about equal to the [rising] amplitude at the moment of sunrise for

this declination and this horizon, as will be clear with a little reflection from what we have already said, such that every degree of declination corresponds to an amount greater than a span on this line on which the marks are to be made. [48] If you make these marks in the way that we mentioned for the meridian, you will have marked, for example, the Sun's place at the moment when it was at the beginning of Cancer and you will also know when it reached the beginning of Cancer, and with remarkable accuracy.

[49] We claim that this procedure may be used to determine the amount of the Sun's equation and the position of its apogee. [50] We explained previously how to find the place of true sunrise from these marks for any position of the Sun on the ecliptic at the time of the observation. [51] Therefore, we can use this for finding the place of true sunrise when the Sun is at the beginning of Cancer, and similarly for the place of true sunrise when the Sun corresponds to any of these marks. [52] Once we know these two places, we measure the distance between them with a ruler and [we also measure] the length of the line[s] from the top of the window, on the [same] side as the image marked [on the opposite wall], to each of these places; thus we have a triangle whose sides are known. [53] Therefore, all the angles can be found, as we have explained above, and in particular the angle at the top of the window which is the rising amplitude (*merḥav*) at this horizon from the degree that the Sun is in when it is at this mark to the beginning of Cancer. [54] Since the obliquity is known and the altitude of the pole for this horizon is also known, the rising amplitude at this horizon for the beginning of Cancer is determined, as we shall soon explain. [55] When this angle at the top of the window is subtracted from the rising amplitude of the beginning of Cancer, the remainder is equal to the rising amplitude of the degree that the Sun is in. [56] We may also use this procedure to determine the degree that the Sun is in at the time of the observation for which this amplitude at this horizon is appropriate, but this needs no explanation for anyone who knows enough of this science to be able to strive for the perfection of this study. [57] Similarly, from this you can find the true place of the Sun at the time it returns to this mark or to another mark. [58] When this is completed, and the times of all these observations are known, there will also be known the time that the Sun enters the beginning of Cancer. [59] We use this procedure, as we did in the case of the figure for the meridian, to determine the amount of the [maximum] solar equation and the position of the solar apogee. [60] We shall explain this procedure further, God willing, when we come to talk about the orb of the Sun.

[61] If you wish to determine easily the length of the lines from the window to the marks on the wall, the walls of the house should be plane and perpendicular [to one another], as we stipulated above; you should measure its [i.e., the wall's] length, the distance from the top of the window to the north

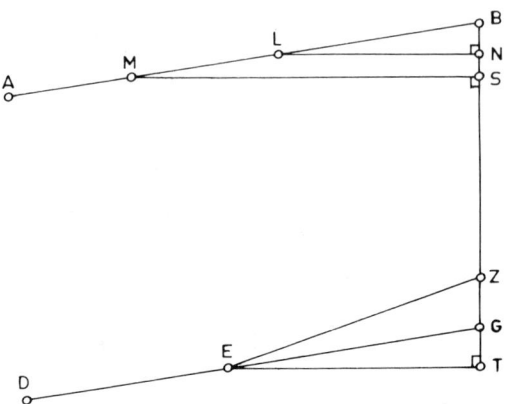

FIGURE 15.2. N 53b; P 26a; Q 17b.

wall, and the distance of the marks from the north wall, from which you will easily know the length of these lines. [62] If the house was not [built] with right angles, there will be some difficulty, but the accuracy found with it will increase as the magnitude of the rising amplitude [increases] there. [63] In this way it happened to us, that we found that the house we prepared for these observations, was not [built] with right angles. [64] It is appropriate for us to inform you how to make the proper adjustments.

[65] To illustrate this, let the western wall with the marks [meet] the floor of the house at line *BA*, the northern wall [meet it at] line *BG*, and the eastern wall with the window on it [meet it at] line *GD* [Fig. 15.2]. [66] Let point *E* be the place that the plumb line reaches on the floor of the house from the top of the window, from which comes the image marked on the western wall. [67] We measure line *BG* with a ruler, and similarly lines *GE* and *AB*. [68] We first mark an arbitrary point on line *BG*, say point *Z*; we draw line *ZE*, and we find the amounts of lines *GZ* and *ZE* with the ruler. [69] All sides of triangle *GZE* are known and so we can find angle *EZG*. [70] We draw *ET* perpendicular from point *E* to line *ZG*, extended if necessary, and we draw line *BGT*. [71] Since angle *TZE* is known, so is angle *ZET* for it is the complement in 90° of angle *TZE*; since line *ZE* is known, line *ZT*, the side subtended by angle *ZET*, is known; therefore, we can also find line *TE*. [72] Since line *ZT* is known and line *ZB* is known by means of the ruler, line *TB* is also known, and it would be the length of the north wall if the north and east walls met at right angles, under the assumption that the north and west walls in fact meet at right angles.

[73] If that assumption does not hold, it is necessary to correct the length of the north wall on that side also, according to the place where the mark is.

[74] Let us assume that the places of the two points marked on the west wall, when a plumb line is dropped from them to the floor of the house, are points L and M. [75] We find angle GBL in the same way that we found angle BGE. [76] Let the perpendicular from point L to line GB, extended if necessary, be line LN—we find line ZN in the same way that we found line ZT. [77] Therefore, we know line TN, the length of the north wall which is appropriate for this mark; in this way we can find line NL, the appropriate distance from point L to the north wall. [78] We draw a perpendicular MS from point M to line GB extended if necessary; we can then find lines TS, the appropriate length of the north wall for this mark, and we can find line SM, the distance from point M to the north wall. [79] In this way you can use this house for this procedure even though its walls are not at right angles; and this is what we sought to explain.

Chapter 16

[1] We cannot make observations of the planets without first knowing the positions of the fixed stars with which they are observed. [2] But there is great uncertainty concerning their positions that arises from the proliferation of opinions we found among our predecessors concerning the motion of the eighth sphere. [3] Clearly, we must first investigate the positions of the fixed stars before we can determine the positions of the planets seen with them. [4] We will use the instrument for observing the Sun and the Moon [together] for we can determine the position of the Sun at the time of the observation without doubt. [5] Since we have found by means of this instrument the elongation between the Sun and the Moon, the apparent place of the Moon is known to us without doubt. [6] From this we can determine the positions of the fixed stars that are seen with the Moon shortly after [*lit.*: close to] the time of the observation, as previously mentioned, but uncertainty still remains here on account of our ignorance of the amount of parallax.

[7] By way of approximation, here we assume that the parallax at conjunction and opposition conforms to what follows from Ptolemy's tables, for it produces close agreement with solar eclipse [observations]. [8] But for the remainder of the month we have noted that, without any doubt, the amount of parallax is no greater than it is at opposition, but for very little. [9] This will become clear in what we shall report, for we have observed the size of the diameter of the Moon by means of the ray that enters the window of the aforementioned instrument, and we did not find it greater, except for a small amount, at quadrature than at opposition, as we shall mention when we come to discuss the Moon. [10] We investigated the size of the lunar diameter with this instrument in all the ways that we mentioned, and with all of them we found that the apparent diameter of the Moon is not greater at quadrature than its apparent size at opposition, but for a small amount. [11] You will find this to be true without a doubt if you investigate it by means of the procedures already described. [12] It follows that the Moon is no closer at quadrature than at opposition, but for a little. [13] With all this [evidence] and despite the general opinion that Ptolemy had determined the amount of

parallax at quadrature with a true proof, we decided to investigate this matter further in the following way. [14] We observed the Moon at quadrature, very nearly, with a fixed star near the ecliptic appropriate for this observation, as we mentioned before; we determined its distance in longitude from [the star] and noted the time of the observation. [15] Then, 4 or 5 hours later, we again observed the Moon with that star, and determined its distance in longitude for that observation; when we have found this, we have determined the apparent motion of the moon for the time interval between the observations. [16] Then we subtract the motion for this time interval according to Ptolemy's model, for in this place there is no harm in depending on a computation based on that model inasmuch as no error would be perceptible in such a short time interval especially at quadrature, as will become clear in what follows. [17] The remainder is an amount which is the surplus of the parallax in the second observation over the parallax in the first observation, if both of them were to be subtracted from the true place of the Moon; or the sum of the two parallaxes, if the first was to be added and the second subtracted from the true place of the Moon; or the amount that the parallax in the first observation exceeded the parallax in the second observation, if both of them are to be added to the true place of the Moon.

[18] It is also possible to determine the amount of parallax for this place by computation according to Ptolemy's tables to see whether the results agree with the remainder in the parallax that we found by [observation]. [19] When we investigated this in this way, we found that it is not appropriate to set the amount of parallax at quadrature greater than at opposition, but for a little. [20] We repeated the observation many times as we shall mention, God willing, when we shall come to speak about the Moon. [21] Once this was clear to us, we again investigated the proof on which Ptolemy based his [excessive] amount of parallax at quadrature. [22] We discovered that his proof was based on the assumption that the maximum lunar latitude north and south of the ecliptic is 5°; but this is not true, as we shall explain without any doubt in the section on the Moon. [23] There it shall become clear that the maximum latitude of the moon is $4\frac{1}{2}°$ as al-Battānī agreed, and with this [change in the parameter] what Ptolemy observed does not disagree with what we assert here, namely, that the parallax at quadrature is not greater than at opposition, as we shall prove later on. [24] It is characteristic of the truth that it is impossible for it to contradict itself; on the contrary, it must agree with itself in all respects.

[25] There is another easier way by which we can determine the position of the fixed stars. [26] We observe the Moon during a lunar eclipse, and we observe the periphery of the shadow on it with some fixed star. [27] We determine the time of the observation, and we find the distance in longitude between the fixed star and the periphery of the shadow by observation.

[28] The semidiameter of the shadow is known approximately, namely 42 minutes, as we shall explain when we come to speak about the Moon, and the amount of parallax for the circle of the shadow seen on the Moon is known very accurately. [29] When we correct the place of the circle of the shadow according to the table of parallax, the true distance between the periphery of the shadow and this fixed star at this time becomes known to us. [30] From this we can easily determine the true distance between the center of the shadow and this fixed star at the time of the observation because the semidiameter of the shadow is known, and the true position of the Sun at the time of the observation which is known yields the position of the center of the shadow, for it is 180° from the Sun. [31] As a result, the place of the fixed star observed with it is known, and the longitude [*lit.*: place] of this fixed star tells us the longitudes of the remaining fixed stars whose longitudes are inscribed in the tables, for we can compute [their longitudes] according to Ptolemy's values for the distance [in longitude] from this star to [each of] them. [32] This is enough in this place for us to begin using observations of planetary positions according to our instructions so that it will become clear whether or not they conform to Ptolemy's models.

Chapter 17

[1] We have mentioned a number of things concerning the Moon which prove without doubt that Ptolemy's model is not correct. [2] For example, according to his model, the diameter of the Moon should be close to twofold greater at quadrature at 180° of anomaly than at opposition at 0° of anomaly.[1] [3] But we did not find much of a difference in the apparent size of the diameter of the Moon, either on account of the elongation [*lit.*: apogee] or on account of the motion of anomaly; we repeated this observation many times with this instrument with which it is impossible for a [perceptible] error to occur, as we explained earlier.

[4] It also became clear to us on the basis of observation that the apparent size of the diameter of Venus is greater at greatest elongation from the Sun than at 0° or 180° of anomaly; on the other hand we did not observe it to be greater at 180° of anomaly than at 0° of anomaly. [5] All this is at variance with what follows from Ptolemy's model, for according to it the diameter of Venus should appear to be greater at 180° of anomaly than at 0° by more than 6 times. [6] We also observed diligently seeking to find the apparent size of Venus at each time relative to the apparent size of the fixed stars of first magnitude (*ʿerekh*) or second magnitude, and in general to determine the variations in the apparent sizes of the planets. [7] For Venus we could determine this by its appearance with the Sun during the day, because when Venus is at its greatest elongation from the Sun, you can see it even in the afternoon sunlight. [8] But when it is closer than 20°, it can not be seen in sunlight. [9] You can also verify this by observing the size of the diameter of Venus at these two places and by observing it at 0° and 180° of anomaly. [10] Another way to verify this is by noting its rays that enter the window of the instrument that we described earlier, and this should be done when the light of the Moon is not shining and it is pitch dark (*ishon ḥoshekh*).

[11] Observations of Mars also do not agree with Ptolemy's model, for it

---

[1] *Lit.*: beginning of anomaly; here and henceforth: 0° of anomaly.

does not appear greater at 180° of anomaly than at 0° of anomaly in the way that it should according to Ptolemy's model. [12] Indeed, it appears to be not more than twice as large at 180° of anomaly than at 0° of anomaly. [13] But according to Ptolemy's model Mars should appear more than 6 times as large at 180° of anomaly than at 0° of anomaly. [14] You should also know that we found this matter somewhat complex for Mars, because when it was retrograde in Leo, we found its size perceptibly greater than that of Saturn, and similarly in Capricorn, but there its size appeared greater than it was in Leo. [15] However, when it was retrograde in Scorpio its size did not seem greater than that of Saturn. [16] This led us to remark on the complexity of this situation, but in any case the [greatest] size that we found did not even reach twice [its least size], as we have mentioned.

[17] For the other planets, this matter is still not clear. [18] It is impossible to determine this for Mercury, because in our geographical latitude (*aqlim*) it is only seen at greatest elongation from the Sun for a few days. [19] It is possible that the apparent smallness in its size when it approaches closer to the Sun is related to the domination (*memshelet*) of sunlight that makes its size smaller, for it is then only seen close to sunrise when the light at the horizon is strong, and all the stars seem smaller then. [20] For Saturn and Jupiter we have not yet succeeded in determining this on account of the small variation found in their sizes by observation, but in any case it is clear that from this point of view Saturn and Jupiter do not conform to what follows from Ptolemy's computation. [21] According to his model the diameter of Saturn should appear to be greater at 180° of anomaly than at 0° of anomaly by close to a fifth of its size; this is definitely not the case when Saturn is observed in such darkness that no sunlight rules (*yishlot*). [22] Indeed, when matters are so arranged, no perceptible difference is found between its apparent size at this place and its apparent size at 180° of anomaly. [23] Moreover, according to Ptolemy's model the diameter of Jupiter should appear to be greater at 180° of anomaly than at 0° of anomaly by close to a third of its size; this is definitely not the case when Jupiter is observed in such darkness that no sunlight rules at all. [24] Indeed, when matters are so arranged, no perceptible difference is found between its apparent size at this place and its apparent size at 180° of anomaly.

[25] You should know that we ascribed what we first discovered concerning the size of Mars on its retrograde arc in Leo to thin clouds through which it was seen at that time. [26] We did this because during its retrogradation in Scorpio its size was not found to be augmented, and because we did not see this increment in the proper order in which it should take place were it due to its closeness to us. [27] You cannot argue that after a short time it [Mars] can appear to be smaller in size than what follows from the appropriate ratio, for we observed its size while retrograde in Capricorn where we found it

somewhat augmented as compared to the size that we found for it in Leo. [28]   We also ascribed the absence of any increment in the size of Mars in Scorpio to the thickness of the vapors (*edim*) through which it was seen at that time. [29]   We then understood that this phenomenon took place because of the comet (*kokhav mezunnav*) that continued to appear for more than 3 months; that vapor came into being under Scorpio and its was drawn from there to somewhat below the north pole: there it burst into flame (*hitlahavut*) and it perished in Scorpio.

[30]   This is what strengthened what we had established at first, namely, that the increment in the apparent size of Mars [in Capricorn?] was due to thin clouds through which it was seen at that time. [31]   We need to investigate this further observationally, but in any event it is clear from what we have said that this does not agree with Ptolemy's model, and we shall expand on this latter, God willing.

CHAPTER 18

[1] It is clear that Ptolemy's models for some of the planets do not agree with their observed sizes, as we have mentioned. [2] Now we should investigate this matter further from the point of view of their observed motions. [3] If their motions in longitude and latitude do not agree with what follows from Ptolemy's models, it will also confirm from this point of view that the Ptolemy's models are inappropriate. [4] Nevertheless, they may lead us to the correct model for each planet such that the motions that follow from it [agree] with the observations.

[5] We decided not to put this investigation here because it would introduce unnecessary duplication—the observations mentioned here would have to be mentioned again inasmuch as we take them as the starting point in order to establish the proper planetary models. [6] Moreover, it is most difficult to introduce this investigation here, for the explanation required is quite long. [7] It would not be possible to use our observations of the five planets to refute Ptolemy's models, on the basis of their disagreement with that follows from his models, without first clarifying the mean position of the Sun. [8] Indeed, one could say that the cause of the discrepancy might have been due to our faulty determination of the Sun's mean position: for Venus and Mercury this would yield a faulty computation because the planet's mean position in longitude is equal to the mean position of the Sun; and for Saturn, Jupiter, and Mars, it would introduce an error in the motion in anomaly, for it is well known that for these three planets, the mean longitude plus the anomaly is equal to the mean longitude of the Sun.

[9] It is also possible to assign this discrepancy in the cases of Venus and Mercury to an error in our determination of their positions in anomaly or to an error in their apogees. [10] Similarly, for Saturn, Jupiter, and Mars, the discrepancy found by the observations could be assigned to an error in the position of the apogee or to an error in their motion in longitude. [11] Further, the discrepancy found by any of the observations could be assigned to an error in our determination of the position of the fixed star with which it was observed. [12] The computation with Ptolemy's models could

lead to a false result for any of these reasons even if these models were correct. [13] But to establish that the discrepancies found by observation in their positions are not the result of any of these causes necessitates a lengthy argument and a large number of observations. [14] Therefore, we shall postpone this investigation until we come to discuss the model appropriate for each of the planets.

CHAPTER 19

[1] It is appropriate here for us to indicate that we must take the true and indubitable observations of the sequence (*seder*) of the celestial motions as the foundation and starting point [for this science], and then we should investigate the model from which this sequence follows. [2] If you find that there is only one model from which this sequence can follow, then it is confirmed that the model is correct without any doubt. [3] If you find more than one model from which this sequence can follow, and also agreeing with the variation in the sizes of the planetary diameters, the investigation confirms that one of these models is correct to the exclusion of any other. [4] Therefore, we must investigate all the models from which [all] varieties (*sugim*) of apparent planetary motions may follow. We shall investigate the characteristics of each of these models, and they will be the mutually exclusive alternatives (*ḥelqey ha-soter*) in this investigation. [5] The model that is ascribed to a planet must produce all the characteristics of its apparent motions in longitude, latitude, and the variation in the apparent size of its diameter, if it does not, it may not be ascribed to that planet.

[6] For this reason, we present the results of our own observations and those of the ancients. [7] We will also present those results of the ancients that we will have not [been able to verify], insofar as there is no contradiction between them, because they may offer us some assistance in perfecting this investigation. [8] This is because the human life span is not sufficient for the observations required for this science. [9] Indeed, a much longer period must transpire before these motions are completed, e.g., the motion of the fixed stars, and the motions of Saturn and Jupiter. [10] We could not determine the planetary mean motions without the aid of quite ancient observations. [11] There is some error [*lit.*: approximation] in any observation and so the determination of the planet's period of rotation will be much closer to the truth if two observations separated by a long time interval are used, and this is quite clear. [12] Therefore, in this science you need the testimony of reliable observers, such as Hipparchus, Ptolemy, al-Battānī, and other careful ancient observers.

[13] Moreover, we have to mention here that the planets, other than the Sun, exhibit two kinds of motion. [14] The first is the motion in longitude that for the Sun and the other planets depends on the position in the ecliptic, but not for the Moon. [15] There is a place where the motion is fastest, and another place opposite it where the motion is slowest. [16] The second is the motion in anomaly, not found for the Sun, which produces direct and retrograde motion, except for the Moon [which has no retrogradation]. [17] This motion [in anomaly] can be seen with all its phenomena [*lit.*: kinds] anywhere in the ecliptic; this is something you can determine observationally in every rotation, and this rotation is completed in a period that is not long. [18] In some places on the ecliptic the correction [for this anomaly] will appear to be greater and at other places smaller, and when it [i.e., the mean longitude of the planet] approaches the place of minimum correction [i.e., the apogee of the deferent], the sum of the corrections for [the anomaly counted in] the direct sense and [the anomaly counted in] the retrograde sense is smaller than when it is far from it. [19] Further, as it approaches the mean place between the place of smallest correction [in longitude] and greatest correction [in longitude], the difference between the correction for [the anomaly counted in] the direct sense and the correction for [the anomaly counted in] the retrograde sense becomes greater. [20] This is the case for all five planets as follows: at intermediate distances which come after the place of smallest correction[1] [symmetric] about 0° of anomaly, the correction for [the anomaly counted in] the direct sense is greater than the corresponding correction for [the anomaly counted in] the retrograde sense, i.e., the correction for 30° of anomaly there is greater than the correction for 330° of anomaly. [21] But [symmetric] about 180° of anomaly one finds the opposite, namely, the correction for [the anomaly counted in] the retrograde sense there will be greater than the corresponding correction for [the anomaly counted in] the direct sense, e.g., the correction for 150° of anomaly there is smaller than the correction for 210° of anomaly. [22] The contrary is the case at intermediate distances which are before the place of smallest correction[2]; in all this there is no doubt except for someone who denies what is perceived with his senses, and this may be seen for most of the planets in a short period of time.

[23] We have determined these things by observation for all the planets except for Mercury, for we have not yet had the opportunity to make numerous observations of this planet. [24] Moreover, we have only observed Saturn and Jupiter on one side of the place of smallest correction [i.e.,

---

[1] I.e., from 0° to 180° in the argument of mean longitude counted from the apogee of the deferent.
[2] I.e., from 180° to 360° in the argument of mean longitude counted from the apogee of the deferent.

apogee]; we found agreement in this matter for this side and it may also be used as evidence for the other side that we did not see, as we shall explain later, God willing. [25] This is easily perceived for Mars on account of the excess between the correction for [the anomaly counted in] the direct sense and the correction for [the anomaly counted in] the retrograde sense, as we mentioned. [26] Further, we determined by our observations of Venus and Mars that their greatest corrections take place at more than 90° of anomaly and at less than 270° of anomaly by an amount about equal to the arc of [maximum] correction due to the motion in anomaly; thus, for Venus the greatest corrections are found at 132° of anomaly and at 228° of anomaly, approximately. [27] We found this by means of the observational instrument that we invented, and we noticed that when it was near this place, its apparent motion in one day is only what is properly assigned to the motion in longitude.

[28] We also reached this [conclusion] in another easier way, for we already found that the effect (*roshem*) of the correction for the motion in anomaly for Venus in the time from (?: *dorekh el*) evening setting to the time of morning setting is equal to the effect of the correction for the motion in anomaly in the time from morning setting to the time of evening setting [?]. [29] We found that the interval from greatest elongation in the evening [*lit.*: the beginning of evening setting] to greatest elongation in the morning [*lit.*: the beginning of morning setting] is very much shorter than the interval from greatest elongation in the morning [*lit.*: the beginning of morning setting] to greatest elongation in the evening [*lit.*: the beginning of evening setting]. [30] The shorter interval does not reach $\frac{2}{5}$ of the other interval, and you can determine this easily for yourself.

[31] For Mars, the results were similar when we observed it before and after it was at the phenomenon (*ᶜinyan*) Ptolemy called the edges (*qiṣwot*) of the night. [32] This is something that is not difficult to determine in one of these ways: it is apparent from observation that the interval from the time when Mars is 90° after the Sun to the time when it is 90° before the Sun is more than twice as long as the time that Mars takes from 90° before the Sun until 90° after the Sun. [33] By the expression "before the Sun," I mean that Mars is to the west [MSS: east] of the Sun, for in that case it rises before sunrise. [34] Similarly for Saturn and Jupiter, although the intervals for them do not differ by very much. [35] This may also be perceived somewhat for the Moon, and from our observations of Mercury it would seem that the same thing holds true for that planet too. [36] All of this is in agreement with the reports of the ancients and there is no difference of opinion in this matter at all.

[37] We should further present here another result of our observations which will be helpful to us in this investigation, but which was only partially

noticed by the ancients. [38] From our observations of the apparent size of the image of the Sun through the window of the instrument that we invented for this investigation we found that its diameter near Cancer 0° in our time is about 0;27,50°, but when it was near Capricorn 0°, its apparent diameter is greater and seems to be about 0;30°. [39] At intermediate distances we found the size of its diameter to lie between these amounts according to the ratio of the distance from these places, and this shows us conclusively that the Sun is fixed to a sphere eccentric to the center of the universe and that the apogee is near Cancer 0° for our time, and its perigee near Capricorn 0°. [40] We will also present here by way of an assumed principle, but later we shall prove it without any doubt, that the motion of the solar apogee is 1° in 43 Egyptian years of 365 days, and about 232 days, $6\frac{1}{2}$ hours, very nearly, and we shall demonstrate this in a later chapter from our own observations as well as those of Ptolemy and al-Battānī. [41] The same seems to hold true for Venus as well. [42] The motion of Saturn's apogee is 1° in about 44 years of $365\frac{1}{4}$ days; that of Jupiter 1° in about 60 years of $365\frac{1}{4}$ days; and that of Mars 1° in about 66 years.

[43] All these results will not be explained here but are presented as assumptions until later when we shall prove them, God willing, and they will be of some help to us in this investigation here. [44] But we decided to introduce these results based on our observations and those of others here so that we may use them as assumptions in our proofs, for in this science you have to rely on sensory perception to find the true models for the celestial orbs, and in this way it is similar to the Science of Physics (*ha-ṭiv'it*) in which proofs are taken *a posteriori* [*lit.*: from the later to the prior]. [45] Nevertheless, someone may argue that in this science the argument should progress as in the other mathematical sciences *a priori* [*lit.*: from the prior to the later]. [46] You will see from what will be proved later that the models (*temuna*) for the orbs and the stars [planets?] are [chosen] according to the motions that follow from them, i.e., they have to produce agreement with the [apparent] motions. [47] Therefore, the motions are prior to these models in the nature of things physically (*eṣel ha-ṭevaʿ*) in the same way that the visual faculty (*ha-koaḥ ha-ro'eh*) is prior to the form (*temuna*) of the eye in the nature of things physically, because the eye is for vision, not vision for the eye; this is self-evident and the Philosopher already mentioned it in his book, *On Animals* (*baʿaley ḥayim*). [48] Be that as it may, this will not be explained here.

CHAPTER 20

[1] After presenting the motions of the planets that are perceived and their apparent variation in size, we should motion all mutually contradictory possible [models] (*kol ḥelqey ha-soter she-efshar*) for the planetary motions. [2] We shall explain the special properties of each of the models, and they will give us preliminary indications for distinguishing them by observation. [3] Before this we must mention two principles that are required for this investigation. [4] The first principle is that the stars [and planets] are affixed to spheres and that they move with the motion of these spheres, the part moving with the whole. [5] The second principle is that the motion of the celestial bodies is necessarily uniform (*shaveh*) in itself, and that the apparent variation is related to the way we see it, not that the motion in itself varies. [6] Indeed it would be impossible for the celestial bodies to be sometimes weak and sometimes strong, as was explained in the Book of the Heavens and the World. [7] Moreover, we found a uniform and regular (*yashar*) order for their motions for long period of time in which a record of events has reached us, and this order is not disturbed (*nitbalbel*) except by what may be ascribed to observational approximation. [8] If the rules governing them are like those governing plants which have youth (*baḥrut*) and old age (*ziqna*) in every year, passing from youth to old age and from old age to youth, their order would be disturbed in a long period of time as is case for plants which in some years are hardy and in other years are weak. [9] In general this principle is quite clear to a careful reader of this book so that a lengthy explanation is superfluous.

[10] Once this has been established, we say that the apparent deviation from the mean motion in longitude may be imagined in one of two ways. [11] Either the center of the sphere on which the motion takes place is located at the center of the earth, and the motion takes place about another point not at its center; or [the center of the sphere] itself is eccentric to the center of the earth, and this may be understood in two ways. [12] Either the eccentric sphere encompasses the earth or it does not, in which case it is affixed to a sphere that encompasses the earth, and this is what Ptolemy called

the sphere of the epicycle (*galgal ha-haqafa*); and Ptolemy proved that these two models are equivalent. [13] They produce the same apparent motion for a planet, where the eccentricity is set equal to the radius of the epicycle, and the deferent is set equal to 60 in both cases. [14] Moreover, when the center of the sphere is put eccentric to the center of the earth, the motion of the sphere may take place either about its center or about another point, and we shall explain, God willing, the properties of each of these models.

[15] We say that [in the models where] the motion takes place on an eccentric sphere, it does not matter whether that sphere encompasses [the center of] the world or does not encompass it; in either case the maximum correction will correspond to an angle greater than 90° of mean motion, counted from the beginning of this motion at the apogee, by the amount of the maximum correction. [16] In other words, if the maximum correction is 11°, it will correspond to 101° from the apogee or 259° from it. [17] But in the model where the planet moves [uniformly] on a sphere whose center is the center of the world about an eccentric point, the maximum correction will take place at 90° of mean motion from the apogee, and similarly at 270° from it.

[18] To prove this, let the eccentric circle on which this motion takes place be circle ABG, and we shall assume that it encompasses the earth [see Fig. 20.1], for what is proved here also holds for the model where the motion takes place on a sphere that does not encompass the earth, as we have mentioned. [19] Let the center of this circle be point D, and the center of the earth point E. [20] The diameter that passes through both these centers is line $ADEG$; point A is the apogee and point G is the perigee. [21] We say that the greatest correction takes place at more than 90° from point A both forward and backward. [22] To find its amount, we draw lines EB and DB. First let the apparent motion, angle AEB, be 90°. [23] Then the mean motion, angle ADB, is greater than angle AEB by the amount of angle DBE, and angle DBE is the correction for this place. [24] I say that angle DBE is the greatest possible angle of correction in this model because line DE is the sine of that angle, and this follows from the assumptions that angle DEB is a right angle and line DB is 60°. [25] If angle AEB were acute (*ḥidda*) or obtuse (*nirḥevet*) the sine of the correction, angle DBE, would be smaller than line DE. [26] This is because the perpendicular from point D to line BE, extended if necessary, which is the sine of angle DBE, either falls between points E and B or outside that interval. [27] Be that as it may, line DE is longer than that perpendicular, because its square is equal to the sum of the squares of that perpendicular and the distance from the foot of the perpendicular to point E. [28] Since the sine of that angle is smaller than line DE, it follows that the angle is also smaller, assuming that the angle in question is less than 90° as is the case here. [29] Thus, it is clear that in this model the

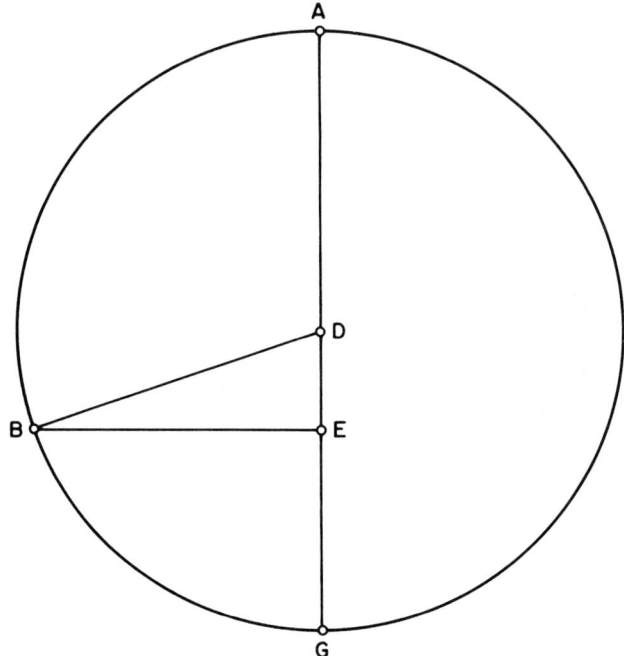

FIGURE 20.1. N 62a; P 30b; Q 21a.

maximum correction occurs when angle *DEB* is 90°, and the mean motion, arc *AB*, is greater than 90° by the amount of angle *DBE*.

[30] Ptolemy explained in the *Almagest* that this follows equally well for the epicyclic model where the radius of the epicycle is set equal to line *DE*, and the motion on the epicycle is set equal to the motion on the eccentric sphere that surrounds the earth [in the eccentric model]. [31] It is clear that the maximum correction in both models takes place when the planet is 90° from its apogee in its apparent motion. [32] Therefore, in both these models there is an apparent change in the size of the planet's diameter as it moves from one place to another in its orbit: the smallest size will be observed at apogee, and the largest size at perigee; and the variation in the apparent size of its diameter in these two places should agree with the ratio of line *AE* to line *EG* in the figure. [33] When the planet is at point *A*, it is seen at a distance equal to line *AE*, and when it is at point *G*, it is seen at a distance equal to *EG* which is shorter than line *AE* by twice the amount of *DE*.

[34] You can also imagine this for the motion on the epicyclic sphere where the sphere which surrounds the earth moves with the planet's mean motion, and [the planet] moves on the epicyclic sphere with a motion equal to

the motion on the sphere that surrounds the earth, i.e., such that the motion on the epicycle and the motion in longitude are completed in the same period. [35] The apparent motion of the planet on the epicycle is sometimes augmented and sometimes diminished with regard to its mean motion, as before. [36] The epicyclic model has a special property, as opposed to any model where the sphere surrounds the earth. [37] In the model where the sphere surrounds the earth we always see the same part of the planet's surface, and this is self-evident to the discerning reader. [38] But in the epicyclic model we do not always see the same part of the planet's surface; indeed, when the planet is truly at its apogee we see the planet's surface that lies on the inner side of the epicyclic sphere, and when it is truly at the perigee we see the planet's surface that lies on the outer side of the epicyclic sphere, and in between we see part of outer surface and part of the inner surface. [39] From this argument it is clear without a doubt that the Moon is not affixed to an epicyclic sphere, as Ptolemy believed. [40] This is because the shading (*ṣel*) seen on the body of the Moon truly belongs to it and does not depend on a mathematical proof, and it is will be made clear later, God willing, that we always see the same side of the surface of the Moon. [41] The proof is that we definitely always see part of the same side of the surface of the body of the Moon, and this would be impossible if the Moon lay on an epicyclic sphere.

[42] Another property of the epicyclic model as distinct from the eccentric model is that the motion of the planet is fastest when it is farthest from the earth, i.e., when the planet passes through the apogee of its epicycle in the direction of the motion in longitude, as Ptolemy assumed for the five planets. [43] But in the eccentric model you will invariably find that the motion is slowest at the farthest point from the earth.

[44] I say that if you put the center of the earth at the center of circle *ABG* in this figure, and let the mean motion take place about a point eccentric to the center of the earth, the maximum correction will take place at 90° of mean motion from the beginning of the motion in longitude. [45] To prove this we may refer to the previous figure, but point *E* is the center of the earth and the center of circle *ABG* as well [see Fig. 20.2]. [46] The mean motion takes place about point *D* and we draw lines *EB* and *DB* as well as *ADEG*. [47] Point *A* is the beginning of the mean motion, and it is clear, as shown before, that the maximum correction in this model takes place when angle *EDB* is a right angle, i.e., 90° before or after point *A*. [48] This is because *EB* is 60 in this model, for it is the radius of the circle *ABG*, and it follows from the previous argument that the maximum correction takes place when angle *EDB* is a right angle.

[49] Another property of this model as distinct from the previous models is that the apparent diameter of the planet always stays the same size without

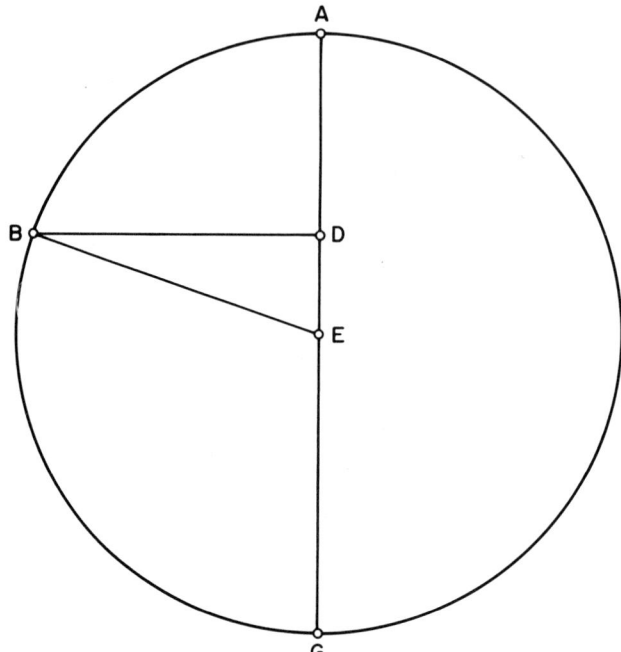

FIGURE 20.2. N 64a; P 31b; Q 21b.

any variation, for it is always seen at the same distance. [50]   Another property of this model as distinct from the previous models is that the amount of line *DE* is made equal in both models, the correction in this model will be greater than in previous models from 0° to 90° of mean motion ahead or behind the beginning of mean motion and in the remainder of the circle it will be the contrary. [51]   The maximum discrepancy will be equal to the arc whose sine is one half of line *DE*.

[52]   To prove this we consider *ABGD* as the circle on which the motion takes place in the second model whose center, point *E*, is also the center of the earth [see Fig. 20.3]. [53]   Let the mean motion take place about point *Z*, and draw line *AZEG*. [54]   Now let us draw, about point *Z*, circle *HBTD* equal to circle *ABGD* intersecting it at points *B* and *D*: line *HAZETG* will be a straight line. [55]   Circle *HBTD* will be the circle of the eccentric sphere whose center is at *Z*, and the mean motion takes place about point *Z* as in the previous model such that circles *ABGD* and *HBTD* illustrate these two models. [56]   We draw lines *ZB* and *EB*, and triangle *ZBE* is isosceles. [57]   We divide line *ZE* in half a point *L* and we draw line *BL*. [58]   It is clear that angle *ZLB* is a

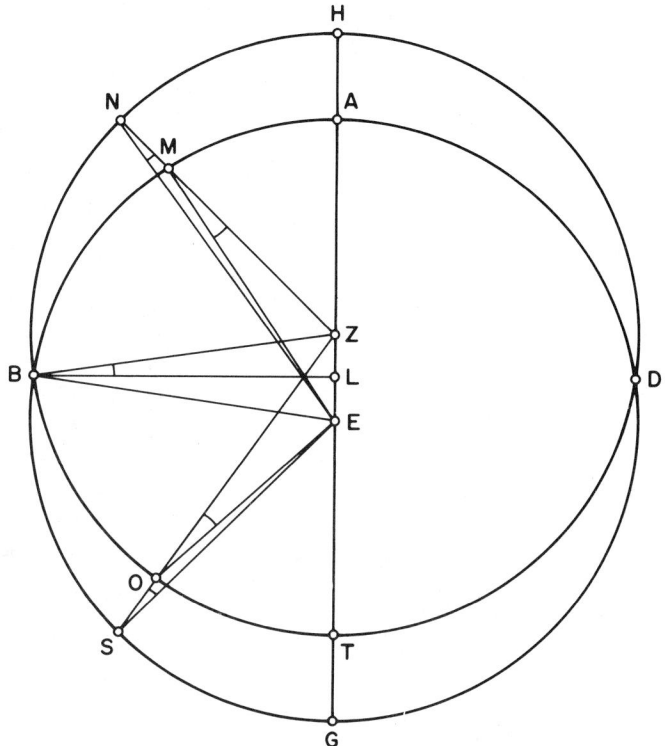

FIGURE 20.3. N 64b; P 32a; Q 21b.

right angle, and that angle *LBZ* is equal to the angle of the arc corresponding to the sine *ZL*. [59] It is clear that angle *AZB*, which is the angle of the mean motion in both models, is greater than 90° by the amount of angle *LBZ*, which is the amount of the angle of the arc corresponding to the sine, half of line *ZE*.

[60] I say that the correction will be greater for the model with circle *ABGD* than it is for the model with circle *HBTD* on both sides of the beginning of the motion until *B* and *D*, and contrariwise for the remainder of the circle. [61] To prove this, let us take an arbitrary point *M* on arc *AB* and extend line *ZM* to point *N* on arc *HB*, and draw lines *EM* and *EN*. [62] Angle *ZME* is the correction for the model with circle *ABGD* for this place, and angle *ZNE* is the correction for the model with circle *HBTD*. [63] But angle *ZME* is greater than angle *ZNE*, and similarly for all points until we reach *B* and *D* respectively on both sides from the beginning of the mean motion. [64] Now we say that the opposite is true for the remainder of

the circle. [65] Let us take an arbitrary point S on arc BG, and draw line ZS intersecting arc BT at point O. [66] We draw lines EO and ES, and we find that angle ZSE, the correction for this place for the model with circle ABGD, is smaller than angle ZOE, the correction for this place for the model with circle HBTD.

[67] It remains for us to consider the models where the mean motion takes place on a sphere whose center is eccentric to the center of the universe, and the motion does not take place about its center. [68] These are the only possibilities: either the mean motion takes place about the center of the earth, or about a point such that the center of the sphere is between it and the center of the earth, or about a point placed between the center of the sphere and the center of the earth, or about a point such that the center of the earth is between it and the center of the sphere, or about a point not collinear with the line through the center of the earth and the center of that sphere. [69] That this exhausts all possibilities is clear from the following argument. [70] In the first instance, it is not excluded that the motion take place either about the center of the earth or about some other point. [71] If it is put at some other point, it may either be on the line from the center of the earth to the center of the sphere or at a point not on the line from the center of the earth to the center of the sphere. [72] If it is put on that straight line, the motion may take place either about a point such that the center of the sphere lies between it and the center of the earth, or about a point that lies between the center of the sphere and the center of the earth, or about a point such that the center of the earth lies between it and the center of the sphere. [73] This set of possibilities is exhaustive (*hekhrahit*) because there is no fourth way to put these three points on one line.

[74] A special property of the model in which the mean motion takes place about the center of the earth is that the motion is observed to the uniform, and the size of the planet appears to vary, i.e., when it is farthest away its size appears smaller than when it is closest; and this is self-evident. [75] Ptolemy set the lunar motion in longitude in this way, but it is clear from what we have said that this fails to conform with the apparent size of the lunar diameter at the various places of the motion in longitude. [76] Indeed, the apparent size of the lunar diameter does not vary in accordance with what follows from this model. [77] According to his model the apparent diameter at opposition should be smaller than its amount at quadrature by about a third. [78] But we found, by observation, that the apparent diameter of the Moon at quadrature is not greater than at opposition except for a little bit, as we have mentioned above.

[79] A special property of the model in which the center of the sphere lies between the center of mean motion and the center of the earth is that the correction at 90° of mean motion is equal to the correction at 90° of apparent

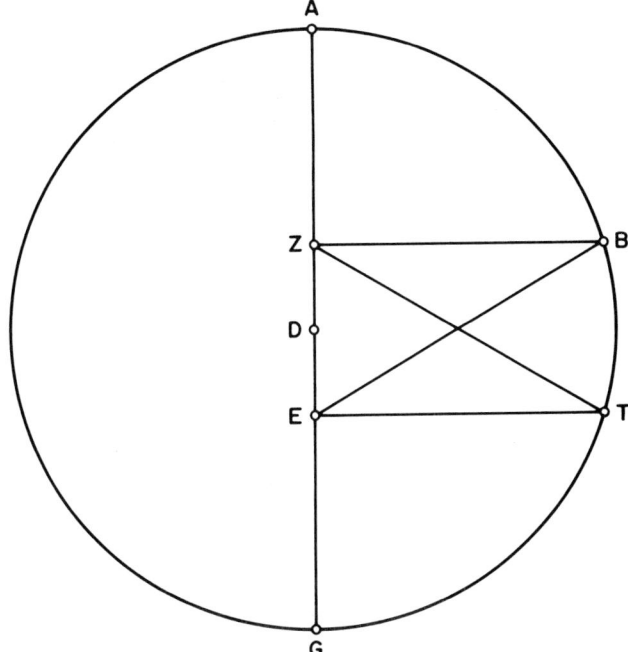

FIGURE 20.4. N 65b; P 32a; Q 22a.

motion. [80] This occurs when the center of the sphere bisects the distance between the center of the mean motion and the center of the earth.

[81] To prove this, let us draw circle *ABG* whose center *D* is eccentric to point *E*, the center of the universe, and let point *Z* be the equant [see Fig. 20.4]. [82] Draw line *AZDEG* and lines *ZB* and *EB*. [83] Assume that angle *AZB* is 90°, and it is the angle of mean motion, and the correction is angle *ZBE*. [84] Then we make angle *AET* a right angle, and it represents the apparent motion. [85] We draw line *ZT*, and it is clear that angle *ZTE* is equal to the correction for this place. [86] I say that angle *ZTE* is equal to angle *ZBE*. [87] Line *ET* is equal to line *ZB* because they lie at the same distance from the center of the circle, line *EZ* is common to both triangles, and angle *BZE* is equal to angle *ZET* because they are both right angles. [88] It follows that line *ZT* is equal to line *EB*, for the triangles are congruent. [89] Moreover, angle *ZBE* is equal to angle *ZTE*, and this is what we sought.

[90] With a little inspection, it is clear from this figure that if line *DE* were longer than line *DZ*, angle *ZTE* would be greater than angle *ZBE*, i.e., that

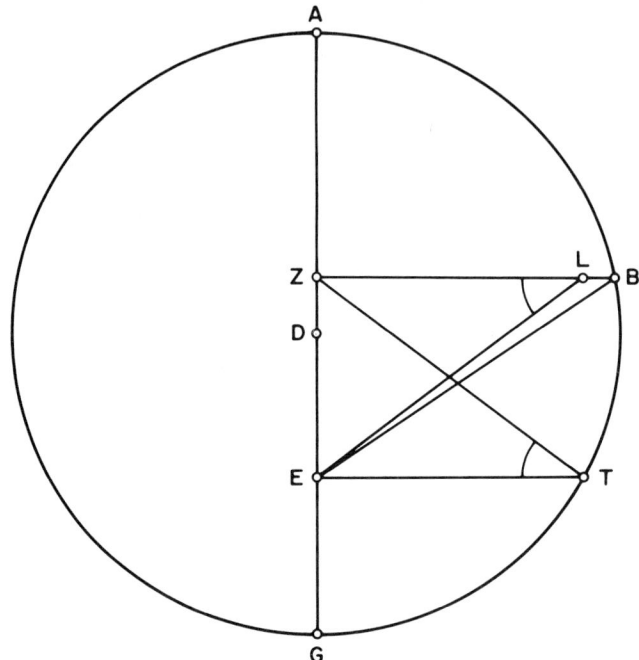

FIGURE 20.5. N 66a; P 32b; Q 22b.

the correction would be greater for 90° of apparent motion than for 90° of mean motion, and vice versa if line *ZD* were longer than line *DE*. [91] To prove this, let us first assume that line *DE* in the previous figure is longer, from which it follows that line *ET* is shorter than line *ZB* [see Fig. 20.5]. [92] On line *ZB* we mark off line *ZL* equal to line *ET*, and we draw line *LE*. [93] It is clear that, as before, angle *ZLE* is equal to angle *ZTE*. [94] Moreover, angle *ZLE* is greater than angle *ZBE*, from which it follows that angle *ZTE* is greater than angle *ZBE*. [95] Similarly, it is clear that if line *ZD* is made longer than *DE*, angle *ZBE* would be greater than angle *ZTE* [see Fig. 20.6], and this what we wished to explain.

[96] Another property that distinguishes this model from the preceding one, in which the mean motion takes place about the center of the eccentric sphere, is that even when the maximum correction in this model is set equal to the maximum correction in that model, the apparent variation in the size of the planet in this model is less than the apparent variation in the preceding model. [97] This is because the distance between the center of the earth and the center of the sphere is smaller in this model than what it was in that

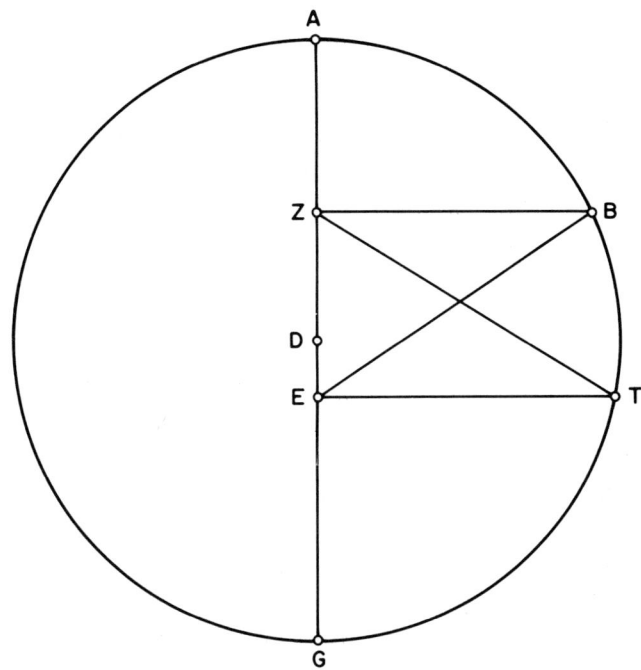

FIGURE 20.6. This figure does not appear in the MSS.

model, and this is self-evident to the careful reader. [98] Similarly, the variation in the distance of the planet from us will be smaller in this model than the variation in that model.

[99] A model in which the correction (*hilluf*) is greater for 90° of mean motion than the correction for 90° of apparent motion is one in which the center of the universe is identical with the center of the sphere and the mean motion does not take place about the center of the universe. [100] But in this model, the excess in the correction between 90° of mean motion and 90° of apparent motion is larger than the corresponding excess in the [previous] model [see Fig. 20.6]. [101] When the distance between the centers in that model is set equal to the distance between points Z and E in this figure, the excess of one line, the radius, over the second line in that model is much greater than the excess of one line over the second line in this model. [102] Clearly this may be explained by an argument similar to the preceding one, i.e., on the radius we mark off a line equal to the second line.

[103] To illustrate this, we draw a figure in which Z is the point about which mean motion takes place, and point E is the center of the earth and the

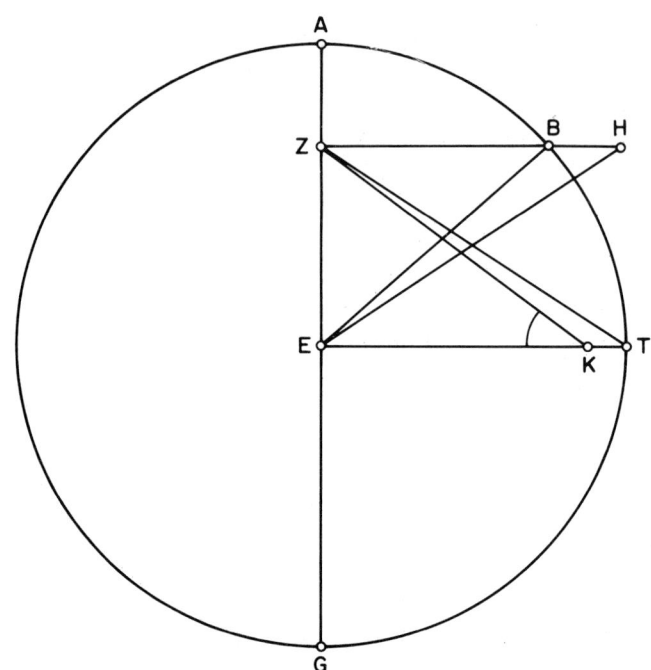

FIGURE 20.7. N 67a; P 33a; Q 22b.

center of the sphere [see Fig. 20.7]. [104] Angle *ZBE* is the angle of correction when the planet is at 90° of mean motion, and angle *ZTE* is the angle of correction when the planet is at 90° of apparent motion. [105] Let circle *ABG* have the same size as before. [106] Since line *ET* in this figure is the radius, it is necessarily greater than the amount of *ET* in the previous figure, for it was distant from the center of the circle by the amount of *DE*, and line *ZB* is smaller than line *ZB* in the previous figure. [107] It follows that angle *ZBE* in the second figure is greater than angle *ZBE* in the previous figure. [108] This will be clear when we extend line *ZB* to *H* such that line *ZH* is equal to line *ZB* in the other figure, and we draw line *EH*. [109] Then the angle *ZHE* is equal to angle *ZBE* in the previous figure, and it is smaller than angle *ZBE* in this figure.

[110] A similar argument shows that angle *ZTE* in this figure is smaller than angle *ZTE* in the previous figure: let us mark off line *EK* on line *ET* equal to line *ET* in the previous figure, and draw line *ZK*. [111] Then angle *ZKE* is equal to angle *ZTE* in the previous figure and greater than angle *ZTE* in this figure. [112] Therefore, the excess of angle *ZBE* over angle *ZTE* in

the second figure will be much greater than the excess of the angle *ZBE* over angle *ZTE* in the previous figure, and this is a special property of this model with which we are now concerned.

[113] Similarly, a model with the property that the correction (*ḥilluf*) for 90° of apparent motion is greater than the correction for 90° of mean motion is one in which the mean motion takes place about the center of the sphere eccentric to the center of the universe. [114] In this model, the excess of the correction between 90° of apparent motion and 90° of mean motion will not be greater than the excess of the correction between these places in the model [to be considered next]. [115] These models may also be distinguished from one another by the property that when the maximum corrections are set equal in them, the body of the planet in the [next] model has a greater variation in apparent size, comparing it at apogee and perigee, than its apparent variation in this model, for the variation in its distance from the earth will be greater in that model, and this will be quite clear to the careful reader.

[116] In the next model let us assume that the center of mean motion lies between the center of the sphere and the center of the earth. [117] I say that a special property of this model, as distinct from the previous models with eccentric spheres, is that if the maximum correction in this model is set equal to the maximum correction in those models, the excess of the apparent size of the planet at perigee over its amount at apogee is greater than it was in any of the previous models, for the excess of the apogee over the perigee is greater in this model, and this is self-evident to the careful reader. [118] I say that in this model the excess of the correction for 90° of apparent motion over the correction for 90° of mean motion is greater than that excess in any of the previous models. [119] Of the previous models the greatest excess was found in the model in which the sphere was eccentric and the mean motion took place about the center of the sphere. [120] We shall explain that this model makes this excess greater than it was in that model, where the maximum correction is the same for both models.

[121] To prove this, we first illustrate the model in which the sphere is eccentric and the mean motion takes place about the center of the sphere. [122] Let the eccentric sphere on which the motion takes place be represented by circle *ABGD* whose center is at *E* and the center of the earth is at *Z* [see Fig. 20.8]. [123] We draw line *AEZG*, and lines *EB* and *ZH* perpendicular to diameter *AG* reaching the circumference of the circle on the same side; then we draw lines *EH* and *ZB*. [124] Angle *EHZ* is the angle of correction that you find when the planet is at 90° of apparent motion and angle *EBZ* is the angle of correction that you find for the planet when it is at 90° of mean motion. [125] Let us mark off line *ET* on line *EB* equal to line *ZH*, and draw line *ZT*. [126] It is clear that angle *ETZ* is equal to angle *EHZ*, because angle *TEZ* is equal to angle *HZE*, and the sides adjacent to angle

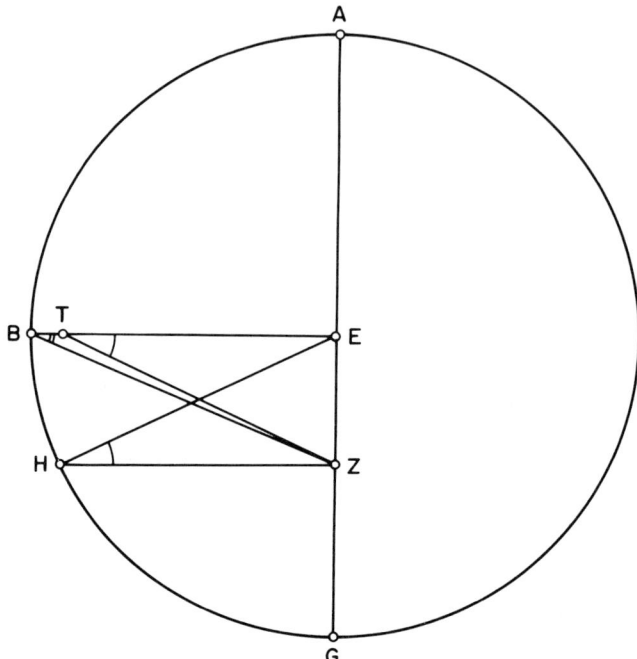

FIGURE 20.8. N 68a; P 33a; Q 23a.

TEZ are equal to the sides adjacent to angle *HZE*, i.e., since the angles and the including sides are respectively equal, the other sides and angles are respectively equal. [127] Therefore, the excess of the correction for 90° of apparent motion over the correction for 90° of mean motion is equal to angle *TZB*, which is the excess of angle *ETZ* over angle *EBZ*.

[128] Now let us consider circle *ABG*, whose center is *E*, as the circle of the eccentric sphere, and the center of the earth at *Z* [see Fig. 20.9]. [129] Let us put point *D*, the center of mean motion for the planet between *E* and *Z*. [130] Draw line *AEDZG*, and then lines *DB* and *ZT* perpendicular to line *AEDZG* reaching the circumference of the circle on the same side. [131] Draw lines *DT* and *ZB*: it is clear, as before, that angle *DTZ* is the angle of correction for 90° of apparent motion and angle *DBZ* is the correction for 90° of mean motion. [132] Let us consider angle *DTZ* in this figure equal to angle *EHZ* in the previous figure [Fig. 20.8] so that the maximum correction is the same amount in both models. [133] Mark off an amount equal to line *DT* on line *DB*; this is possible because line *DB* is longer than line *DT* since *DT* is closer than *DB* to line *DG*, which lies on diameter.

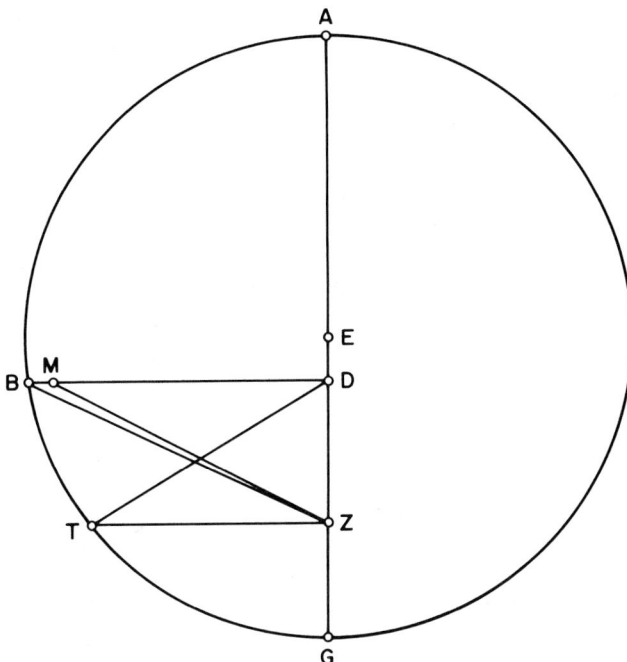

FIGURE 20.9. N 68a; P 33b; Q 23a.

[134] Let the line equal to *DT* be *DM*, and draw line *MZ*: I say that angle *DBZ* in this figure is smaller than angle *EBZ* in the previous figure.

[135] The proof is that since angles *DTZ* and *EHZ* are equal and angles *DZT* and *EZH* are also equal, triangles *DZT* and *EZH* are similar (*mitdamim*). [136] Therefore, the ratio of line *DT* to line *DZ* is equal to the ratio of line *EH* to line *EZ*: since line *DM* is equal to line *DT*, and line *EB* is equal to line *EH*, the ratio of line *DM* to line *DZ* is equal to the ratio of line *EB* to line *EZ*. [137] Since the angle included by *DM* and *DZ* is equal to the angle included by *EB* and *EZ*, for they are both right angles, triangles *DMZ* and *EBZ* are similar, and therefore angle *DMZ* is equal to angle *EBZ*. [138] Since angle *DBZ* is smaller than angle *DMZ*, angle *DBZ* is smaller than angle *EBZ*. [139] Since angles *DTZ* and *EHZ* are equal, the excess of angle *DTZ* over angle *DBZ* is greater than the excess of angle *EHZ* over angle *EBZ*; and this is what we wished to explain.

[140] This model has another special property not found in any of the preceding models: the correction for angles greater than 90° of apparent motion is greater than that for 90° of apparent motion. [141] To prove this we

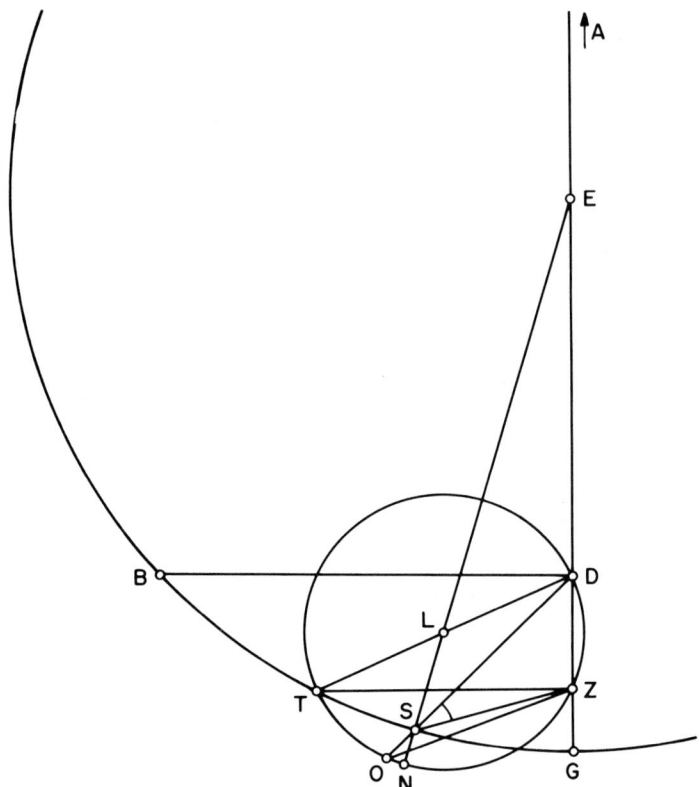

FIGURE 20.10: all MSS display the entire circle. N 68b: *M* is added at the intersection of the small circle and line *DB*, line *ZLM* and line *ZB* are also drawn; P 33b: angle *EZT* is acute; Q 23b.

must consider the previous figure again, and we will explain that there is an angle greater than 90° of apparent motion for which the correction is greater than angle *DTZ*. [142] The proof is as follows: divide line *DT* is half at point *L*, and draw circle *DTNZ* about *L* to circumscribe triangle *DTZ* [see Fig. 20.10]. [143] Draw line *ELN* and call its intersection with circle *ABG* point *S*: it follows that *LS* is the shortest line of those from *L* to the circumference of circle *ABG*, for any *L* not at the center of that circle. [144] Thus *LS* is shorter than *LT* which is equal to *LN*. [145] Draw line *DSO* intersecting circle *DT* at point *O*, and then draw lines *ZS* and *ZO*. [146] It is clear that angle *DSZ* is the angle of correction for this place, and it is greater than angle *DOZ*, which is equal to angle *DTZ*, and this is what we sought to demonstrate.

[147] A model with an eccentric sphere such that the center of the earth is placed between the center of the sphere and the equant point has the special property, as distinct from the previous models, that the swiftest motion takes place when the apparent size of the planet is smallest, and its slowest (*iḥḥur*) motion when it is largest, and this is self-evident. [148] Moreover, if the beginning of this motion is put at the apogee, which lies on the diameter that passes through these centers, the maximum correction corresponds to 90° of mean motion or more than 90° of apparent motion, [and the excess is] equal to the maximum correction; and this is clear with a little reflection. [149] This correction is to be added to the mean motion, whereas in the previous models it was to be subtracted. [150] But if we began the motion from the other end of the diameter, i.e., at the perigee, this model would have the special property, as distinct from the rest of the eccentric models, that the maximum correction corresponds of 90° of mean motion or less than 90° of apparent motion [such that the excess between them] is the maximum correction, and this is clear when taken together with our previous remarks.

[151] A model in which the equant point is not put on the diameter that passes through the center of the sphere and the center of the earth has the distinctive property that the maximum and minimum velocities are not to be found at apogee and perigee. [152] Rather, they are found at the ends of the diameter [*sic*] that passes through the center of the earth and the equant, and this is clear from our previous remarks. [153] From Ptolemy's discussion of Mercury it seems that this model might be appropriate for it, for Ptolemy indicated that the maximum and minimum velocities do not take place at least and greatest distances. [154] Furthermore, he placed the equant in different positions (*maṣavim*) with respect to a point on the diameter that passes through the least and greatest distances, and each of [these positions] with respect to that point on the diameter has special properties, but we will not treat them at length for they are self-evident from our previous remarks.

[155] These are all the possibilities (*ḥelqey ha-soter asher efshar*) that may be imagined to account for the apparent variation in the motion in longitude when the motion in longitude is considered simply. [156] We have merely mentioned the well known epicyclic model which is composed of two motions, because most of its properties have been exhibited for the eccentric model where the mean motion takes place about the center of the sphere, and we did not wish to repeat our remarks.

Commentary

Chapter 1

In this chapter Levi indicates that his goal is to present an astronomical theory that is satisfactory both from a mathematical and a philosophical point of view. He remarks that this had not been achieved in the past, but he does not criticize any of his predecessors by name. In fact, there are very few authors cited anywhere in his astronomical writings, and none of them was his contemporary. We are told that to complete this task in astronomy, the investigator must be skilled in both mathematics and natural philosophy because this task cannot be split up (§§ 14–17). Levi's views are similar to those of Ibn al-Haytham (see Introduction), and I am inclined to believe that Levi knew of them: for another example of similarities between the views of Levi and Ibn al-Haytham, see the commentary to chapter 2.

Chapter 2

In this chapter Levi presents apologetic arguments for the study of astronomy based on its nobility (§ 2), its usefulness for the study of other sciences (§ 3), and for political philosophy (§ 4). The reference to political philosophy is not found in the *Almagest*, Book I, as Levi suggests. Rather, Ptolemy mentions the usefulness of astronomy with regard to "virtuous conduct in practical actions and character" (*Almagest*, I, 1; trans. Toomer, 1984, p. 37). In § 6 he indicates that the creation of the stars and their orbs will be discussed in a later passage: cf. *Wars of the Lord*, VI:2.8; trans. Staub, 1982, especially pp. 261, 273; and Touati, 1973, pp. 277–81. He continues by citing biblical verses (§§ 9,16) in support of the value of astronomy for the understanding of God and His creation. An allusion to astrology may be found in § 10: "God causes diverse actions to emanate from them [the stars] in this sublunary world" (cf.

Wars of the Lord, VI:2.8; trans. Staub, 1982, p. 268). There was considerable debate in the medieval Jewish community concerning the value and legitimacy of astrology: Maimonides was opposed to it, while Levi approved of it. Nevertheless, astrology is not otherwise mentioned in Levi's *Astronomy*, as far as I am aware. For the views of Maimonides on astrology, see Twersky, 1972, pp. 463–73; for those of Levi expressed in *Wars*, Book V, Part 2, see Goldstein, 1976, p. 222.

The only physical consideration in this chapter is that the Moon is opaque and acquires its light from elsewhere (§ 19). In chapter 54 Levi discusses whether the planets acquire their light from the Sun, and argues that only the Moon receives sunlight. If, for example, Venus received sunlight, it should appear in a variety of crescent shapes. But Levi is certain that it never appears crescent-shaped, and so Venus must be self-luminous (P 103b–104a). For Levi's statement that Mercury is self-luminous, see Goldstein, 1969b, p. 54. A similar argument for the self-luminosity of the planets is given by Ibn al-Haytham (ᶜArafat and Winter, 1971).

Chapter 3

Levi is correct in remarking on the state of astronomy in the fourteenth century (§§ 1–3). Although some observations were made by his predecessors, they did not use them to modify Ptolemy's models (cf. Goldstein, 1974a, p. 53). Al-Battānī (*fl.* ca. 900) wrote a standard treatise on Ptolemaic astronomy that was widely used in the Middle Ages by authors who wrote in Arabic, Latin, and Hebrew (cf. Hartner, 1970). On Levi's use of al-Battānī's tables, see Goldstein, 1974a, pp. 28, and 84–85 (nn. 1, 40). On Levi's own observations, see Goldstein, 1979.

In §§ 4–12 Levi is alluding to his own observations where he found discrepancies with what followed from Ptolemy's models (even after allowing for changes in the mean motions): cf. Goldstein, 1974a, p. 28. His modification of Ptolemy's lunar theory based on his own observations is described in Goldstein, 1974a, pp. 53–74. In § 12 Levi notes the long period required to correct certain parameters such as precession which he treats in chapter 61 of his *Astronomy* (see Goldstein, 1975). In §6 the expression "motion of the center" refers to the mean motion in longitude counted from the apsidal line (cf. Pedersen, 1974, p. 280; and Nallino, 1907, vol. 2:335), and in § 7 the expression "inclination of the diameter of the motion in anomaly" refers to the double elongation in the lunar theory (cf. *Almagest*, V, 5; Nallino, 1907, vol. 2:328; and Goldstein, 1974a, p. 54).

In §§ 13–16 Levi introduces some of the problems of observation, and notes that the best kind, for which no instrument is needed (a true conjunction), occurs very rarely. Since true conjunction in longitude is not a rare phenomenon, I take him to be referring to conjunction in both longitude and latitude. In chapter 122 (P 236a) Levi mentions his observation in Avignon on 28 September 1339, about 2 hours before sunrise, of Venus occulting (*histir*) Jupiter "to the appearance of the eye." According to Tuckerman (1964, p. 687) at that time Venus and Jupiter both had longitude 157.8° and latitude 1.0°, rounded to the nearest tenth of a degree; this computation is not sufficiently precise to tell whether this truly was an occultation, or merely a close approach of the two planets to one another at true conjunction. Levi might also be referring to a transit of Venus or Mercury: for his views on this subject see Goldstein, 1969. In § 14 he is alluding to atmospheric refraction as a possible source of observational error, not generally taken into account by medieval astronomers.

Ptolemy's armillary sphere, described in the *Almagest* V, 1 (here called "the instrument of the rings"), is indeed difficult to construct (§§ 17–28: cf. Rome, 1931, pp. 4–5). Levi discusses the problems of the astrolabe more fully in chapter 12, below.

Quite reasonably Levi argues that the positions of the fixed stars should be determined accurately before embarking on correcting the planetary theory, although this is not the procedure used by other medieval astronomers (§§ 26–25). There were many treatises on the astrolabe written in the Middle Ages and Levi does not give sufficient information in § 29 to identify this text. In §§ 30 ff. Levi again emphasizes the sources of error (or "approximation") in making observations, a sensitivity shared by few other medieval astronomers and not stated as sharply by any of them. The alidade with the two pierced plates (in § 33) is a standard part of the astrolabe. However, Levi does not use the standard Hebrew term for alidade (*ma^{c}ṣad*) which may have confused some of his readers (cf. Goldstein, 1977, p. 109; and chap. 12:2).

Here (§ 36) Levi draws attention to the difficulty of finely graduating measuring devices. His elegant solution is given in chapter 12.

The procedure generally used to observe a star or a planet is to find the altitude and then to convert coordinates to longitude and latitude (§§ 37–38). Depending on the circumstances a small observational error combined with an approximation in the conversion procedure can lead to a large error in the longitude of the star. Levi does not give the details here, but in chapter 62, section 8, he tells us how to find the time of day or night from an observed solar or stellar altitude. His procedure is equivalent to the modern formula:

$$t = \text{arc Vers} (\text{Vers } d - [\text{Sin } h \cdot \text{Vers } d]/\text{Sin } h_m]) \qquad (1)$$

where t is the time before or after noon/midnight, d is the arc of daylight/night, h is the observed altitude, and h_m is the altitude at meridian crossing of the observed body on that day (cf. Kennedy, 1973, p. 113; and Goldstein, 1979, p. 127). For the definitions of capitalized trigonometric functions, see the commentary to chapter 4, section 2.

Levi then remarks on the circularity involved in finding stellar longitudes according to accepted procedures (§§ 39–51). Stellar declinations are affected by precession and Levi found that Ptolemy's value for precession failed to account for al-Battānī's and his own observations (cf. Goldstein, 1975). The instrument that Levi invented is generally called the Jacob Staff and it is described in chapter 7, below (cf. Goldstein, 1974a, pp. 21–23). In § 48 and § 52 Levi uses a term here translated "mechanical analogies" which is also found in his discussion of his lunar model (Goldstein, 1974a, p. 67), and in chapter 4:116. The term does not seem to occur in any other medieval text: the closest to it is *mofet tahbuli* in the extant Hebrew version of Averroes's *Epitome of the Almagest* (MS Oxford, Opp. Add. fol. 17 [Neubauer 2011], 122b:7–8: the original Arabic is lost) where it refers to Ptolemy's iterative approximation procedure for finding the eccentricity and apsidal line in the equant model for an outer planet "combining practice (*maʿaseh*) and theory (*hokhma*)"; cf. Neugebauer, 1975, pp. 172ff. (This passage was brought to my attention by Mme J. Lay who is preparing a French translation of Book I of Averroes's text.) In Levi's *Astronomy* the term "mechanical analogy" (*heqqesh tahbuli*) is later used in chapter 46 in the context of preliminary remarks on planetary theory (P 88b:30), in chapter 47 in reference to Ptolemy's method for finding the eccentricity and apsidal line for each of the outer planets (P 93a:3,7), and again in chapter 49 in further remarks on planetary theory: "this belongs to the category of mechanical analogies based on trial (*nisayon*) and reexamination (*hippus*), approximating the correct value until the truth is reached" (P 94b:8–9). Thus, in some contexts "mechanical analogy" clearly refers to Ptolemy's method, but in other contexts (notably in lunar theory) it refers to some other iterative approximation procedure, the details of which are not specified.

In § 51 Levi reaffirms one of his goals which is to account for the observations (both his own and those of his predecessors) by means of geometric models. In this context 'star' may also refer to a 'planet'. In chapter 20, we see the range of models Levi was willing to consider. The discussion of trigonometry (§ 57) is to be found in chapter 4.

CHAPTER 4

There is no claim to originality in this chapter; it is only presented for the benefit of those who are not familiar with the elements of trigonometry. In

Commentary: Chapter 4

section 1 Levi presents some basic definitions. In section 2 he presents a set of theorems from which he will be able to compute his table of sines. This procedure is similar to, but distinct from, Ptolemy's derivation of his table of chords (for Ptolemy's procedure in the *Almagest* see, for example, Pedersen, 1974, pp. 56–65). A translation of the Latin version of this section may be found in Espenshade, 1967. Section 3 includes the derivation of the entries in the sine table based on the theorems in section 2. Section 4 describes the table and the rules for interpolation in it. Finally, section 5 gives the rules for solving plane triangles; note that there is no discussion here of spherical trigonometry.

Where numerical values are given in the text, I have recalculated them by the methods presented in the text, but in those cases where no method is stated I have used procedures that may be assumed to have been available to the author. Where I have found discrepancies I have so noted by adding a parenthetical remark in the commentary in which the recalculated value is preceded by the abbreviation *acc.*, meaning, "accurately computed the value is ..."

SECTION 1. In medieval versions of Euclid's *Elements*, it often happens that the numbering of propositions differs from that adopted by Heiberg in his edition of the Greek text (§ 11: cf. Heath, 1956, vol. 1, pp. 79ff).

SECTION 2. In § 14 we are told that

$$\text{Crd } x = \text{Crd } (360° - x)$$

Hence

$$\tfrac{1}{2}\text{Crd } x = \tfrac{1}{2}\text{Crd } (360° - x)$$

or

$$\text{Sin } (x/2) = \text{Sin } (180° - x/2)$$

In general, I shall use lowercase letters to indicate trigonometric functions with radius 1, and an initial capital letter to indicate trigonometric functions with radius 60, e.g., Sin x = 60 sin x. For the theorem in § 17, see also Euclid, VI.8.

Since triangles ADB and ABG in Fig. 4.1 are similar (§§ 24–26), it follows that

$$\frac{AD}{AB} = \frac{AB}{AG}$$

or

$$AB^2 = AD \cdot AG$$

The author indicated that the purpose of this theorem (§ 27) is to enable him to compute chords and versines from one another:

$$\text{Crd}^2 x = 120 \text{ Vers} x$$

Note that: $\text{Vers} x = 60 - \text{Cos} x$.

We are now informed (§ 33) that this theorem allows us to compute the sine of the arc as well. The notes in MS N do not add any useful information on the text, but the identity of the author of these notes is of interest for determining the readership of Levi's treatise. The notes are introduced by the abbreviation AMP (or AMF) which I interpret to mean: *amar Mordecai Finzi* (i.e., Mordecai Finzi said). For other occurrences of this abbreviation, see MS Bodleian, Lyell, heb. 96, fol. 9a, *et passim*, where Finzi is clearly intended (cf. fol. 8a where his name is written out in full). Thus we can say that Finzi (fifteenth century, Mantua) read at least part of Levi's astronomical text. It is also plausible to assume that the anonymous notes is MS N are due to him. On Finzi, see Steinschneider, 1964, pp. 193–94; and Goldstein, 1985a, 1985b.

Since right triangles *ADB* and *BDG* in Fig. 4.1 are similar (§§ 36–43), it follows that

$$\frac{AD}{BD} = \frac{BD}{GD}$$

or

$$BD^2 = AD \cdot GD$$

i.e., *BD* is the mean proportional between *AD* and *GD* (cf. Euclid, VI. 13). This theorem enables us to compute *BD*, the sine of arc *AB*, from *AD*, the versine:

$$\text{Sin}^2 x = \text{Vers} x \, (2R - \text{Vers} x)$$

where *R* is the radius of the circle.

The theorem discussed in §§ 47–56 states that

$$R = \text{Vers} x + \text{Sin} (90° - x)$$

The theorem discussed in §§ 57–63 states that if an arc, *x*, is greater than 90°.

$$\text{Vers} x = \text{Sin} 90° + \text{Sin} (x - 90°)$$

The theorem discussed in §§ 64–80 states that given the sines and versines of two arcs *x*, *y*, where *x* > *y*, it follows that

$$\text{Crd}^2 (x + y) = (\text{Sin} x + \text{Sin} y)^2 + (\text{Vers} x - \text{Vers} y)^2$$

and

$$\text{Crd}^2 (x - y) = (\text{Sin} x - \text{Sin} y)^2 + (\text{Vers} x - \text{Vers} y)^2$$

The theorem discussed in §§ 81–91 relates the sine and versine of an arc: when $x \leq 90°$

$$R^2 = \text{Sin}^2 x + (R - \text{Vers}\, x)^2$$

and when $x > 90°$

$$R^2 = \text{Sin}^2 x + (\text{Vers}\, x - R)^2$$

The theorem discussed in §§ 92–93 states that

$$\text{Sin}\, x = (\text{Crd}\, 2x)/2$$

The term *metar* is often used for sine as well as chord.

SECTION 3. If, for all $x < 45°$, Sin x and Vers x are given, we can compute the sines and versines of all arcs between $0°$ and $180°$ (§§ 95–101):

1. $\text{Sin}(90° - x) = R - \text{Vers}\, x$
2. $\text{Vers}(90° - x) = R - \text{Sin}\, x$
3. $\text{Sin}(90° + x) = \text{Sin}(90° - x)$
4. $\text{Vers}(90° + x) = 2R - \text{Vers}(90° - x)$
5. $\text{Sin}(180° - x) = \text{Sin}\, x$
6. $\text{Vers}(180° - x) = 2R - \text{Vers}\, x$

Levi then begins the computation of the entries in the table of sines (§§ 102–12). The sines and versines of 90°, 30°, and 18° can be found from theorems in Euclid. Thus, when the radius is set equal to 60,

Sin 90° = 60
Vers 90° = 60
Sin 30° = 30
Vers 30° = 8;2,18,30,46
Sin 18° = 18;32,27,40,15
Vers 18° = 2;56,11,45,58 (*acc.* 2;56,11,47,33)

The half-angle formula is invoked to compute the sines of 45°, $22\frac{1}{2}°$, and $11\frac{1}{4}°$ (on this formula, cf. Pedersen, 1974, p. 60), and from the sines the versines can also be computed as was indicated earlier (§ 81). Similarly, from the sines of 30° and 18°, the sines of their successive half-angles can be found.

In algebraic terms Euclid, XIII.9, states that where D represents the side of a regular inscribed decagon (§§ 107–8):

$$(R + D)D = R^2$$

The statement in § 107 in algebraic terms is:

$$(D + \frac{R}{2})^2 = R^2 + (\frac{R}{2})^2$$

or
$$D^2 + R \cdot D + \frac{R^2}{4} = R^2 + \frac{R^2}{4}$$

Hence
$$D^2 + R \cdot D = R^2$$

or
$$(R + D)D = R^2$$

as above. Before computing the value of D, this equation must be solved for D, and this step is omitted in the text:
$$D = \frac{R(\sqrt{5} - 1)}{2}$$

With $R = 60$, $D = 37;4,55,20,30$ as in the text.

Next (§§ 113–15) the theorem on the chord of the sum of two arcs is invoked (cf. §§ 64–80), and then the relationship between chords and sines. Thus we can compute $\sin 24° = \frac{1}{2} \text{Crd}(30° + 18°)$, and then by successive halvings, the sine and versine of $\frac{3°}{4}$. By the same formulas the sines and versines of all arcs at intervals of $\frac{3°}{4}$ can then be computed.

The goal now is to compute the remaining entries in the sine table with arcs at intervals of $\frac{1°}{4}$ (§§ 116–24). Levi invokes a procedure for finding the sine of $\frac{1°}{4}$ that he calls a "mechanical analogy," a term that he uses on several occasions in his *Astronomy*, but that occurs nowhere else as far as I know: it seems to refer to an iterative computational procedure (cf. chap. 3:48 and my comments *ad loc*; Goldstein, 1974a, p. 67; *idem*, 1974b, p. 285). Here the method involves successive halving of arcs until linear interpolation can be used without disturbing the value sought to the number of places needed. Levi starts with the sine of $8\frac{1°}{4}$ (that depends on the sines of 18° and 30°) and the sine of $3\frac{3°}{4}$ (that depends on the sine of 45°). By successive halving and linear interpolation (omitting the details of his computation), Levi arrives at his fundamental value

$$\sin \tfrac{1°}{4} = 0;15,42,28,32,7 \ (acc. \ 0;15,42,28,29,18)$$

from which he can construct all the entries in his table.

In the next passage (§§ 125–33) Levi motivates his use of intervals of $\frac{1°}{4}$ in his sine table rather than intervals of 1°. In the latter case he tells us linear interpolation can lead to errors as great as $0;15°$ in the arcsine. His test case is to find the arcsine of $59;59,52$. With a sine table arranged at degree intervals one would have to interpolate between $\sin 89° = 59;59,27$ and $\sin 90° = 60;0,0$. Therefore, by this procedure,

Sin 89;45,27° = 59;59,52

and it follows that

Sin 90;14,33° = 59;59,52
Vers 0;14,33° = 0; 0, 8

and thus, by the formula relating sines and versines, i.e.,

$$\text{Sin}^2 x = \text{Vers } x \, (2R - \text{Vers } x)$$

we find

Sin 0;14,33° = 0;30,59,0,33

But according to the table arranged at intervals of $\frac{1}{4}°$

Arcsin 0;30,59,0,33 = 0;29,35°

In this way Levi has shown that an error greater than $\frac{1}{4}°$ was committed, i.e.,

0;29,35° − 0;14,33° = 0;15,2°

SECTION 4. This section concerns the use of the table and linear interpolation.

The table contains only the values of sine of arcs; chords and versines are to be computed from them when needed (§ 135).

The table (§§ 148–51) has been published elsewhere (Goldstein, 1974a, pp. 153–55).

SECTION 5. This section concerns the solution of plane triangles.

In a right triangle, if two sides are known, the remaining side and the two acute angles can be found (§§ 153–61).

Next (§§ 162–72) we are told that if all three sides of a triangle are known, so are its angles. As a first step we have to calculate GD (see Fig. 4.8) as follows:

$$GD = \frac{GB^2 + GA^2 - AB^2}{2GA}$$

To prove this, we may note that by the Pythagorean theorem

$$GB^2 - GD^2 (= BD^2) = AB^2 - (GA - GD)^2$$

Hence

$$GB^2 - GD^2 = AB^2 - GA^2 + 2GA \cdot GD - GD^2$$

or

$$GB^2 + GA^2 - AB^2 = 2GA \cdot GD$$

from which the statement in the text follows (cf. Euclid, II.12,13). If we subtract GD from GA, the result is AD, and from these lines all the angles in triangle ABG can easily be found by solving the two right triangles GDB and ADB. Figure 4.9 presents the case where angle AGD is obtuse, and the proof is analogous to that where the angles are all acute.

We are then told to consider a triangle where two sides and an angle subtended by one of them are known (§§ 173–82). Levi claims that the remaining side and angles may be found, but he has failed to take into account the ambiguous case (cf. Heath, 1956, vol. 1, pp. 306–7). In practice the ambiguity is resolved by noting whether the angles are acute or obtuse (cf. Euclid, VI.7). Now we are told that sines can replace chords in the preceding proof (§§ 183–88). The law of sines is then stated, namely, in any triangle ABG

$$\frac{\operatorname{Sin} A}{\operatorname{Sin} B} = \frac{BG}{AG}$$

The theorem on solving a triangle where the angles and a side are known is also presented without proof but, as Levi remarks, it follows directly from the law of sines.

Finally we are shown how to solve a triangle where two sides and an included angle are given (§§ 189–99). Two cases are considered: Fig. 4.11 displays the case of an acute angle and Fig. 4.12 displays the case of an obtuse angle. The proof is straightforward.

CHAPTER 5

In this chapter Levi discusses the pinhole camera (or camera obscura), and its use in astronomical observations.

Although Levi dies not mention it, the difficulty in accounting for the shape of the pinhole image goes back to an ancient Greek text ascribed to Aristotle, (*Problemata*, Book XV, ch. 6, 911b1): "Why is it that when the sun passes through quadrilaterals, as for example in wickerwork, it does not produce a figure rectangular in shape but circular?" (cf. Lindberg, 1968, p. 158).

In this discussion, Levi assumes the rectilinear propagation of light, a view commonly presented in ancient and medieval treatises on optics (cf. Lindberg, 1976, p. 220, n. 79), but does not state this principle explicitly. He refers to a window (*ḥalon*) of unstated dimensions, although it has to be understood as being small with respect to the distance from the window to the wall on which the image appears (cf. § 22). Levi is clearly aware that the size of the image depends on the angular size of the luminary as well as the size of the opening, and he seems to have been the first astronomer in the West to have realized that for quantitative measurements the size of the opening must be taken into account (Straker, 1971, p. 200). Levi does not cite any sources, and one can only wonder if he was aware, directly or indirectly, of Ibn al-Haytham's *On the Shape of Eclipses*, a treatise that was never translated into Latin or Hebrew (cf. Sabra, 1972, pp. 195–96; Lindberg, 1968, pp. 155–56; for the German translation of this text, see Wiedemann, 1970, vol. 2, pp. 87–101). This chapter of Levi's *Astronomy* has been extensively discussed based largely on the published Latin version (Curtze, 1901; Carlebach, 1910b, pp. 30–34; Straker, 1970, pp. 197–219; Lindberg, 1970, pp. 303–8).

That the size of the image will be greater than the size of the opening is shown by means of an analysis of the formation of the image arising from each point of the opening. The angular diameter of the luminary as seen from the window is not perceptibly different from the angular diameter as seen from the center of the earth (§ 17), and the luminaries are sufficiently far from the earth that the rays from any point on the luminary are effectively parallel (§ 3).

In Fig. 5.1 Levi considers the ray that pass through E, a single point of the opening, and describes the formation of the image LM of the luminary THZ. Point E is the apex of a double cone of rays (TEH and LEM), and so is every other point of the opening. In the Hebrew text point E is at the top of the opening, whereas in the Latin version it is at the center of the opening, but this makes no difference to the argument. It follows that the image is a measure of the size of the luminary provided the size of the opening is subtracted from it, although this condition is not stated until § 23. It also follows that the image is inverted, as noted in §§ 24–25.

To clarify Levi's argument let us consider Fig. 5C-1 (cf. Straker, 1981, p. 275) in which XY is the luminary, AB the opening, and CD the image. The ray from point X that passes through point B of the opening reaches point D on the image, and the ray from point X through point A of the opening is effectively parallel to it and reaches the image at point E. The angular diameter of the luminary is angle XZY which is equal to angle CZD which in turn is equal to angle CAE. To find angle CAE on needs to know AF, the distance from the opening to the image, and CE, the difference between CD, the size of the image, and AB ($= ED$), the size of the opening. Levi presents a version of this argument in chapter 9 (see below), but it seems to have been unavailable to recent scholars.

Another important result is the explanation of the rounding of the corners of images through an opening of polygonal shape (§§ 15–16): cf. Lindberg, 1970, pp. 304–5, and Straker, 1971, pp. 208–9.

Since the image is a measure of the size of the luminary, observing the Sun at different times of the year would give us the variation of its apparent dia-

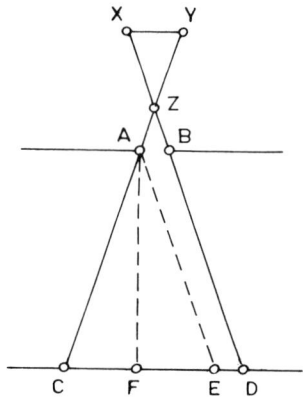

FIGURE 5C-1.

meter, and hence its eccentricity could be determined from the inverse proportion between the distances and the apparent diameters (§§ 18–20). This procedure is not mentioned in the *Almagest*, and I am unaware of anyone before Levi suggesting it. The pinhole image can be used at the time of an eclipse to measure its magnitude (§ 21) by means of a procedure described in 23 that yields the ratio of the diameter of the eclipsed body to the diameter of the luminary (cf. Straker, 1971, pp. 209–91), and it is noted that the image is inverted (§ 24): similar remarks can be found in Ibn al-Haytham's work cited above.

Lindberg (1970, pp. 305–6) argues that though in hindsight it would appear that Levi had a thorough understanding of the formation of pinhole images of the sort Kepler attained; in fact, Levi's interests were more narrowly defined by the astronomical problems for which he intended to use pinhole images, and that his terse remarks cannot be considered a general solution even though there are no flaws in his arguments. Straker (1970, pp. 209–13) takes issue with Lindberg, arguing that Levi could reasonably have expected his readers to draw general conclusions from the discussion of specific cases, and I am inclined to agree with Straker in light of Levi's general style of presentation.

It is difficult to determine the extent of Levi's influence on the subsequent discussions of pinhole images and their use in observing eclipses. I have not found any allusions to these remarks by Levi in later texts, and the studies of Lindberg and Straker suggest that the developments in the Latin west proceeded without reference to Levi's work. To be sure, Levi is rarely cited in the European astronomical literature with the major exception of a passage on the Jacob Staff reported by Commandino in his commentary on Archimedes's *Sandreckoner* published in 1558 (cf. Roche, 1981, pp. 8, 22–23).

In a book published in 1545 Gemma Frisius describes the astronomical use of both the Jacob Staff and the pinhole camera without alluding to Levi's work on both these subjects. He refers to Erasmus Reinhold's commentary on Peurbach's *Theoricae* for the idea of observing a solar eclipse with a camera obscura. Tycho Brahe mentions this passage from Gemma Frisius's book and he also built a Jacob Staff according to Gemma Frisius's instructions (Straker, 1981, pp. 269–70; cf. Roche, 1981, pp. 16–17). In his solar observations (*not* during an eclipse) of 1591 with a camera obscura, Brahe noted the size of the aperture, the distance from the aperture to the screen, and the size of the image, and he may have subtracted the size of the aperture from the size of the image though without any theoretical justification. In 1598 Brahe wrote to Maestlin that the Moon as seen during a solar eclipse with a camera obscura appears smaller than "at other times during full moons when it is equally far away;" thus it is clear that Brahe did not understand the effect of the size of the aperture on the size of the image. In fact, it is the image of the Sun, the luminous source, that appears to be too large and this leads to the illusion that the "image" of the Moon is too small. Straker claims that in 1600 Kepler learned of this method for observing the apparent size of the Sun while he was visiting Brahe in Prague (Straker, 1981, pp. 275–76). Shortly thereafter Kepler established a sound theory for the formation of pinhole images, and he explained the reasons for subtracting the size of the aperture from the image of the luminary noting, among other things, the shortcomings of Gemma Frisius's treatment of the such observations (Straker, 1981, pp. 269 n, 291). Kepler included his analysis of pinhole images in his *Ad Vitellionem paralipomena* published in 1604, a major original contribution to the study of optical phenomena despite the modest title. In this treatise Kepler cites Levi's *Astronomy* in connection with the determination of the center of vision needed to use the Jacob Staff properly, relying explicitly on Commandino's report (see above). But Kepler does not seem to know of Levi's study of pinhole images and does not cite him in this context (cf. Straker, 1971, pp. 218–20).

CHAPTER 6

In this chapter Levi introduces the Jacob Staff (or cross staff) addressing himself first to the problem of locating the center of vision. He considers 3 possibilities: the center of vision lies on (1) the external periphery of the eye, (2) the internal periphery, or (3) in between. He dismisses the first on the grounds that the eye could not grasp the image if it only arrived at a point because there would be no way to determine size or shape (§ 6). It is clear that

Levi accepted a theory of vision in which the rays proceed from the object to the eye (for a discussion of Ibn al-Haytham's intromission theory, cf. Lindberg, 1976, pp. 71–80).

A second argument for determining the center of vision is derived from medical practice: removing a cataract from the eye restores vision. Levi concludes that vision takes place at the center of the eye: "vision takes place in the crystalline lens, and its center is the center of vision" (§ 16). The impression formed by the image is transmitted to the "common sense" in the brain according to Levi, a view that he shared with most medieval writers on this subject (for a discussion of Averroes's treatment of this transmission, cf. Lindberg, 1976, pp. 53–55).

A third argument depends on a mathematical proof using the Jacob Staff, here called simply the staff (*maqel*). It two crosspieces of known ratio to one another are placed on the staff such that the smaller one, closer to the eye, exactly hides the larger one located farther from the eye then, by measuring the distance between the two crosspieces, the center of vision may be determined. In Fig. 6.1 we are given the lengths of the two crosspieces (*text*: plates) GD and EZ, and the distance between them, HT, on the staff, AB. We seek the distance from a known point on the staff to the center of vision at L. The proof is based on the proportionality of the sides of similar triangles LET and LEH:

$$\frac{ET}{GH} = \frac{LT}{LH}$$

Hence

$$\frac{LT}{ET} = \frac{GH}{LH} = \frac{GH - ET}{LH - LT}$$
$$= \frac{GH - ET}{HT}$$

Since ET, GH, and HT are known, LT can be computed (§ 46). Levi then tells us that he performed this experiment many times and found that the center of vision lies at the center of the eye (§ 47). Here he does not say whether he was the only observer on whom the experiment was performed but, when the result is given in chapter 7 (§ 5), we are informed that LA, the distance from the center of the eye to the end of the staff nearest to the eye, is $\frac{1}{20}$ of a span (i.e. about 1 cm) "for most people". Some 250 years later Harriot came up with the same result (Roche, 1981, p. 6; cf. Haasbroek, 1968, pp. 23–28).

Finally, Levi stresses the importance of this result for the accurate deter-

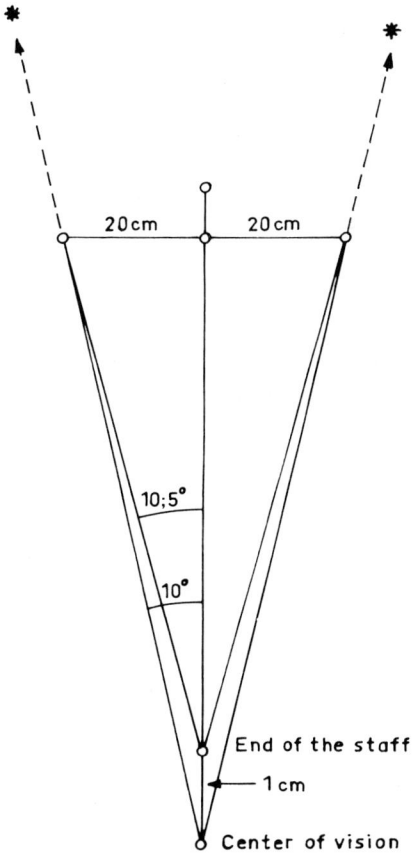

FIGURE 6C-1.

mination of the angular distance between two stars (§§ 48–50). Without this correction, later called the correction for eccentricity, an error of about 10 minutes of arc would be introduced in an angle of 20° (Roche, 1981, p. 6: see Fig. 6C-1, where the crosspiece is assumed to be 40 cm). This instrument was still being used for serious astronomical observations in the seventeenth century; indeed, the observations made by William Crabtree and Jeremiah Horrox from 1635–1640 "constitute the largest surviving body of English astronomical observations made with the cross staff of which I am aware" (Roche, 1981, pp. 25–26). Yet it was only in November 1637 that Horrox realized that he had to correct for the eccentricity to eliminate a significant source of error, although instructions for so doing had already been published in 1537 in England by T. Digges (cf. Roche, 1981, pp. 22, 27).

In his commentary on Archimedes's *Sandreckoner* Commandino presents the arguments from this chapter and explicitly refers them to Levi. Somewhat modified versions of Levi's method were later used by T. Digges (who may have learned of Levi's work form his teacher, John Dee, who met Commandino in 1563), Brahe, Harriot, and others (cf. Roche, 1981, pp. 5, 23, 27, 28). As noted by Commandino, Archimedes's dioptra was similar to, but not identical with Levi's instrument (on Archimedes's dioptra, see Shapiro, 1975). Another commentary on Archimedes's *Sandreckoner* (that escaped Roche's attention) describes the Jacob Staff in detail without mentioning Levi, but with a discussion of the eccentricity of the eye that is reminiscent of Levi's treatment of this subject (Hamellius, 1557, 17v). This author was probably in contact with Ramus in Paris, and it is possible that he learned of Levi's work from Ramus, one of the few scholars to cite Levi in connection with the Jacob Staff.

Chapter 7

This chapter begins with a description of the construction of the Jacob Staff, already introduced in chapter 6. We are told to start with a staff 6 spans long (about 1.5 m) and 1 digit wide (about 2 cm). Pegs are to be placed at the end near the eye so that the line of sight is slightly above the level of the staff (see Fig. 7C-1). Moreover, under these circumstances the center of vision in the crystalline lens is $\frac{1}{20}$ span (about 1 cm) from the end of the staff (§ 5): this is a measured quantity "determined by experiment." Allowance must be made for this "eccentricity" in marking the graduations on the staff. Each span, p, is divided into 8 units, u, such that the first unit begins at the center of vision, i.e., the distance from the end of the staff nearest the eye to the mark for the first span is only $\frac{19}{20}$ of a span or $7;36^u$ (see Fig. 7C-2). Each unit is divided according to the principle of the transversal scale (§§ 9–14), a principle which Levi seems to have discovered although he does not make this claim. The text is perhaps unnecessarily difficult to understand due to the absence of a figure to illustrate this principle. The length of a staff is graduated into 6 spans (1 span is about 25 cm) and each span is graduated into 8 units (1 unit is about 32 mm). We consider a flat surface along the length of the staff whose edges are parallel. Each unit along one edge is graduated into 6 parts (i.e., $0;10^u$ or about 5.2 mm), and each unit along the other edge is graduated into 12 parts (i.e., $0;5^u$ or about 2.6 mm). Transversals are then drawn connecting the points marked on the edges (§§ 10–11), as shown in Fig. 7C-3, and the breadth of the staff is divided lengthwise into 5 strips each that successive intersections with the transversals are $0;1^u$ apart (§ 14).

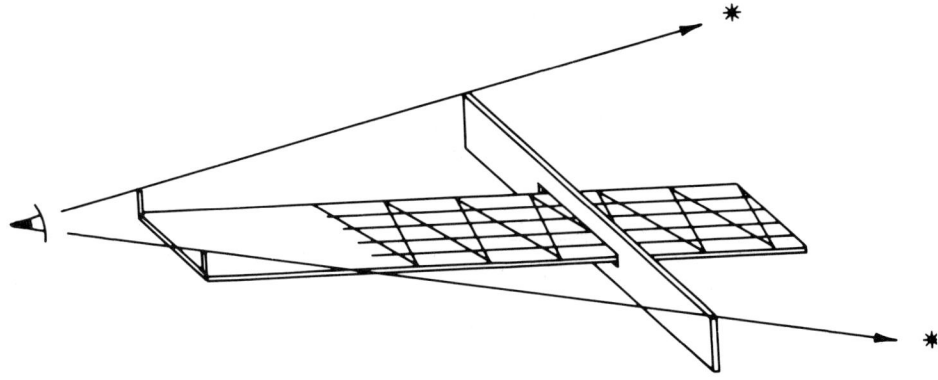

FIGURE 7C-1. The Jacob Staff according to Levi's description.

The crosspieces (here called "plates") each have a round hole, into which the staff is inserted, to allow rotations of the crosspiece without rotating the staff (§ 15). Levi suggests that a number of crosspieces be made so that the crosspiece can be kept far from the eye whatever the angular separation of the stars to be observed (§ 20). It is clear (but unstated) that the error due to the eccentricity of the eye is minimized by keeping the crosspiece far from the eye. The ends of the upper surface of the crosspiece define the lines of sight and so they should be above the surface of the staff by the same amount as the center of eye (§ 16). Levi suggests that crosspieces of the following sizes be made: 24^u, 16^u, 8^u, 4^u, and even 2^u, 1^u, $\frac{1}{2}^u$, and $\frac{1}{4}^u$ in order to observe stars very close to one another (§ 17). Thus the crosspieces vary from 3 spans (about 75 cm) down to $\frac{1}{4}^u$ or $\frac{1}{32}$ of a span (about 8 mm).

We are next told how to hold the staff when making observations: it should be held close to the eye and each of the two stars should be seen touching an end of the appropriate crosspiece (§§ 19–22). We are told that with the largest crosspiece of 24^u, we can observe stars more than 25° apart (§ 19). In fact, the text should read 28° instead of 25°: "5" and "8" look alike in Hebrew numerals, but the manuscripts have 25 written out in words. To demonstrate this, consider Fig. 7C-4 in which line CD represents half the crosspiece, and line AB represents the staff, where point A lies at the center of vision. Let x be angle CAD corresponding to half the observed angle CAE. If line CD is taken to be 12^u (half the length of the largest crosspiece), and AD, the distance from the eye to the crosspiece, to be 48^u (the maximum on the staff), then

$$\tan x = CD/AD = 12^u/48^u$$

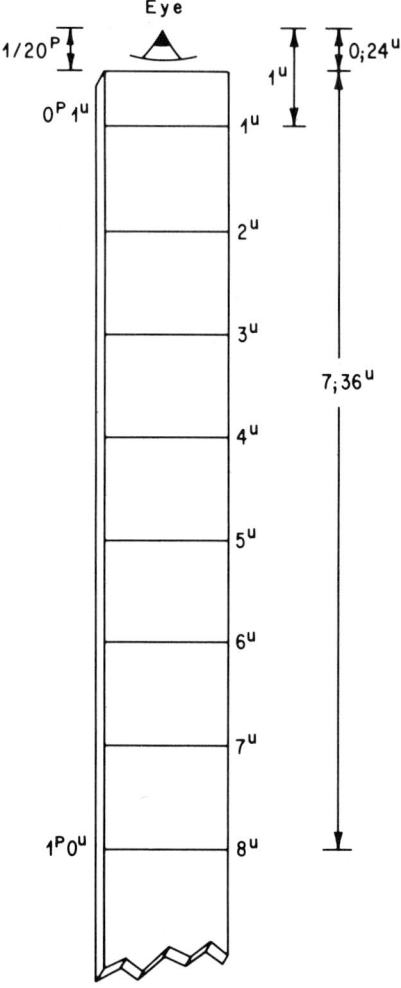

FIGURE 7C-2. The scale on the Jacob Staff.

and

$$x \approx 14;2°$$

If we let x be 12;30°, and leave CD as 12^u, then AD would be 54;8^u, which exceeds the length of the staff. Since the observed angle corresponds to $2x$, it follows that the largest crosspiece can only be used for angles greater than

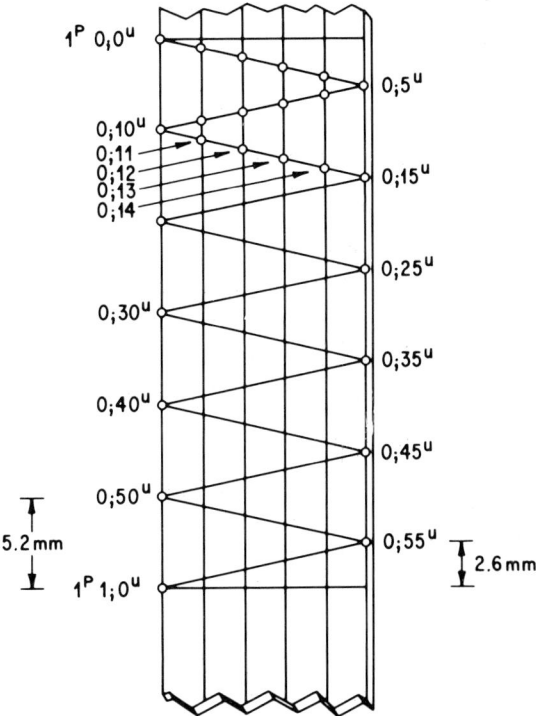

FIGURE 7C-3. The transversal scale for linear measurement drawn according to Levi's description.

about 28°: note that as the crosspiece is brought closer to the eye, the observed angle increases.

Of interest is the remark, not found in any other ancient or medieval text as far as I know, that a light is needed to illuminate the instrument, and that it should be placed behind the observer in order not to interfere with his vision (§ 23): cf. chap. 11 : 1–2). The data to be recorded for the observation are the size of the crosspiece and its distance from the eye (§§ 24–25). An example is then presented, somewhat schematically, for converting the data into the angular separation between two stars where the data are: crosspiece 10ᵘ, distance 40ᵘ (§§ 25–28). First one has to find the distance from the center of vision to one end of the crosspiece, here called the corrected radius (§ 26). In Fig. 7C-4, the crosspiece CE is 10ᵘ and AD, the distance measured on the staff, is 40ᵘ. We find

$$AC = \sqrt{40^2 + 5^2} = \sqrt{1625} \approx 40;19 \tag{1}$$

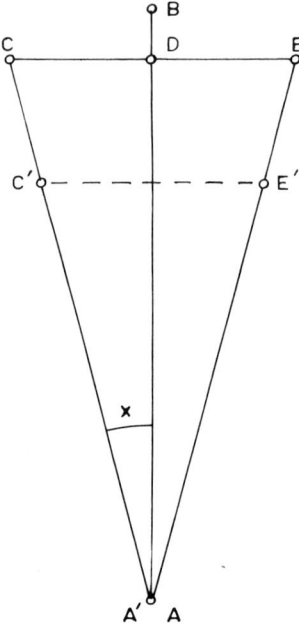

FIGURE 7C-4.

We consider triangle $A'C'E'$ similar to triangle ACE where $A'C'$ is 60. Then

$$\frac{C'E'}{A'C'} = \frac{CE}{AC} \qquad (2)$$

or

$$C'E' = 60 \times 10/40;19$$
$$\approx 14;53$$

and $C'E'$ is called the corrected chord (§ 27). We are then told to look up the corresponding arc in a table of arcs and chords (§ 28), even though Levi has presented a table of arcs and sines (Goldstein, 1974a, pp. 153–55). However, he has already shown that his table can be used for this purpose as well (chap. 4 : 136). The result, unstated in the text, is that the observed angle, CAE, is about 14;15°.

The next task is to convert the observed angular distance between two stars into distances between them in longitude and latitude. A number of cases are considered. If two stars have 0° latitude, i.e., they both lie on the ecliptic, then the observed distance is the difference in their longitudes; similarly, if

Commentary: Chapter 7

both stars have the same longitude, the observed distance is the difference in their latitudes (§ 29): several cases are also discussed (§§ 30–32). A general problem is then presented: to find the longitudes of two stars where the distance between them has been observed and their latitudes are known (§ 33). We are first given a preliminary lemma: the sine of the latitude is perpendicular to the plane of the ecliptic (see Fig. 7.1). Circle $ABGD$ is the ecliptic whose center is at E and whose pole at T. The latitude of a star at Z is arc DZ (on arc DZT); its sine, line ZH, is shown to be perpendicular to line ED and to plane $ABGD$ (§§ 34–42).

In the series of proofs that follow, problems of spherical geometry are solved by means of plane geometry. The first case is that where the latitude of one star, at A, is $0°$, and latitude of the other star is some known quantity, say BZ, where A and B lie on the ecliptic whose center is at E (see Fig. 7.2). We wish to find arc AB where AZ, the observed distance, is known (§§ 43–56). Since arc BZ is known, so is BH, its versine, and it follows that we know EH (the radius minus the versine). Since line AZ, the corrected chord, is known by the observation, and line ZH, the sine of the latitude of the star at Z, is also known, we can solve triangle AHZ for line AH. In triangle AEH, we now have determined all three sides (EA is the radius), and so the angles can be determined (line AT, the sine of arc AB, is drawn perpendicular to line EH and aids in the computation of these angles: cf. chap. 4 : 153–61), in particular we can find angle AEH which is equal to arc AB. Computational rules are stated in §§ 44–46, although the procedure for finding line AT is not mentioned.

The next problem, where both stars have non-zero latitudes, is divided into 2 cases: (1) the latitudes are in the same direction, and (2) the latitudes are in different directions. A verbal description of the procedures (§§ 57–65) is followed by illustrated examples (§§ 66–92). Figure 7.3 illustrates the case where one star, at D, is to the north of the ecliptic, and the other star, at E, is to the south (§§ 66–76). We seek arc AB where line ED, the corrected chord, is known by observation. The goal is to solve triangle HGZ: it is easy to find lines GH and GZ, for they are the differences between the radius and the versines of the latitudes. To find line ZH, called the second chord, we are first told to extend line DZ to T such that line TZ is equal to line EH, then to draw line ET. It follows that line HZ is parallel and equal to line ET, and that line ET can be found by solving right triangle ETD in which ED and DT are known. Hence, line ZH ($= ET$) is also known and triangle HGZ can be solved, for all three sides are known. Thus, we have found angle HGZ which is equal to arc AB, the difference in the stellar longitudes.

The second case is illustrated by Fig. 7.4 where the stars at D and E are both on the same side of the ecliptic, and arc AD, the latitude of one star, is greater than arc BE, the latitude of the other star (§§ 77–83). Again we can easily find lines GH and GZ in triangle HGZ, but we still have to determine line ZH. The procedure is to mark off an amount equal to line EH, the sine of the latitude of arc BE, on line ZD, the sine of the latitude of arc AD, namely line ZT: draw line ET. Note that lines HE and ZT are equal, and that they are perpendicular to plane ABG. It follows that line ZH is parallel and equal to line ET, and that we can solve right triangle ETD for line ET (line ED is the corrected chord found by observation, and line DT is the difference between the sines of the latitudes). Hence, we know line ZH, and triangle HGZ can be solved for angle HGZ which is equal to arc AB, the difference in the stellar longitudes. Another method for finding arc AB is then offered (§§ 84–88): we know all sides of triangle HGZ that lies in the plane of the ecliptic. We can draw perpendiculars HL and BM from H and B to line AG; it is clear that lines HL and BM are parallel. Hence triangles HGL and BGM are similar and that

$$\frac{GH}{HL} = \frac{GB}{BM} \qquad (3)$$

where line GH had already been found, line HL can be found (cf. chap. 4:166 and Fig. 4.8) and line GB is the radius. Hence line BM, the sine of angle BGM can be found, and its arc, AB, is the difference in the stellar longitudes. Figure 7.5 illustrates the special case where the latitudes of the two stars at D and G are equal and in the same direction (§§ 89–92).

After presenting exact solutions to the problem of finding the difference in stellar longitudes, Levi introduces an approximation that can serve for all cases "without harming the observation" (§ 93). Later he adds that this method should only be used for stars within 7° of the ecliptic and whose difference in longitude lies between 10° and 30° (§ 117). The formula proposed (§§ 94–97) is proved with reference to Fig. 7.6, where arc AB is sought, G is the center of the ecliptic, and line ZH the "second chord" (defined in § 58: it is a line in the plane of the ecliptic joining the feet of the perpendiculars from the two stellar positions). Lines DZ and HE are drawn parallel to line AB through points Z and H, whereas line TL is drawn parallel to lines DZ and HE through T, the midpoint of line ZE. Moreover, line ZM is drawn perpendicular to line HE extended, the line ZE is the difference between the versines of the two stellar latitudes, b_2 and b_1, i.e.,

$$ZE = BH - AZ \qquad (4)$$
$$= \text{Vers } b_2 - \text{Vers } b_1$$

It follows that in parallelogram $LTMH$, and right triangle ZMH:

$$TL = MH = \sqrt{ZH^2 - ZM^2} \qquad (5)$$

or, substituting line ZE for line ZM,

$$TL \approx \sqrt{ZH^2 - ZE^2} \qquad (6)$$

Line TL is called the second corrected chord (§ 95). Since lines ZH and ZE are known, line TL can be found, and from it line AB can be computed by appealing to the proportionality of corresponding sides in triangles AGB and TGL:

$$AB = 60 \cdot TL/LG \qquad (7)$$

Note that line LG is called "the remainder of the corrected versine" (§ 96) and it is found as follows:

$$\begin{aligned} LG &= (BG - BH) + ZE/2 \\ &= R - \text{Vers } b_2 + (\text{Vers } b_2 - \text{Vers } b_1)/2 \end{aligned} \qquad (8)$$

From chord AB its arc may be found as before, and this arc is the difference in the stellar longitudes that was sought.

It remains for Levi to show that substituting line ZE for line ZM does not harm the observations when the stellar latitudes are less than 7° and difference in longitude is between 10° and 30° (§ 117). The same figure serves to illustrate this argument in MSS P, Q, with the additional line, GN, drawn perpendicular to AB at point N (see §§ 118–124); MS N displays Fig. 7.7 here. We assume that one star has latitude 7°, the other star latitude 0°, and that the difference in longitude is 10° (one set of extreme conditions according to Levi). Thus $AZ = 0°$,

and
$$BH = ZE = \text{Vers } b = 0;27$$

$$GN = R - \text{Vers } (AB/2) = \text{Cos } 5° = 59;46 \qquad (9)$$

Since triangles ZEM and GAN are similar,

$$\frac{ZM}{ZE} = \frac{GN}{GA} \qquad (10)$$

Therefore
$$\begin{aligned} ZE - ZM &= ZE \, (GA - GN)/GA \\ &= 0;27 \times 0;14/60 \\ &< 0;0,7 \text{ (accurately } 0;0,6,18, \text{ as in N mg.)} \end{aligned} \qquad (11)$$

Thus the use of line ZE instead of line ZM introduces a small difference. Levi does not display all the steps that lead to the conclusion that the effect on the arc sought is negligible. He does indicate, however, that the difference in the squares of lines ZE and ZM is less than $0;0,6$ (§ 124), and this is correct when ZE is $0;27$ and ZM is $0;26,53,42$. He goes on to claim that the effect on line MH is less than $0;0,1$ (§ 124), but offers no proof. To demonstrate this we can compute the value of MH (or TL) from formulas (5) and (6), above, with the values $10;26$ for ZH, $0;27$ for ZE, and $0;26,53,42$ for ZM. The results of the two computations differ by less than $0;0,1$, as stated in the text (§ 124).

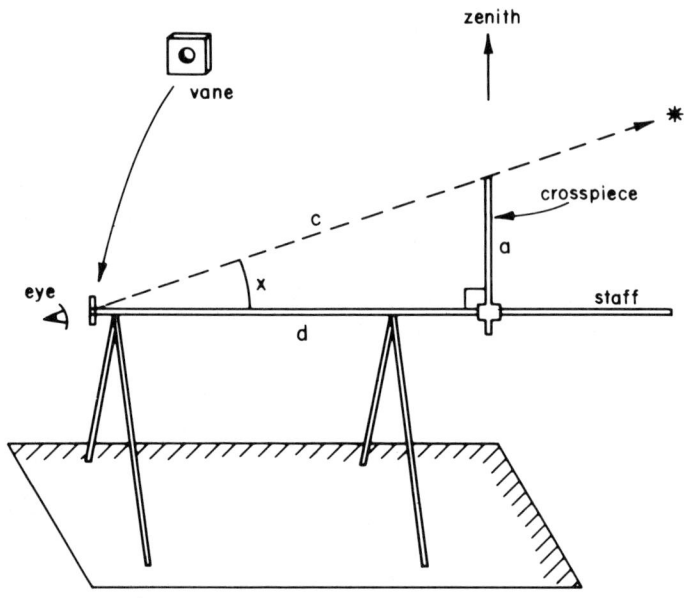

FIGURE 8C-1. The mounted version of the Jacob Staff where the crosspiece is perpendicular to the staff and can slide along it.

A similar argument for the case where the difference in longitude is 30° is discussed briefly in §§ 125–126, and my recomputation agrees with Levi's result.

CHAPTER 8

Levi introduces a second version of the Jacob Staff with fixed legs for observing stellar altitudes. There are two legs in the middle and two at the end near the eye that make the instrument very stable (§ 2): See Fig. 8C-1. This mounted version is fitted with a thin vane pierced by a small hole, near the end of the staff adjacent to the eye, that serves as a sight. In the Latin version, it is claimed that this plate is made of brass (MS Lyon 326, fol. 27b:8), but the Hebrew text does not specify the material (cf. Curtze, 1898, p. 112 [based on a different MS], and Roche, 1981, p. 7). In this way Levi eliminated the necessity of compensating for the center of vision, and the staff is to be graduated from this sight (§ 3). There are six crosspieces (*text*: plates): 60, 40, 30, 20, 15, 10 units (§ 6). The staff is to be kept parallel to the horizon, and the appropriate crosspiece is to be set perpendicular to the staff and checked with a plumb line (§ 7). The distance from the crosspiece to the sight is adjusted until the star is seen grazing the top of the crosspiece (§ 8), and this distance is to be noted as the observational data (§ 9). The size of the staff is not stated, but it may be assumed to be greater than the hand-held version which was 6 spans (about 1½ m); indeed, Levi later refers to a staff that is to be 16 spans or more (chap. 9:13). With the observed distance and the size of the crosspiece, here called the altitude, the angle of this star above the horizon, x, can be computed. In Fig. 8C-1 let d be the observed distance, and a, the observed altitude. Then c, "the corrected semi-diameter", is found as follows (§ 11):

$$c = \sqrt{d^2 + a^2}$$

and

$$\frac{\sin x}{60} = \frac{a}{c}$$

from which x may be found (§ 12). Note that the crosspiece is entirely above the staff in this version. We are reminded to choose the crosspiece that allows the star to be seen at the greatest distance from the sight because that yields the most accurate data, i.e., slight changes in d have little effect on the value of x (§ 13). The observations of stars on the meridian with this instrument yield their declinations because the observed altitude, x, is the sum of the complement of the geographical latitude of the place of the observer and the star's declination (§ 14). With formulas discussed later (chap. 62) the observed altitude of the Sun or of a fixed star can be converted to the time of day or night (§ 15: cf. Levi's solar eclipse observation of 14 May 1333 in Goldstein, 1979, pp. 126–27).

Though Levi's work on the Jacob Staff was known to some European writers who mention this instrument (e.g., Ramus, 1559, p. 62), curiously, not one of them says they used Levi's treatise to design it (Roche, 1981, p. 8).

Chapter 9

In this chapter yet another use is found for the Jacob Staff. Ptolemy presented a set of values for the angular diameters of the planets and the fixed stars, but says little about how they were measured (Goldstein, 1967a, p. 8). Levi claims to have three methods to make the required measurements. The first method is to measure the distance, presumably with the first version of the Jacob Staff, between the near sides of two stars, and then to measure the distance from the far side of one to the near side of the other: the difference between the two measured angles ought to be the apparent stellar angular diameter (§§ 2–6). Clearly this method will not work for the planets and the fixed stars because they are all too small, but it might work for the Moon.

The second method is similar to that ascribed to Hipparchus (cf. *Almagest*, V, 14; and Cohen and Drabkin, 1958, pp. 141–42, for Proclus's description of Hipparchus's dioptra). Observe the star exactly filling a small hole of known size in a plate at a measured distance from the eye, when the hole is above the staff by the same amount as the center of vision (§§ 7–8). From the size of the hole and the distance of the plate from the center of vision, the stellar angular diameter can be computed by the procedures already described (§ 10). Hipparchus's instrument was set up differently. There are two plates on a staff: the plate near the eye has a small sighting hole (as in Levi's instrument described in chap. 8) and the one farthest away from the eye has two small holes through which one is to observe the upper and lower limbs of the luminary. There is no reason to believe that Levi was aware of the relevant passage in Proclus, and Levi is emphatic in claiming that he invented this instrument (§§ 28, 55).

The third method only applies to the luminaries, though Levi claims to have tried it for the bright planets, Venus and Jupiter, as well (§§ 11–12). The instrument described is a combination of the Jacob Staff and the camera obscura, where the staff is to be 16 or more spans (i.e., greater than about 3 m). At one end of the staff is a plate with a circular hole whose diameter is equal to one or two units (where each unit is $\frac{1}{8}$ of a span, as before [1 span ≈ 20 cm]; cf. chap. 7:6) through which the light of the luminary enters (§ 14). At the other end of the staff is a screen parallel to the plate on which the image of the luminary is seen (§ 15). If the distance from the plate to the screen is measured, and the excess of the image over the size of the hole is measured, then the angular diameter of the luminary can be calculated (§§ 16–17). This is shown by means of a worked example that refers to Fig. 9.1. Let the distance from the plate with hole *DG* to the screen with image *ZE* be 100 units, where *DG* is 2 units and *ZE* is 3 units (§ 18). The excess *ZT* of *ZE* over *DG* is 1 when the distance is 100 but 0;36 when the distance is taken to be 60, and thus angle *ZDT* which is equal to angle *AHB*, the apparent diameter of the luminary, is 0;34,22 (§ 20), i.e.,

$$2 \operatorname{Sin} (0;34,22°/2) = 0;36$$

The essential point in the proof that follows is that line *TD* is parallel to line *EH*, where *TD* represents the ray from point *B* of the luminary through point *D* at one side of the hole and *EH* represents the ray from point *B* through point *G* on the opposite side of the hole. This indicated that Levi was well aware of this property of the camera obscura that was discussed earlier in chapter 5 (cf. Curtze, 1901, Lindberg, 1970, pp. 303–8). In the summary of the Latin version this passage is mentioned very briefly (without the figure) and its significance was not noticed (Curtze, 1898, p. 112). Levi reminds us that angle AHB does not differ sensibly from the apparent diameter of AB as seen from the center of the earth (§ 25), except for the Moon (§ 26). We are also told that the longer the staff the more accurate the measurement (§ 27), though no mention is made of the penumbra that may make the observation more difficult. The name that Levi gave to his instrument (§ 29), 'The Revealer of Profundities' (*megalleh ʿamuqot*), appears in the Latin translation as: *revelator secretorum*, but it was not used by anyone other than Levi as a name for this instrument, as far as I know.

There follow two poems that Levi composed in honor of the instrument he invented. In both cases the first letter of the first three lines from Levi's name: L. V. Y. The poems also appear separately in at least two manuscripts, one of which (Bodl. Poc. 280b) has been cited above. In the copy found in Bodleian MS Can. Or. 64 the first line of the page, in Hebrew, may be translated: "These verses were composed by the scholar Maestro Leon de-Bagnoles concerning the staff ... that he called 'The Revealer of Profundities'." The first poem has 8 lines of which the last 4 agree with the text edited here, and the second poem agrees with the text edited here but for the order of the verses which, however, is corrected in the margin to conform with the text edited here. In Bodleian MS Poc. 280b, fol. 222a, the first line of the page, in Hebrew, reads: "These are the verses of R. Levi ben Gershom, of blessed memory, called M. Leon in the secular language." Only the second poem is found here, and it is entirely vocalized. The Hebrew text of the second poem was published (without vowels) by Edelman, 1853, p. 7, apparently on the basis of MS Bodleian Poc. 280b, fol. 222a (cf. Edelman, 1853, p. iii, note b). Both poems were published in Hebrew (Carlebach, 1910a, pp. 152–53), and German translations of them also appeared (Carlebach, 1910b: the second poem begins on p. 27 and the notes to it on p. 103; the first poem is included on p. 34). I am grateful for the assistance of Dr Ray Scheindlin (Jewish Theological Seminary of America) in translating these poems.

The next section of this chapter (§§ 55–58) is devoted to some preliminary remarks on the use of this instrument to observe variations in the apparent diameters of the Sun and the Moon which are understood to depend on the variation in their distances from the earth. The four lunar distances in § 55 refer to the four lunar limits noted by Ptolemy in his table of lunar parallaxes (*Almagest*, V, 18): they are the maximum and minimum distances of the Moon at sygyzy and quadrature evaluated in the *Almagest*, V, 17 (trans. Toomer, 1984, p. 259). The variation in the distances demonstrates that the luminaries lie on eccentric spheres, but not in accordance with Ptolemy's models (§ 57). This investigation, we are told, was motivated by the arguments of physicists against Ptolemy's models (§ 56), an allusion presumably to the discussions in al-Biṭrūjī (which is cited later: see chap. 40, and Goldstein, 1971, vol. 1, pp. 40–43), Maimonides (cf. Goldstein, 1980, pp. 138–39), and Averroes (cf. Carmody, 1952). There is no indication that Levi was aware of Ibn al-Haytham's criticisms of Ptolemy's models; Ibn al-Haytham opposed the use of equant models whereas Levi preferred to depend on them (cf. Goldstein, 1980, pp. 137–42, where references to the earlier literature may be found). In the final section (§§ 59–63), we return to the study of the fixed stars. In a preliminary remark we are told that given the longitude and latitude of one star and the latitude of a second star, its longitude can be determined from the measured distance between the two stars (§ 59). Moreover, given both the longitude of a star and its declination, derived from an observation of the altitude of its meridian crossing, its latitude (here called its inclination from the ecliptic) can be determined (§ 60): no details are offered here.

Levi then notes that stellar coordinates have not been reliably reported and that he had to reinvestigate the positions of the fixed stars by means of many observations (§§ 61–62). Once these positions have been established, planetary positions can be observed with respect to them using the Jacob Staff (§ 64).

Chapter 10

In the first section (§§ 1–7) Levi discusses his method for finding the positions of planets relative to the Sun, whose motion is presumed to have been established. The position of the planets will be observed with respect to nearby stars, and their positions relative to the Sun will be established by observing the elongation of the Moon from the Sun before sunset, and then the elongation of a fixed star from the Moon a little while later. In the *Almagest* (VIII, 2: cf. Pedersen, 1974, pp. 240–45) Ptolemy describes his method for finding the elongation of Regulus from the Sun by means of two observations: one of the elongation of the Moon from the Sun and another, made $\frac{1}{2}$ hour later, of the elongation of Regulus from the Moon. The difficulties in this method arise from the non-negligible motion of the Moon in this time interval as well as the change in its parallax. Apparently Bernhard Walther was the first to introduce an improved method, namely, the use of Venus, rather than the Moon, as the intermediary between the Sun and the fixed star (cf. Beaver, 1970, p. 41; on Walther's observations, see also Kremer, 1980), a method that Brahe used as well (cf. Raeder, 1946, p. 113). In a short time interval the motion of Venus relative to the Sun can be neglected (particularly near greatest elongation), and Venus's parallax is very much smaller than that of the Moon and can also be neglected with respect to the accuracy of ancient and medieval observations. Levi remarks that using Ptolemy's lunar model introduces a small error in the determination of the Moon's motion between the two observations, an allusion to Levi's improved lunar model described in chapter 71 (cf. Goldstein, 1974a, pp. 53–74).

The rest of the chapter is devoted to a description of another modification of the Jacob Staff, this time to enable the observer to find the elongation of the Moon from the Sun. As far as I know this instrument is not mentioned by any subsequent astronomer, and has not been noted in the modern secondary literature. The basic principle is that the Moon is sighted at one end of a crosspiece while the Sun's rays pass through a hole in each of two vanes attached to the staff (§§ 25, 35).

There are two pieces to the instrument (see Fig. 10C-1): a staff and a device consisting of three plates that form an isosceles right triangle with one leg extended (§§ 10–11). This triangle is to be attached to the staff in one of two ways, depending on the elongation of the Moon from the Sun (§ 30). If the elongation is less than or equal to 90°, one leg of the triangle is set perpendicular to the staff through which it is inserted, and farther from the eye than the hypotenuse through which the staff is also inserted (§ 33), but when the elongation is between 90° and 180° the leg lies between the eye and the hypotenuse (§ 34). The triangle is rigid and can slide along the length of the staff. Attached to it are vanes with holes (§ 25), through which the Sun's rays can pass, set perpendicular to the plane of the triangle (§ 26) at vertices D and Z (see Fig. 10.1) as well as at points along EZ and ED such that the path of the Sun's rays will form angles with respect to EZ or ED of 15°, 30°, 45°, 60°, 75°, and 90° (§§ 16–21). Apparent elongation has to be corrected for (lunar) parallax to find the true elongation (§§ 36–37): for Levi's parallax table, see Goldstein, 1974a, pp. 116–122.

Figure 10.2 (redrawn as Fig. 10C-1) illustrates the use of this instrument when the elongation is 90° and the triangle is set for elongations less than or equal to 90°, i.e., the hypotenuse is closer to the eye (§ 42). The Sun's rays are assumed to pass through vanes at D and L such the angle EDL is 15° (§ 41). The first plate, DE, extends beyond the staff, ABG, to point M and the triangle is set such that at the moment of the observation the center of the Moon is seen in the direction AM, where point A represents the center of vision (§ 38). Since the distance BM is fixed, and the distance AB is measured along the staff, angle BAM in right triangle ABM can be computed and so can its complement, angle AMB (§ 44). If we assume that angle BMA is 75° and that angle EDL is 15°, it is claimed that the elongation is 90°. The proof requires us to extend line DL to N on line AM forming triangle DMN. It follows that angle MND is the supplement of the sum of the other two angles;

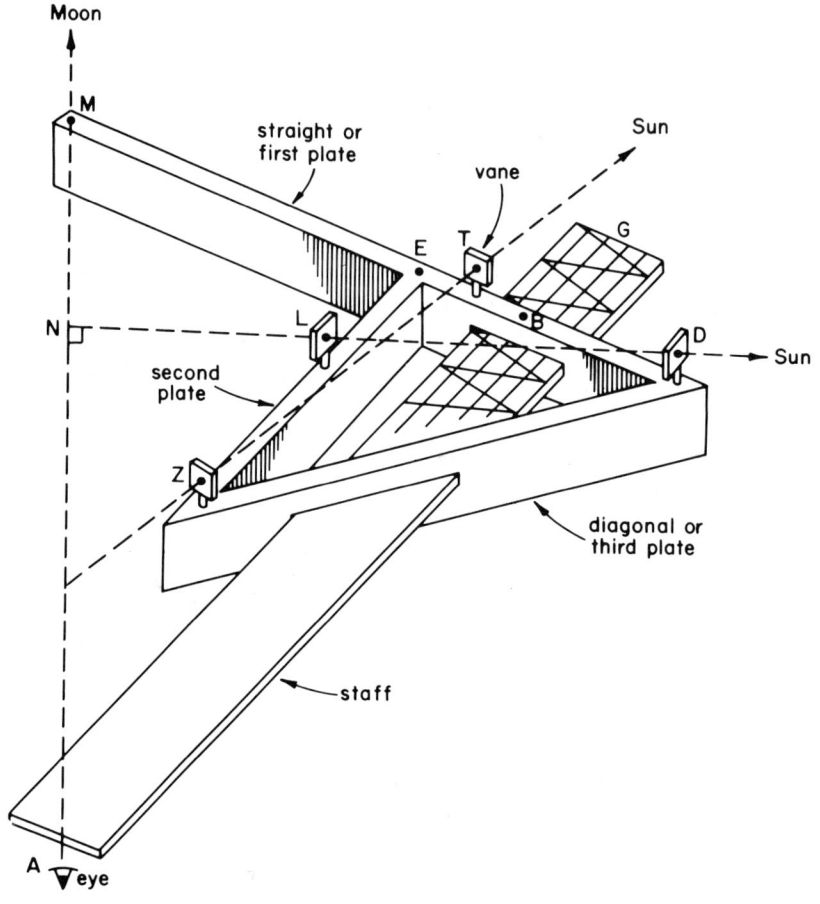

FIGURE 10C-1. The modified Staff for observing the Moon's elongation from the Sun when it is less than 90°.

hence it is 90° (§§ 46–47). We are told that the angle formed at N is not significantly different from the angle formed at A and so we can consider it to be the apparent elongation (§§ 48–49), i.e., the Sun's rays would reach A on a line parallel to line DL. If the periphery of the Moon was seen in the direction AM, rather than its center, one must correct the apparent elongation for the lunar radius (§ 50).

The second case concerns an elongation less than 90° and also appeals to Fig. 10.2. It is now assumed that the Sun's rays pass through vanes at T and Z such that angle EZT is 15° (or angle ETZ is 75°), and that angle BMA is again 75°. If we extend line TZ to line AM (not shown in the MSS), a triangle is

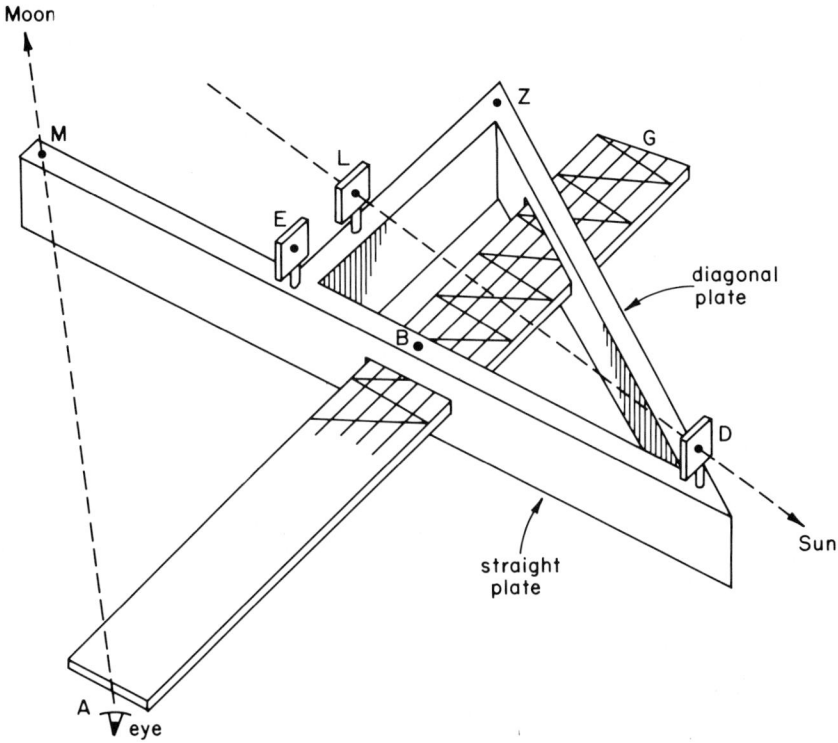

FIGURE 10C-2. The modified Staff for observing the Moon's elongation from the Sun when it is greater than 90°.

formed, and the elongation sought is the supplement of angles BMA and ETZ, i.e., 30° (§ 52).

The third case concerns an elongation greater than 90° and Fig. 10.3 is introduced to illustrate it (see Fig. 10C-2). Now the diagonal plate is farther from the eye than the straight plate through both of which the staff has been inserted. The staff is represented by line AG, leg DE is perpendicular to it at B, the Sun's rays pass through vanes at D and L, and the Moon is seen in the direction AM where point M lies on leg DE extended (§ 55). To find the elongation sought, line NA is drawn parallel to line DL, and line SA is drawn parallel to line DM: the elongation sought is claimed to be angle NAM (§§ 57–59). Here we are explicitly told that the Sun's rays along DL are parallel to rays that would reach point A because of the small distance between them compared to the distance to the Sun (§ 60). Since angle NAS is equal to the observed angle LDM (in the rigid triangle), and angle SAM is the supplement of angle *BMA* also known from the observation, if follows that angle *NAM*, the sum of angles *NAS* and *SAM*, has been determined (§§ 61–63).

A fourth case is considered (Fig. 10.3): the Sun's rays are now assumed to pass through vanes at *D* and *E* at the same time that the Moon is observed in the direction *AM* (§ 64). The elongation is thus the supplment of angle *BMA*, i.e., angle *SAN* is zero (§§ 65–67).

Chapter 11

The "instrument" in this chapter is the Jacob Staff described in chapter 7, rather than the modified version described in chapter 10. A number of instructions are introduced to aid an observer, and to me they indicate that Levi actually used this instrument for astronomical observations. First he remarks on the need to illuminate an instrument at night so that measurements can be taken and that the ends of the crosspiece be visible (§§ 1–2: cf. chap. 7:23). Note that the light is to be put behind the observer, for if it were in front of him, it would interfere with his ability to see of the stars. We are next told that, to avoid introducing error, the two stars whose distance is being observed should lie in the plane of the instrument (§ 3).

The next section (§§ 4–8) concerns distortions due to refraction resulting from clouds and from the thickness of vapors near the horizon that affect both apparent positions and apparent sizes. The remark that in sunlight a star close to the Sun appears smaller than otherwise (§ 8), presumably refers to Venus, the only "star" ordinarily seen during the day (star and planet are often designated by the same word in Hebrew). For the visibility of Venus, and even of Jupiter, in the daytime, see Minnaert, 1954, p. 100.

Levi also mentions that the Jacob Staff is not suited for observing stars more than 40° apart (§ 9) because they could not be seen simultaneously (Brahe describes a sextant for observing angular distances up to 60° by a single observer: Raeder, 1946, p. 78). If the latitudes of the stars are not known precisely, the distance between them (in longitude?) should not be less than 20° (§§ 10–12; cf. chap. 7:117).

Chapter 12

In this chapter Levi discusses three sources of error in the determination of stellar altitude using an astrolabe. The first is that the "zero-point" (here designated the diameter of the instrument) may not lie in the direction to the

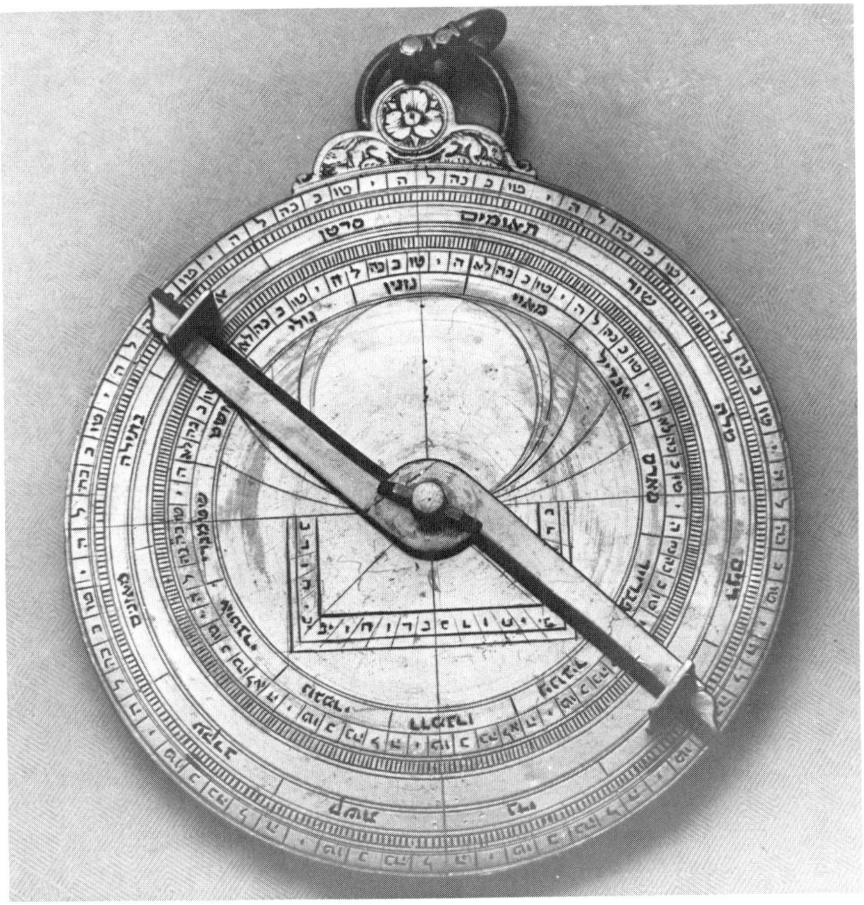

FIGURE 12C-1. The back of an astrolabe inscribed in Hebrew with the alidade directed towards a point in the sky whose zenith distance is 47°: Chicago, Adler Planetarium, No. M-20.

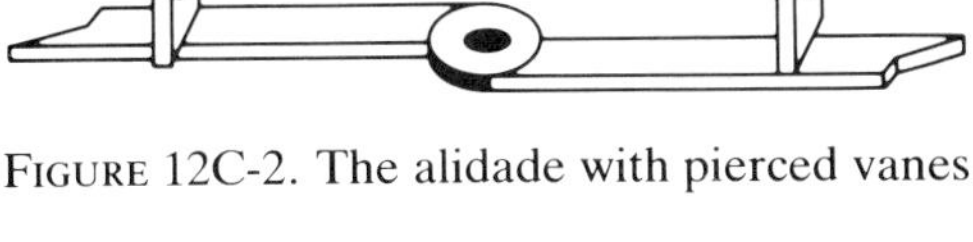

FIGURE 12C-2. The alidade with pierced vanes.

zenith, thus introducing a systematic error in all observations. Second, the line of sight may not be in the same direction as the alidade's reading on the scale, again causing a systematic error. Third, a random error is introduced by reading the scale to minutes whereas it is only graduated in degree intervals. Levi suggests a special procedure to deal with the first source of error, care in construction for the second, and a new way to mark the instrument for the third. His approach to the last problem is to construct a transversal scale, a device usually associated with Tycho Brahe's mural quadrant. The determination of fine subdivisions of a degree for astronomical observations was widely discussed in the sixteenth and seventeenth centuries, and ultimately was solved by the introduction of the vernier scale and the micrometer (cf. Grant, 1852, pp. 449 ff., 461 ff.; King, 1955, pp. 84 ff.).

To facilitate the subsequent discussion let us consider the back of a typical medieval astrolabe (see Fig. 12C-1: Chicago, Adler Planetarium, No. M-20; for a description of this fifteenth century instrument inscribed in Hebrew, see Goldstein, 1976b). Of interest to us here is the ring on top for suspending the astrolabe such that its vertical diameter may be directed towards the zenith, and the scale graduated in degree intervals on the rim. An alidade with pierced vanes (see Fig. 12C-2) such that the holes are aligned along a diameter of the instrument is used to sight an object in the sky whose zenith distance can then be read on the scale.

According to Dreyer (1963, pp. 329 ff.), Brahe mentioned that he got the idea for the transversal scale to subdivide a straight line from Homilius (d. 1562) or from Scultetus. Scultetus "stated that the method was already known to Purbach and Regiomontanus" (Dreyer, 1963, p. 330), but Dreyer found nothing to support this in their published works. He further informs us that Digges (1573) ascribed the introduction of the transversal scale on the cross staff to an English instrument maker, Richard Chancellor (d. 1556: cf. Taylor, 1956, pp. 195, 205; Taylor, 1954, pp. 19 f., 170). However, the transversal scale for linear measurements was discussed by Levi ben Gerson in chapter 7, above, and in the commentary to that chapter we have already considered the influence of the Latin version of Levi's work on the subsequent history of the cross staff.

It can now be seen that the transversal scale for angular subdivisions was also invented by Levi ben Gerson, although he does not specifically take credit for it. It has been claimed that the earliest example of a simple transversal scale is found on the face of the Merton Astrolabe ca. 1350 (see Gunther, 1923, pp. 208 ff., Gunther, 1932, p. 297), but in fact the "transversals" on it were intended to account for the four-year calendar cycle, as the inscription on the rim indicates. Thus it is clear that the Merton Astrolabe and Levi's treatise are completely unrelated to one another. On the other hand, an astrolabe dated 1483, now in Florence (Gibbs et al., 1973, p. 492), clearly

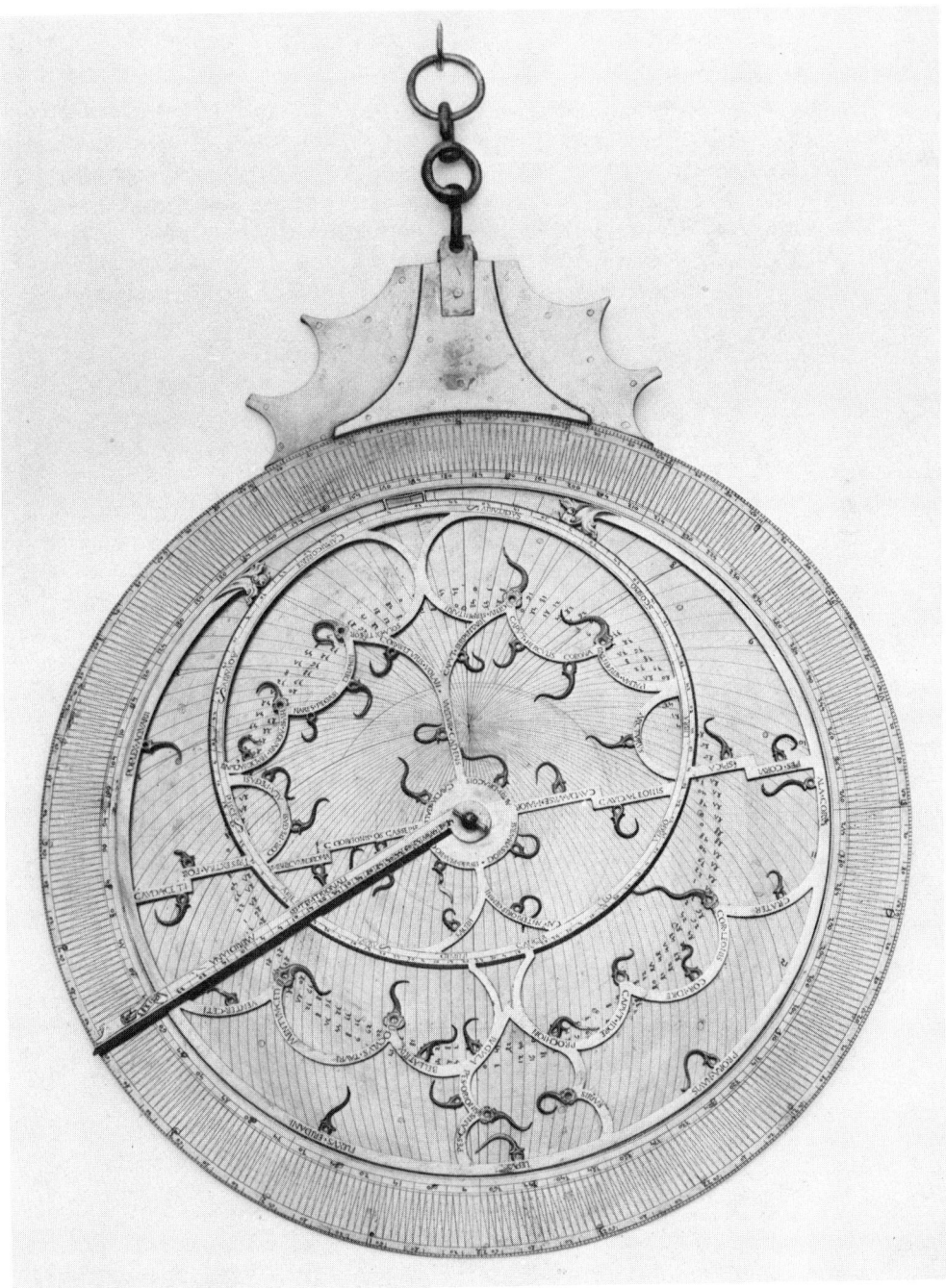

FIGURE 12C-3. The front of an astrolabe, dated 1483, with a transversal scale on the rim: Florence, Istituto e Museo di Storia della Scienze, No. 1096.

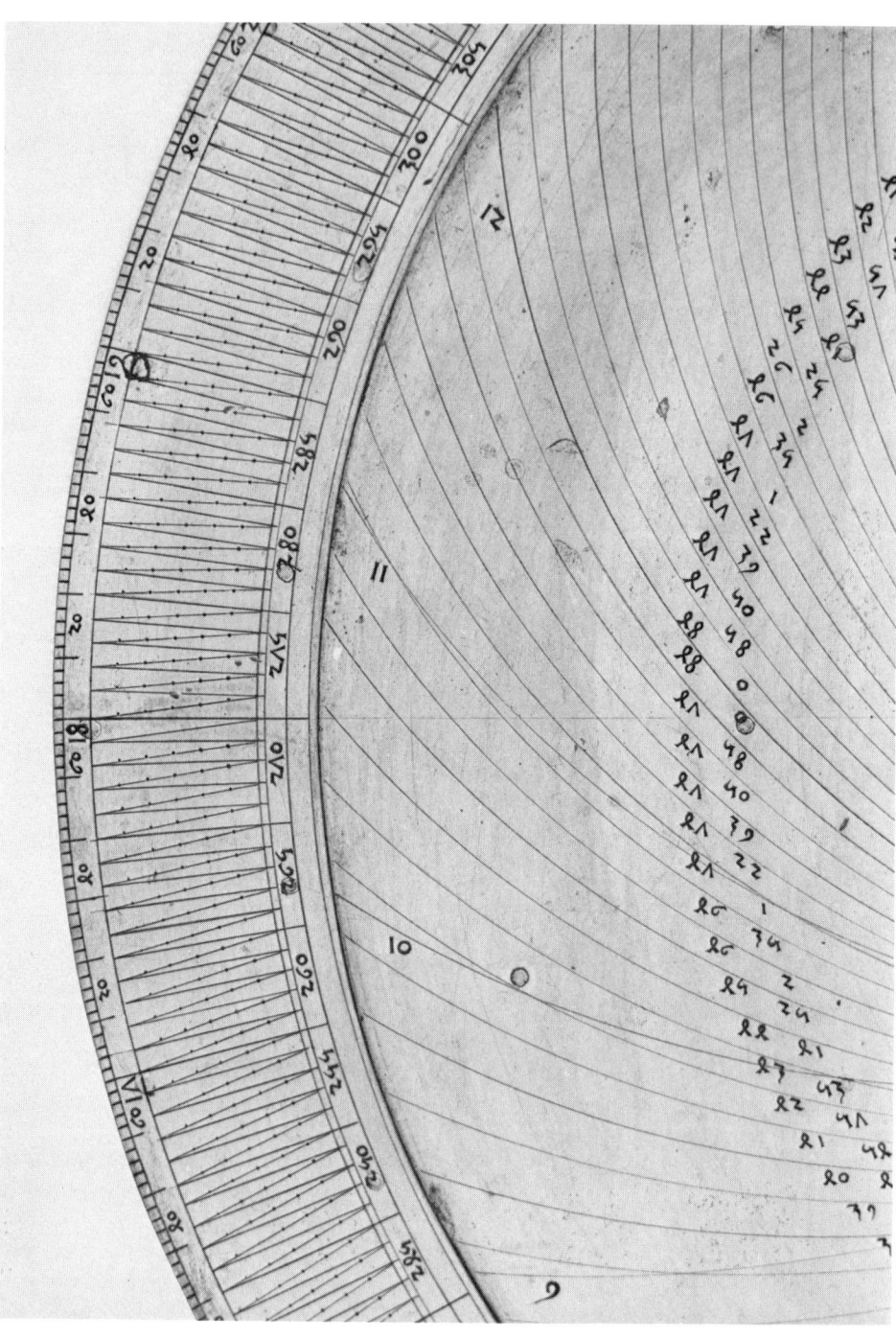

FIGURE 12C-4. Detail of the rim of the astrolabe in Fig. 12C-3.

Commentary: Chapter 12

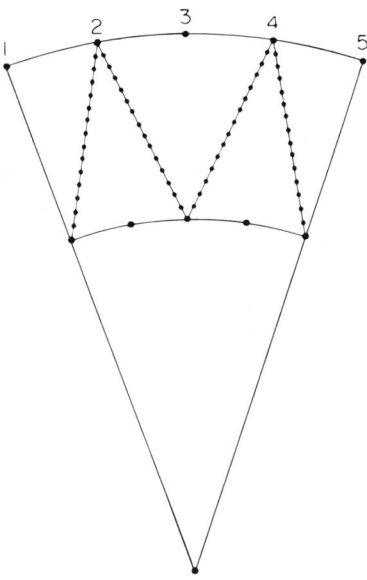

FIGURE 12C-5. The transversal scale for angular measurement drawn according to Levi's description.

displays a transversal scale where each degree is divided into sixths, i.e., 10 minutes of arc (see Fig. 12C-3, and Fig. 12C-4); as far as I know this is the only extant example of a transversal scale for angular divisions before the time of Tycho Brahe. Although the scale on this astrolabe does not conform precisely to Levi's description, it seems reasonable to assume that the maker was influenced by Levi's treatise, directly or indirectly (cf. Introduction).

Levi's procedure is to draw two circles concentric with the center of the instrument such that the radius of the inner circle is $\frac{5}{6}$ of the outer circle, and to mark degrees on both of them (§§ 15–16). Then he tells us to draw diagonal lines from every other point on the inner circle to the two points on the outer circle 1° away from it in either direction (§§ 17–18: cf. chapter 7, where a similar procedure is invoked for dividing a line into fine subdivisions). Instead of simply marking points on each transversal dividing it into 15 equal parts, Levi actually computed the distances from the center to each of these points on the transversal (§§ 20 ff.; see Fig. 12C-5). Theoretically this is a more accurate procedure, but in fact these two sets of points probably could not be distinguished on an instrument of the size he had in mind.

For finding the direction to the zenith accurately Levi suggests taking two observations of solar altitude in a short time interval near noon such that the solar altitude will not have changed, turning the instrument 180° about its vertical axis between the observations so that the altitude readings with the alidade can be marked at two places on the rim symmetricaly about the zenith direction. The interval between the two marks is then bisected and this yields the "zero-point" for the altitude scale (§§ 7–10). To keep the vertical axis stable during the observations, Levi tells us to suspend a plumb-bob from the instrument (§ 6). Curiously, it is to be suspended from a ring at the lower part of the instrument whose width is equal to the thickness of the instrument: I know of no medieval astrolabe fitted with such a ring with the possible exception of one at the Adler Planetarium (No. M–26: on the back of this astrolabe Aries 0° corresponds to March $12\frac{1}{2}$, and this implies a date *ca.* 1275). If the alidade was on the eastern side of the astrolabe for the first observation of the Sun, the instrument should be turned 180° about a vertical axis so that the alidade is on the western side for the second observation (§ 8). After the instrument has been rotated, the alidade is to be redirected to the Sun. If the astrolabe is suspended from a fixed object by a cord, such a rotation would introduce a torque in the cord that might interfere with the second observation. If the cord were detached and then attached again and time allowed for the instrument to become motionless, a longer interval would elapse between the two observations. However, Levi does not discuss this aspect of the observational technique. There are other texts on the astrolabe that deal with sources of error in the construction of the instrument, but I have not found any other text that describes this procedure. In *The Treatise on the Astrolabe*, traditionally ascribed to Masha'allah (but now see Kunitzsch, 1981), there is a passage that comes close to Levi's discussion. However, this text suggests the use of a plumb-line to make certain that the astrolabe is properly set up, not to aid in its construction (Gunther, 1929, pp. 143–44):

> And know that the rings by which an astrolabe is suspended are bevelled on their spinae for as far as one runs upon the other, as if upon the edge of a sword, lest their movement be impeded, and there might perhaps be some leaning towards one side or other. That this may not be, the hole in which the allidada is, and that [second hole] in which the bevelled ring is, register most accurately on the middle line on the Mother, yet from this cause there may be some torsion [or error] in the taking of an altitude. And this you ought to test in this way. Pass a thread through the hole and hang something heavy from it; then hang up the astrolabe by another thread from the same hole. Then if thread falls on thread and if there is no divergence, the astrolabe will be true. But if there is divergence, study to adjust it by moving the hole towards that side to which the thread diverges, [if God be willing].

A set of calculations is introduced to determine the distance from A, the center of the instrument, to point Z on the transversal GD, for each angle GAZ from 0° to 1° at intervals of 0;4° (§§ 21 ff., cf. Fig. 12.1). All calculations in §§ 23–49 were checked with a modern sexagesimal sine table, where appropriate, and the results are correct but for some uncertainty in the last place. The first step is to calculate the lengths of GE and ED (§ 23), and then to find the length of GD (§ 24):

$$GD^2 = GE^2 + ED^2 \qquad (1)$$
$$= (1;2,50)^2 + (9;59,27)^2$$
$$= 1;5,48 + 1,39;49,0$$
$$= 1,40;54,48 \qquad (2)$$

Hence $GD = 10;2,44$. The variant readings for GD (§ 24) are: 10;2,44 (N mg), 10;2,46 (N), and 10;2,47 (P, Q). In § 28 GD is said to be 10; 2,47 in all three manuscripts with N mg again reading 10;2,44. It seems likely that the annotator of N (Finzi) correctly recomputed the square root and that the text tradition is best represented by the reading 10;2,47.

The next step is to solve triangle GED for angle GDE (§§ 26–29):

$$\text{Sin } GDE = (GE/GD)\cdot 60 \qquad (3)$$
$$= (1;2,50/10;2,47)\cdot 60$$
$$= 6;15,16$$

Therefore

$$\text{angle } GDE = 5;59° \qquad (4)$$

Since angle GDE is an angle exterior to triangle GAD, it is equal to the sum of angles GAD and AGD. But angle GAD is given as 1°; hence angle AGD (= angle AGZ) is 4;59° (§ 29). Then the law of sines is applied to triangle GAZ to find line AZ:

$$\frac{\text{Sin } AGZ}{\text{Sin } GZA} = \frac{AZ}{AG} \qquad (5)$$

or, substituting 4;59° for angle AGZ:

$$\frac{\text{Sin } 4;59°}{\text{Sin } GZA} = \frac{AZ}{60} \qquad (6)$$

where angle GZA is equal to $180° - (4;59° + n\cdot 0;4°)$, for $n = 0, \ldots, 15$. Table 12.1 displays Levi's results (§§ 32–50). Presumably, for purposes of locating the points on the transversal, the last place in the value for AZ was only meant to aid in the rounding to minutes.

Some comment on terminology is appropriate here. In § 2 the expression "the diameter of the instrument" refers to the vertical axis of the astrolabe that is supposed to point toward the zenith, but may be inclined to that direction. The Hebrew word here translated alidade (§ 2) is *luaḥ* which means table or plate: my rendering of it is based on the context. One would expect *maʿṣad* (from the Arabic ʿidāda) but, as far as I can determine, Levi does not use *maʿṣad* at all and one is forced to conclude that he did not know the word. Gandz (1970, pp. 258 ff.) discussed the Hebrew terminology for the parts of the astrolabe: *luaḥ* is used for the mater or for the plates that are placed within it. For alidade Gandz (p. 261) gives *beriaḥ* and *maʿṣar*; the letters *resh*

Table 12.1. Distance of Points on the Transversal Scale

	Angle GAZ	AZ
1	0; 0°	60
2	0; 4	59;12,31
3	0; 8	58;26,45
4	0;12	57;41,41
5	0;16	56;57,50
6	0;20	56;14,59
7	0;24	55;33,20
8	0;28	54;52,53
9	0;32	54;13,14
10	0;36	53;34,22
11	0;40	52;56,26
12	0;44	52;19,31
13	0;48	51;43,28*
14	0;52	51; 8,13
15	0;56	50;33,45
16	1; 0	50

*The variant 51;43,58 is slightly better.

and *dalet* are difficult to distinguish in most medieval Hebrew scripts, but in a manuscript that I checked, the form is clearly *maʿṣad* (cf. Goldstein, 1967b, Hebrew section, pp. 58–59: MS Bodleian heb. Michael 400, 61v:29,31,32). For the corresponding Arabic terminology, see Hartner, 1968. For a description of an astrolabe inscribed in Hebrew, see Goldstein, 1976b.

Chapter 13

We are told that this chapter will be devoted to two methods for finding the local meridian. The first method is called "approximate": we note the moment when the Sun appears to be highest in the sky, i.e., at noon, and we draw a straight line on the floor of a house following the rays of the Sun that enter along the side of the window (facing south, to be sure) and that is our meridian (§ 3).

The second method is rather complicated and we are told that our geographical latitude (or the height of the pole) must first be determined (§ 5). Levi suggests the use of the astrolabe for measuring noon solar altitudes at the two solstices, a method similar to that of Ptolemy (cf. Britton, 1969). If we call the noon solar altitude at winter solstice h_w, the noon solar altitude at summer solstice h_s, and ϕ our local latitude, then (§§ 7–8)

$$\phi = 90° - [h_w + \tfrac{1}{2}(h_s - h_w)]$$
$$= 90° - \tfrac{1}{2}(h_w + h_s)$$

and ε, the obliquity, is also determined (§§ 9–10):

$$\varepsilon = \tfrac{1}{2}(h_s - h_w)$$
$$= 23;33°, \text{ very nearly.}$$

The strategy is then to find a point on the western wall of the room, at sunrise on winter or summer solstice, that represents the image of the center of the Sun as seen through a window on the eastern wall of the same room. The direction from the window (more precisely the upper northern corner of the window) to the point on the western wall is marked on the floor. From the obliquity and the local latitude the ortive amplitude can be calculated (i.e., the arc from the east point to the solar rising point on the horizon), and hence an east-west line can be drawn on the floor. Finally, the meridian is drawn perpendicular to the east-west line passing through a point directly below an edge of a window facing south. I am not aware of this procedure in any earlier astronomical text. However, al-Bīrūnī (eleventh century) described a similar method for deriving the meridian making use of observations of the shadow of a gnomon at sunrise from which the east-west line is found by means of the ortive amplitude: the meridian is then drawn perpendicular to the east-west line [Kennedy, 1976, vol. 1, p. 167, vol. 2, p. 94; cf. commentary to chap. 15, eq. (7), below]. There are many subtleties in Levi's description, many of which are not explained, and I will attempt to clarify them here.

We begin by considering a room whose floor lies in the plane of the horizon, and whose opposite walls are parallel such that the wall facing east has a window whose upper edge is parallel to the floor and whose sides are perpendicular to the floor (§§ 11–12). A line is drawn on the western wall parallel to the floor about 5 spans (approximately 1 m) below the height of the top edge of the window on the eastern wall (§ 13). Since we are told (§ 84) that the room may be as large as 40 or 50 spans (about 8 to 10 m), in effect Levi has marked an artificial horizon that the Sun will cross when it reaches an altitude of about 5° to 7°. No explanation is offered so it is unclear if this was done because of obstacles (e.g., another building) that concealed the eastern horizon, or because of his awareness that atmospheric refraction would affect the observation most seriously at the true horizon. But, in the absence of a specific statement on the problems of horizontal refraction, it does not seem appropriate to ascribe this knowledge to him. A second line is drawn parallel to the line on the western wall about 1 span below it (§ 14). The image of the window on the western wall is marked when its upper edge reaches the upper line, and the mark is placed on the upper line directly above the place where it crosses the lower line (§ 15). No explanation is given, and it took me a long time to discern the reasons for this procedure. In Figure 13C-1 the room is

172   The Astronomy of Levi ben Gerson

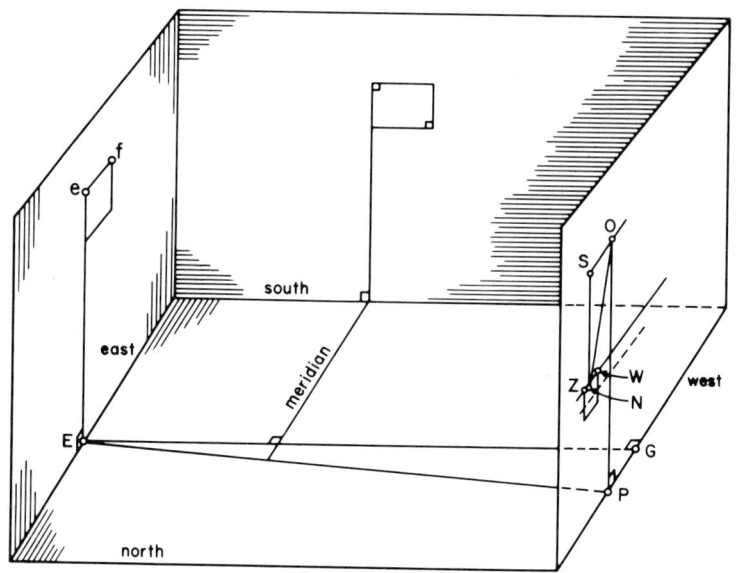

FIGURE 13C-1.

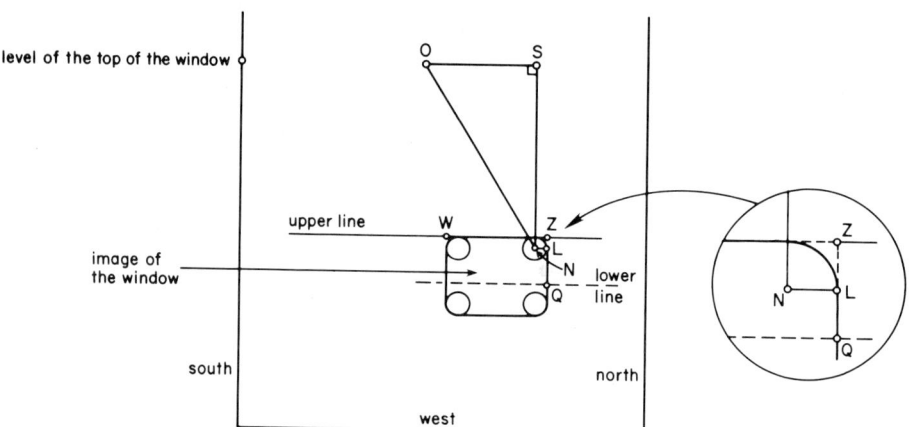

FIGURE 13C-2. The western wall in Figure 13C-1, as seen from inside the room.

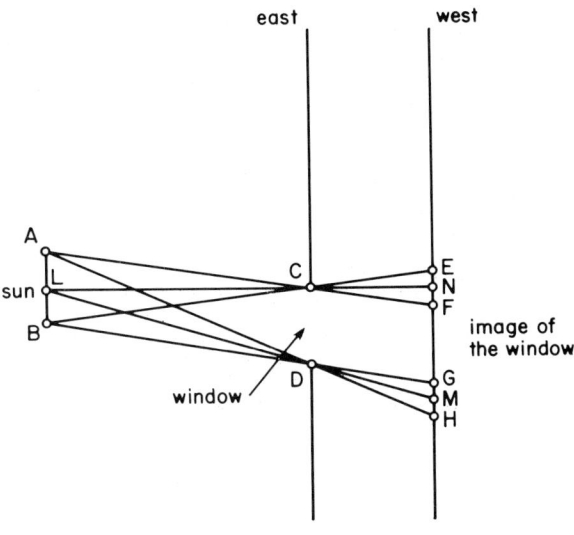

FIGURE 13C-3.

shown with a window whose upper edge is *ef*, on the eastern wall, and whose image on the western wall is *ZW*: note that for a window of considerable size the orientation of the edges is conserved. In Fig. 13C-2 we see only the western wall: the image of the window has rounded corners because each point on the window generates a circular image corresponding to the Sun, and the images overlap (see Fig. 13C-3 and chap. 5:16; cf. Lindberg, 1970, pp. 304–7). When the upper edge of the image reaches the upper line (the image descends as the Sun rises), the northern side of the image crosses the lower line at *Q*. If *QZ* is drawn perpendicular to the upper line, point *Z* satisfies the conditions stated by Levi in the ensuing discussion. Line *QZ* is greater than *ZL*, the solar radius, when the lower line is set at one span or more below the upper line, i.e., the angle formed by *QZ* and a point on the upper edge of the window is a little more than 1° where the angle formed by *ZL* and a point on the upper edge of the window is taken to be 0;15° (§ 23).

Levi introduces Fig. 13.1 to help us understand the subsequent steps in his procedure. The difficulty is that the same plane represents both the south wall and the west wall. We consider only the point at the upper edge on the northern side of the window, point *E*. The image of the Sun *TH*, where *T* is the center of the Sun and *H* its lower limb, is seen on the western wall as line *ZL*, where the image is inverted, i.e., *Z* is the image of the lower limb and *L* is the image of the center of the Sun (§ 20). Angle *ZEL* is the angular diameter of the Sun taken to be 0;15° at winter solstice as determined by observation

(§ 23), using an instrument—presumably the one described above in chapter 9 (cf. note to chap. 9:11 ff). From angle *ZEL* and the size of the room, the length of *ZL* is computed (§§ 24–27), because it is very difficult to measure the rounding of corners of the image on the western wall (cf. Fig. 13C-2). Point *Z* is determined by the observation and point *L* fixed by computation. Then from *L* a line is drawn parallel to the horizon and we mark the point *N* such that *ZL* is equal to *LN* (§ 29), i.e., point *N* is the image of the center of the Sun that passes through point *E* on the window (Fig. 13C-2). Clearly point *N* is to the south of point *L* (§ 28). From point *N* we draw *NS* on the western wall perpendicular to the horizon such that the height of point *S* above the floor is equal to the height of point *E* above the floor (§ 30), i.e., line *ES* lies in a horizontal plane. The next step is to measure line *NS* whose position and length serve as our observational data. Levi remarks that yet another correction is needed in order to locate the point on the western wall corresponding to sunrise (§ 31). Point *S* is at the appropriate height on the wall, but the point we seek is to the south of it on the western wall, as we shall see.

The next section (§§ 32–39) is concerned with preliminary matters related to the day-circle of the Sun. This circle is parallel to the equator and hence its inclination to the horizon in the same (§ 34). Moreover, the projected image of the Sun formed by rays that pass through the center of the equator follows a circular path parallel to the equator but on the other side of it, and the earth can be considered as a point lying at the center of the equator (§§ 36–37). Thus, when the Sun lies to the north of the east point, its image lies to the south of the west point by an equal amount (§ 38). The next remark (§ 39) introduces the proof that follows for the computation of the distance from point *S* to point *O* in Fig. 13.1. We are told that it is necessary, but not sufficient, to know the angle between the equator and the horizon, because we have to find the angle between a different plane that meets the horizon at a different angle. This plane is defined as the great circle passing through the Sun's rising point and its position at the time of the observation. A new figure is introduced (§ 40: Fig. 13.2, redrawn as Fig. 13C-4) in which the day-circle of the Sun is *ZLH*, meeting the horizon at line *ZH*, and parallel to the equator whose center is point *E*. The horizon is circle *ABDG* where point *B* is the east point and point *G* is the west point. At the moment of observation, the center of the Sun is at point *L* and its image on the western wall is at point *R* (the ray passes through point *E*, the center of the equator). The angle between the plane of the Sun's day-circle and the horizon (the complement of the local latitude) is here represented by angle *LKN* where *LK*, which lies in the plane of the day-circle and *NK*, in the plane of the horizon, are both perpendicular to line *ZH*. Point *N* is defined as the foot of the perpendicular drawn from point *L* to the horizontal plane. The line *LN* is equal to the sine of the

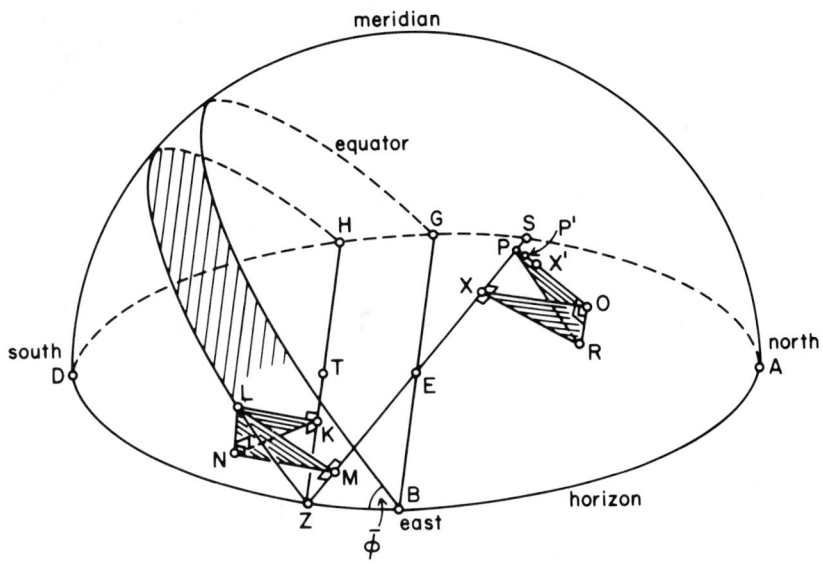

FIGURE 13C-4.

altitude of the Sun at point L and that angle can be determined with an astrolabe (§ 45). Levi now remarks that if we were observing at point T, the center of the Sun's day-circle, angle LKN would be appropriate, but since we are at point E, we must find a different angle shown as angle LMN in the figure where point M lies on the line from the true solar rising point to the center of the equator. In other words, since we project both Z and L through point E, we must consider the plane defined by those three points and its inclination to the horizon, i.e., angle LMN (§§ 47–50). Clearly angle LMN is not equal to angle LKN that was previously considered.

The next section (§§ 51–56) explains how angle LMN is computed. We are given the altitude of point L by observation, from which arc ZL on the day-circle can be computed (the method to be explained later) and hence its chord, line ZL with respect to ZT, the radius of the day-circle. But the ratio of the radius of the day-circle to the radius of the equator is known: hence, the ratio of line ZL to the radius of the equator (to be set at 60) is also determined (§ 53), and from it the amount of arc ZL on the great circle can easily be found. Line LM is the sine of arc ZL on the great circle and so it too is known (§ 55). Since we had already found line LN, right triangle LMN is determined and we can solve it for line NM and angle LMN, the inclination to the equator of the great circle passing through points Z and L (§ 56).

The next step is to draw the image of triangle LMN on the western wall

(§§ 57–60). We first assume that the western wall is perpendicular to the direction ZES in Fig. 13C-4. In that case, corresponding to right triangle LMN is triangle RXO, where LER is a straight line and point R, the center of the observed image of the Sun, corresponds to point L, the center of the Sun at the time of the observation. Plane RXO is parallel to plane LMN, and both of them are perpendicular to line ZES. Triangle ROX is marked on the western wall as triangle NSO in Fig. 13.1 such that angle SNO (i.e., angle ORX in Fig. 13C-4) is equal to angle NLM (in Fig. 13C-4). If the western wall were perpendicular to line ZES, we would be finished with this part of the argument, because point O, to the south of point S in Fig. 13.1, would represent the image of the center of the Sun at the time of its true rising. But there is no reason for the western wall to be perpendicular to line ZES in Fig. 13C-4 because point G is the west point and a wall truly on the west side of the room would be perpendicular to line BEG.

We are now told how to make the relatively small correction for the difference between the west wall and plane XOR in Fig. 13C-4. Point R is still the image of the center of the Sun at L and point R shall be considered to lie on the west wall (corresponding to point N in Fig. 13.1). Point O in Fig. 13C-4 corresponds to point S in Fig. 13.1, and the plane of the west wall is represented by plane ROP in Fig. 13C-4 such that line PO is in the horizontal plane. Hence, point P is the projection on the western wall of point Z, the true rising of the Sun, and we seek the amount of line OP, i.e., we have to solve right triangle OXP in Fig. 13C-4. Clearly line OP is always greater than line OX (§ 68), and line OX may be found from the similarity of triangles LMN and ROX where RO corresponds to the measured quantity NS in Fig. 13.1.

Though for all practical purposes there is no significant difference between OX and OP (§ 68), Levi goes on to describe an iterative procedure for arriving at a more accurate determination of OP (§§ 69–74). The line from the window to the place of apparent rising (§ 70) is line EX' where X' lies in the horizontal plane on the west wall such that OX' = OX, i.e., we have rotated triangle ROX about OR into the plane of the west wall. We are then asked to find angle EX'O, i.e., to measure this angle (§ 71). The next step is to locate a point P' closer to P than X' where P corresponds to the place of true rising. If we let a be angle EX'O, we can compute the length of OP' ($\approx$ OP) as follows:

$$\sin a = OX/OP'$$

Since angle a and line OX are known, the length of OP' (called OP in § 72) is determined. But as we can see from Fig. 13C-5, which illustrates a portion of the horizontal plane in Fig. 13C-4:

$$\sin b = OX/OP$$

where b is angle EPO. If angle a is not a right angle, sin a is less than 1, and

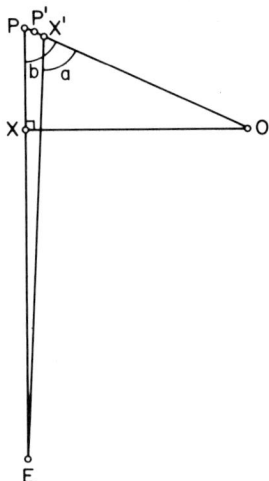

FIGURE 13C-5.

OP' is greater than OX ($= OX'$). Moreover, since angle a is greater than angle b, sin a is greater than sin b, and thus

$$OP > OP' > OX'$$

In other words point P' is closer to P than was our point X'. Hence, we can iterate the procedure starting with a measurement of angle $EP'O$ and this will get us even closer to point P (§ 74). Levi says that, in fact, there is little difference between X' and P and it will have hardly any effect on the meridian whose determination is the ultimate goal (§ 68–69).

Once the point on the western wall corresponding to true sunrise has been marked, we are ready to find the meridian in the room (§§ 75–84). First we must find the points on the floor that correspond to the upper northern edge of the window on the eastern wall and the image of the center of the Sun at true sunrise on the western wall. This is done by means of a plumb line (§§ 75–76). The line connecting these two points corresponds to line ES in Fig. 13C-4 (§ 77). Angle SEG in Fig. 13C-4 is equal to angle BEZ, the ortive amplitude, and this angle is to be found by means of the appropriate table (§ 78). No such table has been found among those of Levi but, in any event, the ortive amplitude can be computed from the obliquity and the local latitude [see comments to chap. 15, eq. (7)]. With this angle a line, corresponding to line EG in Fig. 13C-4, is to be drawn on the floor (see Fig. 13C-1)—this is our east-west line (§§ 79–81)—and a line perpendicular to it will be the meridian. If we have a window on the southern wall, we are told to drop a plumb-line from one of its upper edges to a point on the floor, and from that point to draw our meridian (§§ 82–83). When the ray from the center of the Sun passes the edge of this window on the southern wall to the meridian line on the floor, it is noon (§ 83). Finally the size of the room is given: 40 or 50 spans (about 8–10 m), presumably in the east-west direction (§ 84; see chap. 15 : 44).

A room 8–10 m long with a window about 5 m above the floor (chap. 15 : 11) seems to be too large for an ordinary household, and it seems reasonable to suppose that Levi was referring to the synagogue in Orange. No information is available about that building, but the surviving eighteenth century synagogue in Cavaillon (not far from Orange) has these dimensions, approximately. Moreover, the former synagogue in Avignon, described in 1765, was even larger: there were 4 windows on the west wall and 6 on the south wall measuring about 2.5 m in height by 1.25 m in width (10 *pans* by 5 *pans*), and the room was about 14.5 m from north to south (7 *cannes* 2 *pans*), about 12 m from east to west (6 *cannes*), and about 11.5 m high (5 *cannes* 6 *pans*: Moulinas, 1980, pp. 18, 26). It seems most unlikely that the synagogues in these towns were enlarged subsequent to the fifteenth century because of the general decline in the size of the Jewish population as well as the difficulties in obtaining permission for any modifications in these structures. It is generally understood that synagogues are oriented to the east, but I know of no study of actual synagogue orientations in general or of those in Provence in particular.

CHAPTER 14

In this chapter Levi emphasizes the difficulties to be resolved before embarking on planetary theory. The observed positions of the planets depend on the positions of the fixed stars, which in turn depend on the position of the Moon, which in turn depends on the position of the Sun. We are told that Ptolemy's lunar model is not adequate even at syzygy (§§ 2–4), and elsewhere the error may reach $\pm 1\frac{1}{2}°$ in longitude (§5). Another difficulty in the lunar model concerns parallax (§§ 6–8), i.e., Levi does not accept Ptolemy's lunar distances because, as he states elsewhere, the Moon is not seen to be twice as large in diameter at quadrature than at opposition, as Ptolemy's model requires (cf. chap. 17:2–3). As for the Sun, Levi is aware of the difficulties in observing equinoxes cited by Ptolemy in *Almagest*, III, 1, and calls attention to discrepancies in the solar parameters used by later astronomers (§§ 9–15). Of special interest is his remark that a small error in the observed solar altitude may produce a large error in solar longitude (§ 11: cf. *Almagest*, III, 1; trans. Toomer, 1984, p. 134), e.g., at equinox 1° in solar longitude corresponds to about 0;24° in solar declination (the co-latitude plus the solar declination is equal to the noon solar altitude). Ptolemy's maximum solar equation was indeed 2;23° (§ 12; cf. *Almagest*, III, 6), and al-Battānī used a value close to 2;0°: in fact, 1;59,10° (Nallino, 1907, vol. 2, p. 81; cf. Nallino 1903, vol. 1, p. 214, for a list of other early Islamic astronomical texts which used values close to 2;0° for this parameter). Ptolemy's value for the solar apogee was indeed Gemini 5;30° (§ 13: cf. *Almagest*, III, 4) whereas al-Battānī used the value Gemini 22;15° for the year 1194 Seleucid Era that began on 1 Sept. 881 (Nallino, 1903, vol. 1, p. 214), with a precession of 1° in every 66 solar years (Nallino, 1903, vol. 1, p. 114, cf. Goldstein–Sawyer, 1977, p. 167). With al-Battānī's parameters, the solar apogee reached Gemini 29;15° (§ 14) in 1343 (the year prior to Levi's death), i.e., Gemini 22;15° + 7° = Gemini 29;15°, and 7° of precession corresponds to 462 years at 1° in 66 years. If we add 462 years to the date used by al-Battānī, i.e., 881 A.D., the result is 1343, which is about 12 centuries after the observations of Ptolemy (§ 13). Later on, Levi describes his determination of the solar apogee, the motion of the solar apogee, and the motion of precession (see chapters 15, 57, and 61; cf. Goldstein, 1974a, p. 94; and Goldstein, 1975).

The value Gemini 17;30° for the solar apogee (§ 13) is close to the value in the Hindu astronomical text, the *Suryasiddhanta*: Gemini 17;15° (cf. Nallino, 1903, vol. 1, p. 218). It is also close to the value Gemini 17;50° in the *Toledan Tables* which, according to Toomer (1969, pp. 321–22), is probably due to al-Zarqāl and intended to be the sidereal longitude of the solar apogee at the epoch of the Hijra, 622 A.D.

CHAPTER 15

This chapter is concerned with the determination of the solar parameters: the obliquity, the maximum equation, and the apogee (§§ 1–3). The procedure depends on the use of a room in a house with a window on the south side, presumably the same room mentioned in chapter 13. We are first given an idealized account where the walls and floor are perpendicular, and then a procedure to make adjustments for the inclination of the walls to the floor, as was necessary for the room Levi actually used (§ 63).

The meridian is drawn on the floor from a mark directly below the western edge of the window on the south wall according to the procedure described in chapter 13 (§§ 4–7). To the west of this meridian by about half a span (i.e., about 10 cm) another line is drawn parallel to the meridian (§ 8). The function of this line is to aid in determining when the image of the center of the Sun reaches the meridian (see Fig. 15C-1). The distance between the two lines

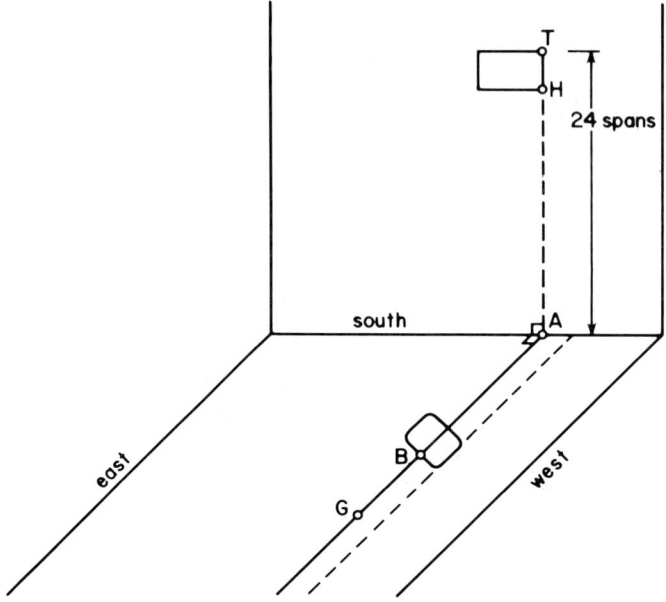

FIGURE 15C-1.

would correspond to the size of the image of the solar radius ($\frac{1}{4}°$) if it were $\frac{1}{8}$ span, for we are correctly told that near summer solstice, a change of 1° in solar declination corresponds to about $\frac{1}{2}$ span where the height of the top of the window is 24 spans and the geographical latitude is 44° (§ 11). In Fig. 15C-2 let the distance AB on the meridian be s, and the angle at B be α. Then

$$s = \frac{24}{\tan \alpha} \quad (1)$$

If $\bar{\phi}$ is the co-latitude, and δ the solar declination, then

$$\alpha = \bar{\phi} + \delta \quad (2)$$

If

$$\alpha_1 = 46° + 23;33°, \text{ then } s_1 = 8;57^p$$

$$\vec{\alpha}_2 = 46° + 22;33°, \text{ then } s_2 = 9;25^p$$

Thus

$$s_2 - s_1 \approx 0;30^p$$

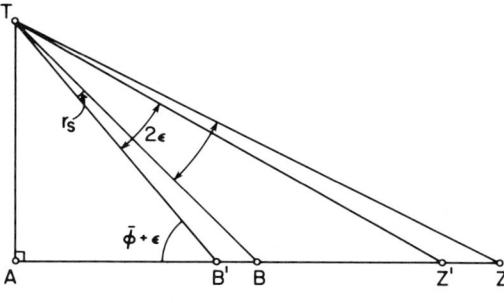

FIGURE 15C-2.

When the image of the western edge of the window reaches the more westerly line on the floor, we are to make a mark at point B where the image crosses the other line (farthest from the south wall) and that is the image of the top of the window (§ 9). The Sun reaches its highest noon altitude at summer solstice (Cancer 0°), and hence the image of the window will be closest to the south wall at that time (§ 10). The two parallel lines on the floor can be used to form a transversal scale (cf. chapter 7) to allow the distance of the image from the south wall to be measured with greater precision (§ 12).

It is difficult to determine the day on which solstice takes place because of the small change in solar declination at that time. Therefore, Levi indicates that he started observing the image of the Sun on the floor about 10 days before the solstice and each day thereafter until the mark was closest to the south wall (§ 13). He then continued to mark the place of the image on the floor until it returned to the mark made 10 days prior to the solstice (§ 15). This would appear to be a good way to determine the time of the solstice because it takes advantage of the symmetry about that moment. In Levi's time this procedure was particularly simple because the solar apogee was very near the summer solstice (§ 21), and so the velocity of the Sun was symmetric about it.

The next section describes a method for finding the moment the solstice from observations made at noon before and after the day on which the solstice took place. In the first case it is assumed that the image returned to the same mark 19 days later: hence, the solstice took place 9 days 12 hours after the moment the Sun reached the mark the first time (§§ 16–21). In the second case the image does not return exactly to the same mark so that interpolation is required to find the moment corresponding to the image reaching that mark (§§ 22–27). In Fig. 15.1, line TH is the western edge of the window, A is the point on the floor directly below the western edge of the window, and ABG the meridian on the floor. The Sun reached point G 10 days before solstice and point E 9 days before solstice, but after the solstice the closest to point G that it reached at noon was point D. We then use linear interpolation to find x, the appropriate fraction of a day:

$$\frac{GD}{GE} = \frac{x}{1 \text{ day}} \qquad (3)$$

Once the moment after the solstice corresponding to point G is determined, the same bisection procedure used in the first case allows us to find the moment of the solstice. The same process can be used to find winter solstice (§ 27). If the solar apogee was at Cancer 0°, the times between the solstices would be equal (§ 28); if they are not equal, the apogee lies on the arc that the Sun took longer to traverse, i.e., the solar velocity is slowest at apogee (§ 30). In the preceding section there was no need to correct the intersection of the image with the meridian for the solar radius because its effect is very nearly the same for all the observations near a single solstice, i.e., the point on the meridian corresponding to the center of the Sun is about 0;7 spans closer to the south wall near summer solstice than the image of the top of the window, corresponding to the lower limb of the Sun. In Fig. 15C-2 point T represents the top of the window on the south wall, and AZ is the meridian on the floor. Point B' is the image of the center of the Sun at summer solstice, and point B is the image of its lower limb. Similarly, point Z' is the image of the center of the Sun at winter solstice, and point Z is the image of its lower limb. We wish to find BB' where:

$$AB = \frac{24}{\tan(\overline{\phi} + \varepsilon - r_s)} \qquad (4)$$

and

$$AB' = \frac{24}{\tan(\overline{\phi} + \varepsilon)} \qquad (5)$$

Thus

$$BB' = \frac{24}{\tan(\overline{\phi} + \varepsilon - r_s)} - \frac{24}{\tan(\overline{\phi} + \varepsilon)} \qquad (6)$$

where r_s is the radius of the Sun at summer solstice. If we let $\phi = 44°$, $\varepsilon = 23;33°$, $r_s = 0;15°$, and line $AT = 24$ spans, then line BB' will be 0;7 spans.

In the next section we are told how to find the obliquity from observations of the two solstices (§§ 32–39). In Fig. 15.1, point B represents the image of the top of the window on the floor at summer solstice, and point Z is its image at winter solstice. Thus, angle BTZ is twice the obliquity and it can be found from measurements of AB and AZ. The text tells us to measure TB, TZ, and BZ (§ 35), but it is clearly difficult and unnecessary to measure TB and TZ.

We are also told to correct for the solar radius (§ 36), but it is not clear that Levi was aware that the effect of the solar radius is noticeably different at the two solstices. The procedure in the text (§ 36) is to add half of angle BTZ to angle AZT and then to add the solar radius to the sum in order to get $\bar{\phi}$. This procedure assumes that the apparent solar radius at summer solstice is equal to the apparent solar radius at winter solstice, i.e., that angle $B'TB$ is equal to angle $Z'TZ$ in Fig. 15C-2. However, according to Levi they are slightly different and this might produce a small error in the determination of geographical latitude. Half of angle BTZ is assumed to be the obliquity, rather than half of angle $B'TZ'$, using the same approximation (§ 35). With the latitude and the obliquity we can find the noon altitude of the Sun on any day (§ 37), and similarly from an observed noon altitude, the true longitude of the Sun can be determined (§ 38), i.e., if h_n is the noon solar altitude, and δ its declination, then

$$h_n = \bar{\phi} + \delta$$

Given a solar longitude, we can find the declination by means of a table of declinations (cf. Goldstein, 1974a, p. 159) and hence h_n can be computed. If h_n is observed, we subtract $\bar{\phi}$ from it which gives us δ, and from the declination table we can find the corresponding solar longitude. Finally, we are told to observe solar longitudes by means of noon solar altitudes 40 or 50 days before and after summer solstice in order to determine the solar apogee and maximum equation. The procedure is not presented here, but it is to be found in chapter 57 (P 105b–107b). There we are told of observations when the Sun had the same declination on 24 April 1334 and 4 August 1334, and the interval between them is given as 101 days and 18;26 hours. Levi's observation of summer solstice on 13 June 1334 is reported in chapter 55 (P 104a:30). From these three observation he computed the solar apogee and eccentricity using Ptolemy's method for finding the lunar apogee and eccentricity from three lunar eclipses (chap. 57; cf. Neugebauer, 1957, pp. 210–14) that Ptolemy also used to find the apogee and eccentricity of each outer planet from observations of three oppositions (§ 39; cf. Neugebauer, 1975, pp. 172–79). In al-Bīrūnī's *Chronology of Ancient Nations* (trans. Sachau, 1879, p. 167) we are told that his teacher, Abū Naṣr Ibn ʿIrāq (flourished ca. 1000), had recognized that the solar apogee and eccentricity could be determined from any three observations of the Sun, not necessarily at solstices and equinoxes as Ptolemy had done, or at the mid-points between them as the Muslim astronomers of the ninth century had done. Levi's results in chapter 57 are that the solar apogee lies at Cancer 3° and that the solar eccentricity is 2;14 (where the radius is 60).

The next section (§§ 40–42) tells us that the apparent diameter of the Sun would vary according to its distance from the center of the world, and that observations of these apparent diameters would allow us to determine the solar eccentricity. With the staff arranged as a camera obscura, as described in chapter 9, Levi suggests that such a test could be performed. His actual determination of the solar eccentricity, based on observations of the solar diameter at summer and winter solstices in 1334, is given in chapter 56 (P 104b), but he later changed the solar eccentricity based on an analysis of solar eclipses (cf. Goldstein, 1979, p. 104).

This section ends (§ 43) with a note on what can be determined with the completed solar theory: the true position of the Sun at any time, and the moments of the solstices that can be derived from the position of the apogee and the maximum solar equation (provided that an initial position is known).

The following section (§§ 44–48) describes the experimental conditions for observing the ortive amplitude. The room is specified as having a north wall of length 40 spans (8 m) or more (§ 44). Two lines are drawn parallel to the floor on the west wall, 6 spans below the level of the top of the window on the east wall for the first line, and another line about 1 span below it (§ 45). This arrangement is essentially the same as that described in chapter 13, except that there the upper line on the west wall was only 5 spans below the level of the top of the window (chap. 13:13). The procedure for marking the upper line at the time of the observation (§ 46) is the same as that described in chapter 13 (cf. Fig. 13C-2, where point Z represents the appropriate mark). We are then told to note the solar declination from which the ortive amplitude can be computed and that each degree of declination corresponds to more than a span on the line on the western wall to be marked (§ 47). To explain this, consider the modern formula for ω, the ortive amplitude (cf. Nallino, 1903, vol. 1, p. 178):

$$\sin \omega = \frac{\sin \delta}{\cos \phi} \qquad (7)$$

If $\phi = 44°$

$$\sin \omega = 1;23 \sin \delta \qquad (8)$$

When $\delta = 0°$, $\omega = 0°$; when $\delta = 1°$, $\omega = 1;23°$. But when $\delta = 22;33°$, $\omega = 32;13°$; when $\delta = 23;33°$, $\omega = 33;44°$. Thus a change of 1° in declination corresponds to more than 1° in ortive amplitude. At a distance of 40 spans from the window to the line on the west wall, 1° of ortive amplitude corresponds to about 0;42 spans. Thus, in the interval where the declination changes from 0° to 1°, the difference in ortive amplitude is 1;23° corresponding to 0;58 spans, and in the interval where the declination changes from 22;33° to 23;33°, the difference in ortive amplitude is 1;31° corresponding to 1;4 spans. We can make marks on the line on the west wall from which the moment of the solstice can be derived by a precedure analogous to that used on the meridian (§ 48). This procedure ought to be more precise than that on the meridian (§ 48) because here 1° of declination near solstice corresponds to 1;4 spans on the line on the west wall, whereas the same change in declination corresponds to only half a span on the meridian line (§ 11).

This experimental arrangement is now invoked to determine the solar apogee and maximum equation (§§ 49–60). We are reminded that the way to find the true solar position at sunrise has already been given (in chapter 13), and this method can be used on any day (§ 51). Since the ortive amplitude can be converted from a measured length to an angle (with the vertex at the top corner of the window), and the geographical latitude is known, we can find the solar declination from formula (7), above, or its equivalent. Since we know the maximum declination, we can convert this declination angle to solar longitude (for example, by means of a table: cf. Goldstein, 1974a, p. 159). From a set of correspondences of times and solar longitudes, the eccentricity and apogee can be found, but the procedure is not described here (§§ 59–60).

There follows a section (§§ 61–64) in which additional elements of the experiment are described. The walls should be plane and perpendicular to each other as well as to the floor. Measurements of the distance of the marks on the line on the west wall should be made from the north wall and the distance from the top corner of the window on the east wall to the north wall should also be measured (see Fig. 13C-1). These measurements then allow us to determine the distance from the top corner of the window on the east wall to the marks on the west wall. If the walls are not perpendicular, adjustments must be made, and they are described in the next section (§§ 65–79). In § 62 we are told that the inaccuracy due to non-perpendicular walls will be reduced when the ortive amplitude is sufficiently large (i.e., near the solstices). Levi's remark that in the room he used the walls were not perpendicular indicates that for him this was not a theoretical exercise but that he actually did the experiments (§ 63).

In Fig. 15.2, all lines are on the floor: AB is the west edge, BG the north edge, and GD the east edge (§ 65). Point E is directly below the top corner of the window on the east wall (§ 66). Lines GE, AB, and BG are measured (§ 67). From point E we draw a line to an arbitrary point Z on line BG, and a perpendicular line to BG meeting it at point T. By a simple combination of measured quantities and the solution of triangles, the length of line TB can be found (§ 68–72). If the west wall were perpendicular to the north wall (§ 72), line BT would be the length of the north wall needed for the computations mentioned above (§ 61). But if the north wall does not meet the west wall at right angles, another adjustment must be made (§§ 73–78). Points L and M on line AB correspond to two marks on the line on the west wall (§ 74). From L and M perpendiculars are drawn to line BG (the north edge of the floor) meeting it at points N and S, respectively. The lengths of TN and TS can be found, and so can the lengths of *LN* and *MS* (again by combining measurements with the solutions of triangles). Thus the room can be used despite the absence of perpendicularity of its walls to one another (§ 79).

Chapter 16

Since the positions of the planets will be observed with respect to the fixed stars, we are reminded that the positions of the fixed stars are not known precisely, and there are a variety of opinions to be found in the antecedent literature (§§ 1–3). From what follows it is clear that our concern here relates to the positions of the fixed stars relative to the ecliptic rather than relative to one another (§ 31). To find the position of a fixed star relative to the ecliptic, one can measure the elongation of the Moon from the Sun before sunset and then the elongation of the star from the Moon after sunset on the same day (§§ 4–6). The difficulty remains that one must correct the position of the Moon for parallax in order to convert the observed elongation of the star from the Moon to its elongation from the true lunar position, and Levi remarks that there remains some uncertainty in the value for lunar parallax (§ 6). As a first approximation Levi accepts Ptolemy's values for parallax at the syzygies, i.e., he accepts the Ptolemaic values for the distance of the Moon from the Earth at syzygy (§ 7). But since Levi is convinced that Ptolemy's values for the lunar distances at quadrature are seriously in error, he does not intend to use Ptolemy's parallax values for the remainder of the synodic month (§ 8). We are now informed that Levi measured the apparent lunar diameter at opposition and quadrature with the instrument he described earlier (i.e., the combination of the staff and the camera obscura), and found the lunar diameter at quadrature only slightly greater than at opposition (§§ 9–12: in contrast to the doubling of the size of the Moon that follows from Ptolemy's model). The conditions of Levi's observation are now specified more clearly, but no date is given. Near quadrature the Moon's elongation from a fixed star is observed twice separated by an interval of 4 or 5 hours (§§ 13–15). The computed true motion of the Moon according to Ptolemy's model in this interval is subtracted from its observed motion in the same interval, and the remainder is the difference in the lunar parallax at the times of the two observations (§§ 16–17). Levi remarks that whatever its deficiencies Ptolemy's model is sufficiently accurate over an interval of 4 or 5 hours for the present purpose (§ 16). This difference can then be compared to the computed parallaxes from Ptolemy's tables and, repeatedly, Levi found a noticeable discrepancy (§§ 18–20).

Levi does not accept Ptolemy's claim that the parallax is very much greater at quadrature than at opposition (§ 19: cf. *Almagest*, V, 13). He argues that the error may be due to Ptolemy's parameter for the inclination of the lunar orb from the ecliptic: Levi prefers $4\frac{1}{2}°$ (claiming erroneously that this is al-Battānī's value) to Ptolemy's 5° (§§ 22–24: cf. Goldstein, 1974a, pp. 132–34).

The next section describes an alternative method for determining the positions of the fixed stars relative to the Sun (§§ 25–32). At the time of a lunar eclipse we can see the shadow of the Sun on the surface of the Moon, and the center of the shadow must be diametrically opposite the center of the Sun. Levi tells us to measure the distance from the edge of the shadow to a fixed star during a lunar eclipse and then to correct this measurement for the radius of the shadow, here given as 0;42° (§§ 27–28: cf. Goldstein, 1974a, pp. 123–128). Thus the elongation of the fixed star from a point diametrically opposite the Sun has been found, and hence its position can be precisely stated, provided one takes into account the small parallax that applies to the shadow (§§ 29–30). Once the position of one star has been determined, we are told we can use Ptolemy's star catalogue to find the positions of the other fixed stars relative to it (§ 31). Subsequently the fixed stars can be used in the observations of the planets to test whether the planetary positions conform to Ptolemy's models (§ 32).

CHAPTER 17

In this chapter Levi emphasizes the observations of apparent planetary and lunar diameters that fail to conform with the consequences of Ptolemy's models. He first considers the Moon (§§ 2–3) and tells us that the twofold variation in the lunar diameter that follows from Ptolemy's model does not agree with his observations. Surprisingly, no one before the fourteenth century seems to have paid attention to this obvious difficulty. Independently, Ibn al-Shāṭir (Syria, 14th century) was also aware of this and produced a model in which he eliminated this effect on the size of the Moon (cf. Roberts, 1957). Levi goes on to remark that Venus is not seen to vary sixfold in apparent diameter (§§ 4–5) as he believes follows from Ptolemy's model: in fact, according to Ptolemy, the ratio of maximum to minimum distance for Venus is 104 to 16 (cf. Goldstein, 1967a, p. 7). Rather, Levi claims the diameter of Venus is greater at maximum elongation than at either superior or inferior conjunction. As we now know, the reason for the relatively slight variation in the apparent brightness of Venus (confused with apparent diameter in ancient and medieval sources) is that the phases of Venus more or less compensate for the change in distance from the earth (cf. Price, 1959, pp. 212–14). Levi remarks on the visibility of Venus during daylight (§ 7), but does not use Venus as an intermediary to find the elongation from the Sun to a fixed star, a method introduced by Bernhard Walther in the late fifteenth century (cf. Beaver, 1970, p.

41). He also suggests observing the apparent diameter of Venus with a camera obscura on a moonless night (§ 10).

Mars is considered next: Levi claims that it varies in size only twofold as opposed to sixfold according to Ptolemy (§§ 11–13). In fact, Ptolemy claims that Mars varies in distance sevenfold (cf. Goldstein, 1967a, p. 7). He then alludes to observations of Mars near opposition, i.e., when it is brightest, and found that under some circumstances it was greater in size than Saturn (§§ 14–16). The surprisingly small size of Mars in Scorpio is reconsidered later (§§ 25–29) and an explanation is offered that depends on the theory of comet formation. Levi argues that a comet seen for three months was formed from vapors in the sublunary sphere directly below Mars and that the associated turbulence made Mars appear smaller than it would otherwise. From other contemporary reports in Europe and China, this comet can be dated to 1337 (cf. Goldstein, 1972, p. 45). Levi's account is of special interest because he introduced a cometary observation in an astronomical context; no earlier example is known to me. A final remark on Mars (§ 30) is intended to explain the slight augmentation of its apparent size in Capricorn (noted in §§ 14 and 27) and here refraction due to the intervention of thin clouds is invoked.

The remaining planets are given short treatment (§§ 17–24). Mercury is only seen near greatest elongation and that does not provide sufficient data. Moreover, it is only seen close to sunrise (or sunset) near the horizon and "all stars seem smaller then" (§ 19). There are two effects of which Levi mentions only one: stars appear dimmer near the horizon during twilight. The other effect is that owing to the absorption of light by the air the apparent magnitude of a star diminishes as the star approaches the horizon (cf. Minnaert, 1954, p. 75). For Saturn and Jupiter, Levi did not succeed in determining the variation in apparent diameter, but he was nevertheless certain that it is smaller than that which follows from Ptolemy's model (§ 20). In the *Planetary Hypotheses*, Ptolemy gives the ratio of maximum to minimum distance of Saturn as 7 to 5 (Goldstein, 1967a, p. 7): Levi considers a ratio of 6 to 5 (§ 21) for reasons he does not explain. He goes on to say that this is not supported by observation in total darkness (i.e., when there is no moonlight, as in § 10, despite the text which has "no sunlight"). According to Ptolemy the ratio of maximum to minimum distance for Jupiter is 37 to 23 (Goldstein, 1967a, p. 7): Levi states that this ratio is close to 4 to 3 (§ 23). The same argument mentioned for Saturn is also applied to Jupiter (§§ 23–24).

Chapter 18

In this chapter Levi announces his intention to reconsider the planetary models in order to account for the observed sizes, mentioned earlier, that do not agree with what follows from Ptolemy's models. However, he does not present his other planetary observations here because he argues that the discrepancies between observation and computation based on Ptolemy's models might be due to other causes. In fact, Levi's observed planetary positions are presented in a much later section of this book (cf. chapters 109, 113, 117, and 122). The planetary theory is not complete in any of the surviving manuscripts and it is possible that Levi did not finish writing that section.

Levi indicates that the mean solar positions are fundamental to the determination of the planetary models and they must be established on a firm basis before proceeding to the planetary theory. Similarly, the positions of the fixed stars are used in the planetary observations, and so they should be established beforehand. In sum, Levi argues that a discrepency between observation and computation is insufficient to disconfirm a planetary model until all other possible sources of error have been eliminated and the best possible parameters have been used. This careful procedure was not generally employed in the 16th and 17th centuries on those occasions when astronomers compared different models or sets of tables. Rather, they often compared models with values for the same parameter that differed from one another (cf. Thoren, 1967, p. 158; and Wilson, 1970, p. 100).

CHAPTER 19

This chapter begins by defining the problems to be resolved by the planetary models: they must account for the apparent motions in longitude and latitude as well as the variation in the apparent planetary diameters. A series of models will be described (in chapter 20) and the characteristics of each model will be examined. Then the model from which the observed characteristics follow will be accepted as true to the exclusion of other possible models for that planet. Hence, it is necessary to collect observations including those of the ancients because "the human life-span is not sufficient for (all) the observations needed for this science" (§ 8), particularly for the determination of the mean motion parameters. The advice to use observations separated by a long interval of time for determining mean motion parameters (§ 11) is derived from Ptolemy (*Almagest*, III, 1; trans. Toomer, 1984, p. 137).

In the next section (§§ 13–22) the characteristics of planetary motion are described in a general way. There is an equation that depends on the motion in longitude, and another that depends on the anomaly. The motion in anomaly for the planets produces direct and retrograde motion, and the effect of the motion in anomaly is different on different parts of the ecliptic.

Levi then tells us that he has had few opportunities to observe Mercury (§ 23) and that all his observations of Saturn and Jupiter were made while they

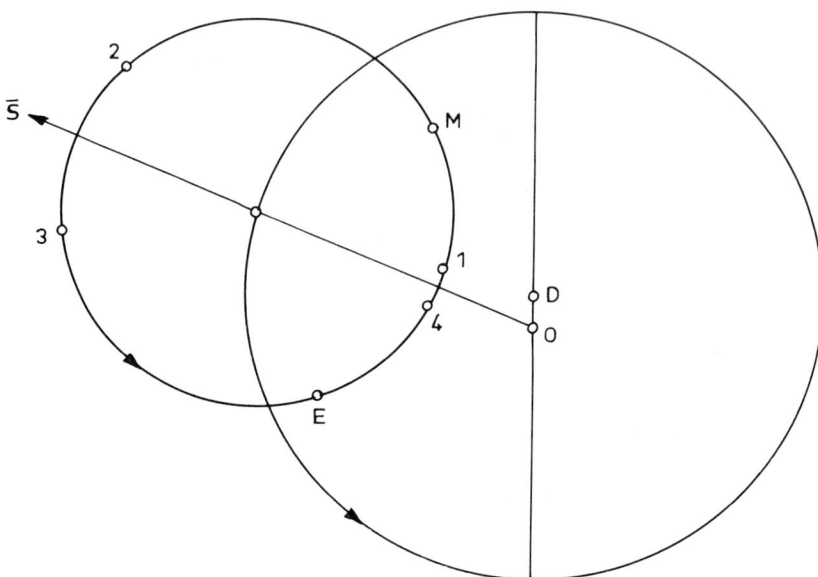

FIGURE 19C-1. Venus: (1) first rising in the morning; (M) greatest elongation in the morning; (2) last rising in the morning; (3) first setting in the evening; (E) greatest elongation in the evening; (4) last setting in the evening; the observer is at O, and $O\bar{S}$ is the direction to the mean Sun.

were only on one side of their respective apogees (§ 24). For Venus and Mars he was able to verify that the greatest corrections took place at more than 90°, and less than 270°, of anomaly (§ 26), in conformity with calculations based on Ptolemy's model. To do this he made observations with the Jacob Staff, here called: "the instrument that we invented" (§ 27). Levi correctly remarks that at maximum correction the apparent daily motion is equal to the motion in longitude (§ 27), and for Venus this means that its apparent motion at that place is equal to the motion of the Sun.

The remarks in §§ 28–30 concerning Venus are not clear to me unless one interprets them to refer to greatest elongation in the morning and in the evening in which case the statements are true (see Fig. 19C-1): the arc EM, from greatest elongation in the evening (E) to greatest elongation in the morning (M), is less than arc ME, from greatest elongation in the morning to greatest elongation in the evening. Otherwise, one should refer to (1) first rising in the morning, (2) last rising in the morning, (3) first setting in the evening, and (4) last setting in the evening. In this case the time from last setting in the evening to first rising in the morning is very much less than the time from last rising in the morning to first setting in the evening, i.e., the period of invisibility near

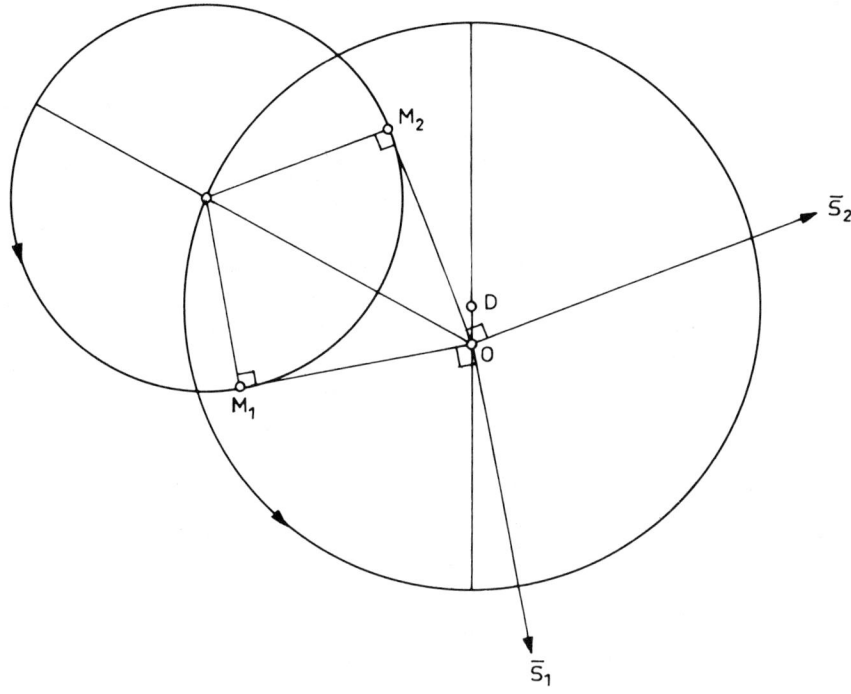

FIGURE 19C-2. When Mars is at points M_1 and M_2, the mean Sun lies in the directions $O\bar{S}_1$ and $O\bar{S}_2$, respectively; the observer is at O.

inferior conjunction is much less than the period of invisibility near superior conjunction.

In the next section (§§ 31–36) the other planets are considered. When the apparent elongation of Mars from the Sun is 90°, Mars lies at the point on its epicycle where the radius of the epicycle is perpendicular to the line from that point to the observer (see Fig. 19C-2): this corresponds to greatest elongation for Venus and perhaps supports the interpretation of the previous section. The expression "edges of the night" (§ 31) clearly corresponds to acronychal rising/setting that originally referred to the planetary phase when the planet rises in the east as the Sun sets in the west, or when the planet sets in the west as the Sun rises in the east. In the *Almagest*, *akronychos* simply means opposition (cf. Neugebauer, 1975, p. 1091; *Almagest*, X, 6 [trans. Toomer, 1984, pp. 483, 484 n. 29], Arabic trans. al-Ḥajjāj: *aṭrāf al-layl* [MS Leiden, Or. 680, fol. 145b:9]). Neugebauer (1975, p. 792) notes a passage in Pliny (*Nat. Hist.* II, 60) where reference is made to the "sensitivity" of Mars to the rays of the Sun at 90° elongation, and perhaps Levi's remark depends on a tradition from astrological sources. Levi makes use of the fact that for an outer planet the radius from the center of the epicycle to the planet is always parallel to the direction of the mean Sun, and then claims that the interval from M_1 to M_2 is less than half the interval from M_2 to M_1. Note that at M_1 Mars rises before sunrise, and at M_2 Mars sets after sunset.

In the next paragraph (§§ 37–42) we are told that the observations of the apparent solar diameter prove that the Sun lies on an eccentric sphere because at summer solstice (Cancer 0°) its apparent diameter was 0;27,50°, and at winter solstice (Carpricorn 0°) it was 0;30°. In a later passage (chap. 56: P 104b) we find descriptions of dated observations of the apparent solar diameter at the two solstices taken in 1334 with a staff that had a pierced plate at one end and a screen at the other end (cf. chap. 9:11–17). Parameters for the motion of the solar and planetary apogees are presented here without proof which, we are told, will be forthcoming in a later chapter.

Finally, we are reminded that Astronomy is like Physics in that one must often use *a posteriori* arguments, rather than *a priori* arguments, as in mathematics. Attention is drawn to an Aristotelian passage in the book *On Animals* in which it is stated that the visual faculty is prior to the form of the eye physically because the eye is for vision, not vision for the eye (§ 47: cf. Aristotle, *Parts of Animals* I, 1, and *De anima*, 412b:18 ff). Similarly, the planetary motions are prior to the models physically.

Chapter 20

In this chapter Levi presents a qualitative discussion of the kinematics of circular motion in which he considers a number of models. In some cases the center of the orb coincides with the position of the observer, and in others it coincides with the equant point (about which mean motion takes place); but arbitrary positions for these three points inside the orb are also considered. Levi is particularly interested in the angular argument corresponding to maximum equation, and in the variation in velocity under these different conditions; I am not aware of a comparably exhaustive discussion in any other ancient or medieval treatise.

Levi's lunar models are more complicated than any discussed here, but similar principles are invoked (cf. Goldstein, 1974a, pp. 53–74; Goldstein, 1974b). Moreover, he seems to have used a "skew equant model" in constructing a table of lunar velocities (see §§ 90–95: cf. Goldstein, 1974a, p. 114). By "skew equant model" I mean a model in which the observer and the equant point are set at different distances in opposite directions from the center of the orb.

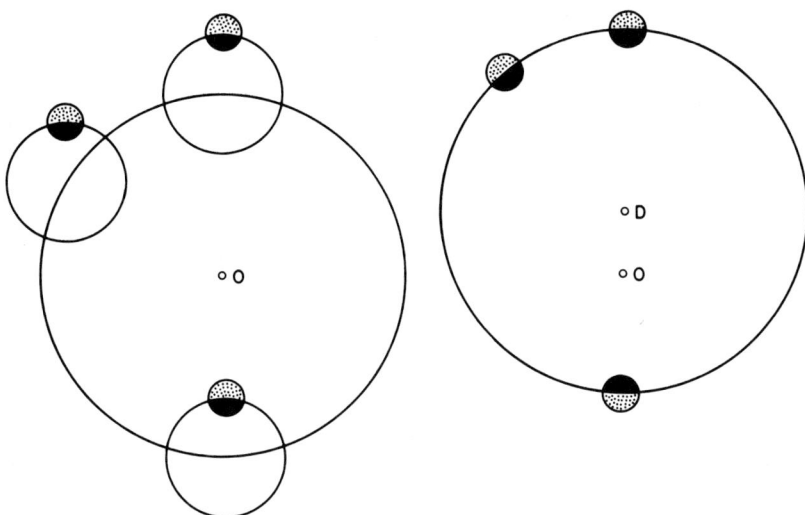

FIGURE 20C-1. The apparent disk of the Moon according to the epicyclic and eccentric models: point O represents the observer, and point D the center of the deferent. Note that the Sun may shine on both the solid and the shaded portions of the Moon.

At the beginning of this chapter Levi alludes to Aristotle, *De caelo*, II, 6 and 8 (§§ 4–6). He then turns to Ptolemy for his demonstration that the epicyclic and eccentric models are equivalent (§ 12: *Almagest*, III, 3; cf. Neugebauer, 1959). Levi illustrates the simple eccentric model (see Fig. 20.1) where angle DBE is the correction corresponding to angle ADB, its argument (§§ 18–29). Ptolemy proved that the maximum correction takes place when the angle at the observer (point E) is 90° (*Almagest*, III, 3). Levi distinguishes the epicyclic model from the eccentric model on physical grounds, namely, the epicyclic model implies that we should see both sides of the moon, contrary to the observed phenomena. Hence, he concludes that the Moon cannot lie on an epicycle (§§ 34–41: see Fig. 20C-1): it is a most unusual argument (see the Introduction).

Levi goes on to describe an epicyclic model where the motions on the epicycle and on the deferent take place in the same sense (§§ 42–43). In this case, swiftest motion takes place at the apogee of the epicycle, and this conforms to the way Ptolemy constructed his planetary models. For an epicyclic model to be equivalent to an eccentric model, the motion on the epicycle must take place in the sense opposite to that on the deferent, and Ptolemy constructed his models for the Sun and the Moon accordingly.

Levi then introduces a concentric model where the observer is put at E, the center of the orb, and the equant at another point, D (§§ 44–51: see Fig.

20.2). The maximum correction corresponds to 90° of mean motion, and the apparent size of the planet does not vary.

Levi wishes to compare the corrections for the same argument in the eccentric and concentric models (§§ 52–66). In Fig. 20.3 point Z, the equant, corresponds to point D in Figs. 20.1 and 20.2. Circle $HBTD$ about Z corresponds to the eccentric orb in Fig. 20.1, and circle $ABGD$ about E corresponds to the concentric orb in Fig. 20.2. If point M lies on arc AB and point N on arc HB such that ZMN is a straight line, then the correction at M on the concentric orb is greater than the correction at N on the eccentric orb. At point B, where the orbs intersect, the corrections are equal, and the argument AZB is equal to the sum of angle ZBL and right angle ZLB. When the argument is greater than angle AZB, e.g., angle AZO, the correction on the eccentric orb is the greater one, and this is what Levi sought to prove. Levi correctly remarks that $\sin ZBL = ZE/2$.

Levi now turns to models where the three points are distinct, i.e., the center of the orb, the equant, and the observer (§§ 67–73). In the first instance he assumes that all three points lie on a line; later, he considers the general case. Before continuing this discussion, he restates his objections to Ptolemy's lunar model (§§ 75–78) in terms similar to those found elsewhere in this treatise (see, for example, chap. 17:2–3). In the first of this set of models point Z is the equant, point E is the observer, and the eccentricity is bisected at D, the center of the orb, as in Ptolemy's planetary models (§§ 79–89: see Fig. 20.4; cf. Neugebauer, 1957, pp. 198 ff). Levi proves that the correction for 90° of mean motion is equal to the correction for 90° of apparent motion. A small, but significant, alteration is introduced in the model so that the eccentricity is no longer bisected (§§ 90–95). If DE is greater than ZD, the correction for 90° of apparent motion is greater than the correction for 90° of mean motion (see Fig. 20.5), and contrariwise if DE is less than ZD (see Fig. 20.6). This skew equant model is compared with the eccentric model where the maximum corrections are set equal to one another. Levi notes that there is less variation in the planet's size here because the observer lies closer to the center of the orb.

In §§ 99–112 Levi reverts to the model illustrated by Fig. 20.2 where the observer lies at E, the center of the orb (see Fig. 20.7): it has the property that the correction for 90° of mean motion is greater than for 90° of apparent motion, a property it has in common with the skew equant model illustrated in Fig. 20.6. Levi then proves that the excess between the two corrections in this model is greater than the excess between them in the skew equant model.

Levi reconsiders the simple eccentric model (see Fig. 20.8), to note that the correction for 90° of apparent motion is greater than that for 90° of mean motion. It is then compared with a model with the same maximum correction where the equant lies between the observer and the center of the orb (see

Fig. 20.9), in order to show that the latter model produces a greater variation in the apparent size of the planet, due to the greater distance between the observer and the center of the orb (§§ 113–139). Levi also demonstrates that the model illustrated by Fig. 20.9 produces a greater difference between the corrections at 90° of mean and apparent motions than is produced by the simple eccentric model. The essence of his proof is as follows. Angle DTZ in Fig. 20.9 is set equal to angle EHZ in Fig. 20.8, which is greater than angle EBZ in Fig. 20.8. But angle EBZ is Fig. 20.8 is greater than angle DBZ in Fig. 20.9. Therefore

$$\text{angle } DTZ - \text{angle } DBZ > \text{angle } EHZ - \text{angle } EBZ$$

where the angles on the left side of the inequality refer to Fig. 20.9, and those on the right side to Fig. 20.8.

The model where the equant lies between the center of the orb and the observer (§§ 140–146) has another property that Levi calls to our attention: the maximum correction corresponds to an angle greater than 90° of apparent motion. To prove this he introduces Fig. 20.10 and uses the following lemma (§ 143): the shortest line from a point inside a circle to its circumference lies on the radius passing through that point (cf. Euclid, III.7).

The proof of this theorem may be summarized as follows. In Fig. 20.10, angle DTZ is the correction corresponding to 90° of apparent motion and angle DSZ is a correction that corresponds to an angle greater than 90° of apparent motion. Angle DSZ is greater than angle DOZ because it is an exterior angle of triangle SOZ. But angle DOZ is equal to angle DTZ because they are both inscribed in the same arc of the circle whose center is L. Hence

$$\text{angle } DSZ > \text{angle } DOZ = \text{angle } DTZ$$

Levi next considers a model where the observer at O lies between C, the center of the orb, and E, the equant point (§§ 147–150: see Fig. 20C-2). In this case the swiftest motion takes place at A, the apogee, where the planet has the smallest apparent size. He then discusses the corrections that apply to this model.

Only the general case remains (§§ 151–154: see Fig. 20C-3); in this model the equant point does not lie on the diameter passing through the center of the orb and the observer. Levi claims, incorrectly, that maximum and minimum velocities occur at the ends of the chord that passes through the equant and the observer. No proof is offered, only the phrase: "and this is clear from our previous remarks" (§ 152). I have checked the velocities in a number of cases: for small eccentricities, Levi's suggestion is a fair approximation, i.e., the maximum and minimum velocities are assumed at points on the circle not more than a few degrees away from the places where he believed they occur.

196 The Astronomy of Levi ben Gerson

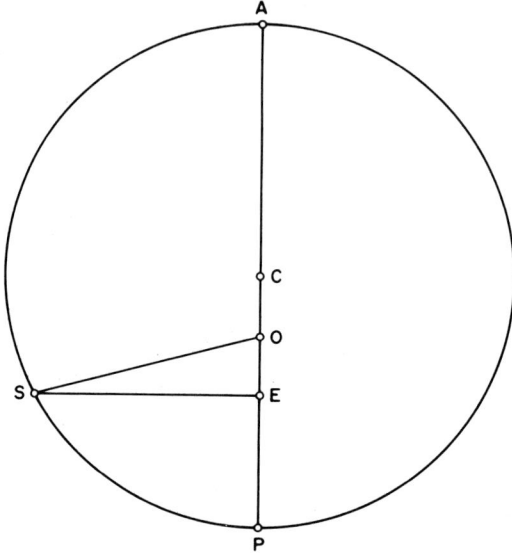

FIGURE 20C-2.

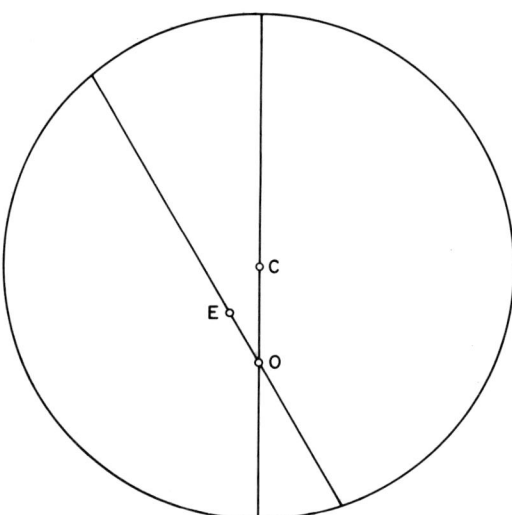

FIGURE 20C-3.

But for large eccentricities, the discrepancy between them can reach 45° or more. Levi refers to Ptolemy's Mercury model because, in general, the equant does not lie on the diameter passing through the center of the orb and the observer (cf. Neugebauer, 1957, pp. 200 ff). Ptolemy was not at all interested in the velocity of a point on Mercury's deferent because the planet lies on the epicycle, and the center of the epicycle (which moves uniformly about the equant point) is fixed on the instantaneous deferent. Levi attempted to solve a difficult problem in kinematics, but the mathematical techniques at his disposal were inadequate for the purpose.

References

Aaboe, A., 1963. "On a Greek Qualitative Planetary Model of the Epicyclic Variety", *Centaurus*, 9:1–10.

Aaboe, A., 1964. *Episodes from the Early History of Mathematics*. New York: Random House.

Aaboe, A., 1974. "Scientific Astronomy in Antiquity", *Phil. Trans. Royal Society of London, A.* 276:21–42.

Aiton, E. J., 1981. "Celestial Spheres and Circles", *History of Science, 19*:75–114.

cArafat, W., and Winter, H. J. J., 1971. "The Light of the Stars—A Short Discourse by Ibn al-Haytham", *British Journal for the History of Science*, 5:282–88.

Barker, P., and Goldstein, B. R., 1984. "Is Seventeenth Century Physics Indebted to the Stoics?", *Centaurus*, 27:148–64.

Beaujouan, G., 1974. "Observations et calculs astronomiques de Jean de Murs (1321–1344)", in *Actes du XIVe Congrès international d'histoire des sciences*, 2:27–30. Tokyo-Kyoto.

Beaver, D. deB., 1970. "Bernard Walther: Innovator in Astronomical Observation", *Journal for the History of Astronomy*, 1:39–43.

Britton, J. P., 1969. "Ptolemy's Determination of the Obliquity of the Ecliptic", *Centaurus*, 14:29–41.

Cantera Burgos, F., 1931. "El Judio Salmantino Abraham Zacut", *Revista de la Academia de Ciencias*, 27:63–398.

Carlebach, J., 1910a. "Selections from the Writings of Rabbi Levi ben Gerson", in M. Stern (ed.), *Festschrift zum vierzigjährigen Amtsjubiläum des Herrn Rabbiners Dr. Salomon Carlebach in Lübeck*, 151–78 [Hebrew Section]. Berlin: Verlag "Hausfreund".

Carlebach, J., 1910b. *Lewi ben Gerson als Mathematiker*. Berlin: L. Lamm.

Carmody, F. J., 1952. "The Planetary Theory of Ibn Rushd", *Osiris*, 10:556–86.

Cohen, M. R., and Drabkin, I. E., 1958. *A Source Book in Greek Science*. Cambridge, MA: Harvard University Press.

Constable, G., 1975. "Horologium stellare monasticum (saec. XI)", in *Corpus Consuetudinum Monasticarum. Tomus VI: Consuetudines Benedictinae Variae*, 1–18. Siegburg: Franciscum Schmitt Success.

Curtze, M., 1898. "Die Abhandlungen des Levi ben Gerson über Trigonometrie und den Jacobstab", *Bibliotheca Mathematica, NS 12*:97–112.

Curtze, M., 1901. "Die Dunkelkammer", *Himmel und Erde, 13*: 225–236.

Curtze, M., 1902. "Der Briefwechsel Regiomontan's mit G. Bianchini, J. von Speier und C. Roder", in *Urkunden zur Geschichte der Mathematik im Mittelalter und der Renaissance*, 185–336. Leipzig: Teubner.

de Santillana, G., 1955. *The Crime of Galileo*. Chicago: University of Chicago Press.

Drake, S., 1957. *Discoveries and Opinions of Galileo*. New York: Doubleday Anchor.

Dreyer, J. L. E., 1953. *A History of Astronomy from Thales to Kepler*. New York: Dover (reprint ed.).

Dreyer, J. L. E., 1961. "Mediaeval Astronomy", in R. Palter (ed.), *Toward Modern Science*, 2 vols., 1:235–56. New York: Noonday Press.

Dreyer, J. L. E., 1963. *Tycho Brahe*. New York: Dover (reprint ed.).

Edelman, H., 1853. *Dibrey Hephez: "Acceptable Words"*. London: Abraham Pierpoint Shaw and Co.

Espenshade, P., 1967. "A Text on Trigonometry by Levi ben Gerson", *The Mathematics Teacher, 60*:628–37.

Feldman, S., 1984. *Levi ben Gershom (Gersonides): The Wars of the Lord, Book One*. Philadelphia: The Jewish Publication Society of America.

Friedländer, M. (trans.), 1956. *Maimonides: The Guide for the Perplexed*. New York: Dover (reprint ed.).

Gandz, S., 1970. "The Astrolabe in Jewish Literature", in *Studies in Hebrew Astronomy and Mathematics*, 245–62. New York: Ktav.

Gibbs, S., Henderson, J., and Price, D., 1973. *A Computerized Checklist of Astrolabes*. New Haven: Department of History of Science and Medicine, Yale University.

Goldstein, B. R., 1967a. "The Arabic Version of Ptolemy's *Planetary Hypotheses*", *Transactions of the American Philosophical Society, NS 57,4*:1–55.

Goldstein, B. R., 1967b. *Ibn al-Muthannā's Commentary on the Astronomical Tables of al-Khwārizmī*. New Haven: Yale University Press.

Goldstein, B. R., 1969a. "Preliminary Remarks on Levi ben Gerson's Contributions to Astronomy", *Proceedings of the Israel Academy of Sciences and Humanities*, 3:239–54.

Goldstein, B. R., 1969b. "Some Medieval Reports of Venus and Mercury Transits", *Centaurus, 14*:49–59.

Goldstein, B. R., 1971. *Al-Biṭrūjī: On the Principles of Astronomy.* 2 vols. New Haven: Yale University Press.

Goldstein, B. R., 1972. "Theory and Observation in Medieval Astronomy", *Isis*, 63:39–47.

Goldstein, B. R., 1974a. *The Astronomical Tables of Levi ben Gerson.* Hamden, CT: Archon Books.

Goldstein, B. R., 1974b. "Levi ben Gerson's Preliminary Lunar Model", *Centaurus*, 18:275–88.

Goldstein, B. R., 1975. "Levi ben Gerson's Analysis of Precession", *Journal for the History of Astronomy*, 6:31–41.

Goldstein, B. R., 1976a. "Astronomical and Astrological Themes in the Philosophical Works of Levi ben Gerson", *Archives Internationales d'Histoire des Sciences*, 26:221–24.

Goldstein, B. R., 1976b. "The Hebrew Astrolabe in the Adler Planetarium", *Journal of Near Eastern Studies*, 35:251–60.

Goldstein, B. R., 1977. "Levi ben Gerson: On Instrumental Errors and the Transversal Scale", *Journal for the History of Astronomy*, 8:102–12.

Goldstein, B. R., 1979. "Medieval Observations of Solar and Lunar Eclipses", *Archives Internationales d'Histoire des Sciences*, 29:101–56.

Goldstein, B. R., 1980. "The Status of Models in Ancient and Medieval Astronomy", *Centaurus*, 24:132–47.

Goldstein, B. R., 1981. "The Hebrew Astronomical Tradition: New Sources", *Isis*, 72:237–51.

Goldstein, B. R., 1985a. "Scientific Traditions in Late Medieval Jewish Communities", in G. Dahan (ed.), *Les Juifs au regard de l'histoire: Mélanges en l'honneur de M. Bernhard Blumenkranz*, 235–47. Paris: Picard.

Goldstein, B. R., 1985b. "Descriptions of Astronomical Instruments in Hebrew", in D. A. King and G. Saliba (eds.), *Essays in Honor of E. S. Kennedy*, in *Annals of the New York Academy of Sciences* (forthcoming).

Goldstein, B. R., 1985c. "Levi ben Gerson: On the Relationship between Physics and Astronomy", in *Proceedings of the International Conference on the Interrelations between Physics, Cosmology, and Astronomy: 1300–1700* (Tel Aviv–Jerusalem 1984). Jerusalem (forthcoming).

Goldstein, B. R., 1985d. "Star Lists in Hebrew", *Centaurus* (forthcoming).

Goldstein, B. R., and Bowen, A. C., 1983. "A New View of Early Greek Astronomy", *Isis*, 74:330–40.

Goldstein, B. R., and Sawyer, F. W., III, 1977. "Remarks on Ptolemy's Equant Model in Islamic Astronomy," in Y. Maeyama and W. G. Saltzer (eds.), *Prismata*, pp. 165–81. Wiesbaden: Franz Steiner Verlag.

Grant, R., 1852. *History of Physical Astronomy*. London: Henry G. Bohn.

Gunther, R. T., 1923. *Early Science in Oxford*, vol. 2. Oxford: University Press.

Gunther, R. T., 1929. *Chaucer and Messahalla on the Astrolabe*. Oxford: University Press.

Gunther, R. T., 1932. *The Astrolabes of the World*. 2 vols. Oxford: University Press.

Haasbroek, N. D., 1968. *Gemma Frisius, Tycho Brahe and Snellius and their Triangulations*. Delft: Netherlands Geodetic Commission.

Hamellius, P., 1557. *Commentarius in Archimedis, librum de numero arenae*. Paris: G. Cavellat.

Hartner, W., 1968. "The Principle and Use of the Astrolabe", in *Oriens-Occidens*, 287–311. Hildesheim: Georg Olms Verlag.

Hartner, W., 1969. "Naṣīr al-Dīn al-Ṭūsī's Lunar Theory", *Physis*, 11:287–304.

Hartner, W., 1970. "Al-Battānī", in *Dictionary of Scientific Biography*, 1:507–16. New York: Scribners.

Heath, T. L., 1913. *Aristarchus of Samos*. Oxford: Clarendon Press.

Heath, T. L., 1956. *The Thirteen Books of Euclid's Elements*. 3 vols. New York: Dover (reprint ed.).

Hirschfeld, H., 1904. *Descriptive Catalogue of The Hebrew MSS. of the Montefiore Library*. London: Macmillan.

Jervis, J. L., 1977. "Toscanelli's Cometary Observations: Some New Evidence", *Annali dell'Istituto e Museo di Storia della Scienza di Firenze*, 2:17–20.

Kellner, M. M., 1979. "R. Levi Ben Gerson: A Bibliographical Essay", *Studies in Bibliography and Booklore*, 12:13–23.

Kennedy, E. S., 1966. "Late Medieval Planetary Theory", *Isis*, 57:365–78.

Kennedy, E. S., 1973. *A Commentary upon Bīrūnī's Kitāb Taḥdīd al-Amākin*. Beirut: American University of Beirut.

Kennedy, E. S., 1976. *The Exhaustive Treatise on Shadows by al-Bīrūnī: Translation and Commentary*. 2 vols. Aleppo: Institute for the History of Arabic Science.

Kepler, J., 1959. *Gesammelte Werke: Briefe 1620–1630*. Vol. 18. M. Caspar (ed.). Munich: C. H. Beck'sche Verlagsbuchhandlung.

King, H. C., 1955. *The History of the Telescope*. Cambridge, MA: Sky Publishing Corp.

Kremer, R. L., 1980. "Bernard Walther's Astronomical Observations", *Journal for the History of Astronomy*, 11:174–89.

Kunitzsch, P., 1981. "On the Authenticity of the Treatise on the Astrolabe Ascribed to Messahalla", *Archives Internationales d'Histoire des Sciences*, 31:42–62.

Langermann, Y. T., 1985. "An Unknown Astronomical Treatise by Levi ben Gerson", *Kiryat Sefer* [in Hebrew] (forthcoming).

Lindberg, D. C., 1968. "The Theory of Pinhole Images from Antiquity to the

Thirteenth Century", *Archive for History of Exact Sciences*, 5:154–76.

Lindberg, D. C., 1970. "The Theory of Pinhole Images in the Fourteenth Century", *Archive for History of Exact Sciences*, 6:299–325.

Lindberg, D. C., 1976. *Theories of Vision from Al-Kindi to Kepler*. Chicago and London: University of Chicago Press.

Loewinger, D. S. and Weinryb, B. D. 1965. *Catalogue of the Hebrew Manuscripts in the Library of the Juedisch-theologisches Seminar in Breslau*. Wiesbaden: Harrassowitz.

Manitius, K., 1963. *Ptolemäus: Handbuch der Astronomie*. 2 vols. Leipzig: Teubner (reprint ed.).

Margoliouth, G., 1915. *Catalogue of the Hebrew and Samaritan Manuscripts in the British Museum, Part III*. London: British Museum.

Minnaert, M., 1954. *The Nature of Light and Colour in the Open Air*. New York: Dover.

Monfasani, J., 1976. *George of Trebizond*. Leiden: Brill.

Monfasani, J., 1984. *Collectanea Trapezuntiana: Texts, Bibliographies, and Documents of George of Trebizond*. Binghamton: SUNY Press.

Moulinas, R., 1980. "Les vieilles synagogues d'Avignon et du Comtat Venaissin", *Archives juives*, 16:14–26.

Munk, S., 1859. *Mélanges de philosophie juive et arabe*. Paris: A. Franck.

Murr, C. T. de, 1801. *Notitia trium codium autographorum Iohannis Regiomontani in bibliotheca Christophori Theophili de Murr*. Nuremberg: In Bibliopolio Wolfio-Penkeriano.

Nallino, C. A., 1903–07, 1899. *Al-Battānī sive Albatenii Opus Astronomicum*. 3 vols. Milan: Il Reale Osservatorio di Brera in Milano.

Neugebauer, O., 1957. *The Exact Sciences in Antiquity*, 2nd ed. Providence: Brown University Press.

Neugebauer, O., 1972. "Planetary Motion in P. Mich. 149", *Bulletin of the American Society of Papyrologists*, 9:19–22.

Neugebauer, O., 1975. *A History of Ancient Mathematical Astronomy*. Berlin, New York: Springer-Verlag.

Nolhac, P. de, 1887. *La Bibliothèque de Fulvio Orsini*. Paris: F. Vieweg.

North, J. D., 1977. "The Alfonsine Tables in England", in Y. Maeyama and W. G. Saltzer (eds.), *Prismata*, 269–301. Wiesbaden: Franz Steiner Verlag.

Novak, B. C., 1982. "Giovanni Pico della Mirandola and Jochanan Alemanno", *Journal of the Warburg and Courtauld Institutes*, 45:125–47.

Pedersen, O. 1974. *A Survey of the Almagest*. Odense: University Press.

Petz, H., 1888. "Urkundliche Nachrichten über den literarischen Nachlass Regiomontans und B. Walters 1478–1522", *Mitteilungen des Vereins für Geschichte der Stadt Nürnberg, 7*:237–62.

Pico della Mirandola, G., 1952. *Disputationes adversus astrologiam*. Vol. 2. E. Garin (ed.). Florence: Vallecchi Editore.

Pines, S., 1964a. "Ibn al-Haytham's Critique of Ptolemy", in *Proceedings of the Tenth International Congress of the History of Science in Ithaca 1962*, 547–50. Paris: Hermann.

Pines, S., 1964b. "La dynamique d'Ibn Bajja", in *Mélanges Alexandre Koyré I*, 442–68. Paris: Hermann.

Price, D. J., 1959. "Contra-Copernicus: A Critical Re-estimation", in M. Clagett (ed.), *Critical Problems in the History of Science*, 197–218. Madison: University of Wisconsin Press.

Raeder, H., Strömgren, E., and Strömgren, B., 1946. *Tycho Brahe's Description of his Instruments and Scientific Work*. Copenhagen: E. Munksgaard.

Renan, E., 1893. "Les écrivains juifs français du XIVe siècle", in *Histoire littéraire de la France, 31*:351–789. Paris: Imprimerie Nationale.

Righini Bonelli, M. L., and Settle, T. B., 1979. "Egnatio Danti's Great Astronomical Quadrant", *Annali dell'Istituto e Museo di Storia della Scienze di Firenze, 4*:3–13.

Roberts, V., 1957. "The Solar and Lunar Theory of Ibn al-Shāṭir", *Isis, 48*:428–32.

Roche, J. J., 1981. "The Radius Astronomicus in England", *Annals of Science, 38*:1–32.

Rome, A., 1931. *Commentaires de Pappus et de Théon d'Alexandrie sur l'Almageste. Tome I: Pappus d'Alexandrie, Commentaire sur les livres 5 et 6 de l'Almageste*. Rome: Biblioteca Apostolica Vaticana.

Rose, P. L., 1975. *The Italian Renaissance of Mathematics*. Geneva: Librairie Droz.

Rosen, E., 1965. "Copernicus on the Phases and the Light of the Planets", *Organon, 2*:61–78.

Roth, C., 1959. *The Jews in the Renaissance*. Philadelphia: The Jewish Publication Society.

Sabra, A. I., 1972. "Ibn al-Haytham", in *Dictionary of Scientific Biography, 6*:189–210. New York: Scribners.

Sabra, A. I. 1976. "The Physical and the Mathematical in Ibn al-Haytham's Theory of Light and Vision", in *The Commemoration Volume of Bīrūnī International Congress in Tehran*, 439–78. Tehran: The High Council of Culture and Art.

Sabra, A. I., and Shehaby, N. (eds.), 1971. *Ibn al-Haytham: al-Shukūk ᶜalā Batlamyus*. Cairo: National Library Press.

Sachau, C. E. (trans.), 1879. *The Chronology of Ancient Nations of Albīrūnī*. Frankfurt: Minerva (reprint ed. 1969).

Settle, T. B., 1978. "Dating Toscanelli's Meridian in Santa Maria del Fiore", *Annali dell'Istituto e Museo di Storia della Scienza di Firenze*, 3:69–70.

Sezgin, F., 1974. *Geschichte des arabischen Schrifttums. Band V: Mathematik*. Leiden: E. J. Brill.

Shapiro, A., 1975. "Archimedes's Measurement of the Sun's Apparent Diameter", *Journal for the History of Astronomy*, 6:75–83.

Shatzmiller, J., 1972. "Gersonides and the Jewish Community of Orange in his Day", in B. Oded et al. (eds.), *Studies in the History of the Jewish People and the Land of Israel*, vol. 2, pp. 111–26. Haifa: University of Haifa [in Hebrew].

Shatzmiller, J., 1974. "Further Information about Gersonides and the Orange Jewish Community in his Day", in B. Oded et al. (eds.), *Studies in the History of the Jewish People and the Land of Israel*, vol. 3, pp. 139–43. Haifa: University of Haifa [in Hebrew].

Staub, J. J., 1982. *The Creation of the World According to Gersonides*. Chico, CA: Scholars Press.

Steinschneider, M., 1893. *Die hebraeischen Uebersetzungen des Mittelalters*. Berlin: Kommissionsverlag des bibliographischen Bureaus.

Steinschneider, M., 1964. *Mathematik bei den Juden* (2nd ed.). Hildesheim: Georg Olms Verlag.

Straker, S., 1971. *Kepler's Optics: A Study in the Foundations of 17th Century Natural Philosophy* (unpublished dissertation). Bloomington: Indiana University.

Straker, S., 1981. "Kepler, Tycho, and the 'Optical Part of Astronomy': the Genesis of Kepler's Theory of Pinhole Images", *Archive for History of Exact Sciences*, 24:267–93.

Swerdlow, N., 1972. "Aristotelian Planetary Theory in the Renaissance: G. B. Amico's Homocentric Spheres", *Jounral for the History of Astronomy*, 3:36–48.

Taton, R., 1971. "G. D. Cassini (Cassini I)", in *Dictionary of Scientific Biography*, 3:100–04. New York: Scribners.

Taylor, E. G. R., 1954. *The Mathematical Practitioners of Tudor and Stuart England*. Cambridge: University Press.

Taylor, E. G. R., 1956. *The Haven-finding Art*. London: Hollis & Carter.

Thoren, V., 1967. "Tycho Brahe's Discovery of the Variation", *Centaurus*, 12:151–66.

Thorndike, L., 1934. *A History of Magic and Experimental Science*. Vols. 3, 4. New York: Columbia University Press.

Toomer, G. J., 1969. "The Solar Theory of az-Zarqāl: A History of Errors", *Centaurus*, 14:306–36.

Toomer, G. J., 1984. *Ptolemy's Almagest*. New York: Springer-Verlag.

Touati, C., 1973. *La pensée philosophique et théologique de Gersonide*. Paris: Les éditions de minuit.

Tuckerman, B., 1964. *Planetary, Lunar, and Solar Positions: A. D. 2 to A. D. 1649 at Five-day and Ten-day Intervals*. Philadelphia: The American Philosophical Society.

Twersky, I., 1972. *A Maimonides Reader*. New York: Behrman House.

Weil, G. E., 1980. "Sur une bibliothèque systématiquement pillée par les Nazis (Breslau)", in G. Nahon et C. Touati (eds.), *Hommage à Georges Vajda*, 579–604. Louvain: Editions E. Peeters.

Westman, R., 1977. "Magical Reform and Astronomical Reform", in *Hermeticism and the Scientific Revolution*, 5–91. Los Angeles: W. A. Clark Memorial Library.

Wickersheimer, E., 1929. *Recueil des plus celebres astrologues et quelques hommes doctes faict par Symon de Phares*. Paris: Librairie Champion.

Wiedemann, E., 1970. "Ueber die Camera obscura bei Ibn al Haitam", in *Aufsätze zur arabischen Wissenschaftsgeschichte*, 2 vols., 2:87–101. Hildesheim: Georg Olms Verlag.

Wilson, C. A., 1970. "From Kepler's Laws, So-called, to Universal Gravitation: Empirical Factors", *Archive for History of Exact Sciences*, 6:89–170.

Index

Aaboe, A., 3
Adler Planetarium, 163, 168
Afendopolo, Kaleb, 10
Aiton, E. J., 5
Akronychos, 112, 191
al-, *see* the next part of the name
Alidade, 28f, 82, 133, 164, 169
Almagest, 1, 3, 6, 24, 30, 131, 133, 142, 157f, 178f, 186, 189, 191, 193
Amico, G. B., 14
Angulus Mantuanus, 12; *see also* Finzi
Anomaly, 31; *see also* Planets, models for
Aperture, polygonal, 141; *see also* Camera obscura
Apian, P., 13
Apogee, 31; *see also* Sun
motion of, 113
Apollonius, 2
Apparent sizes of planets, *see* Planets, sizes of
ᶜArafat, W., 8, 132
Archimedes, 142, 146
Aristotle, 2ff, 6f, 140, 192f
Armillary sphere, 9, 28, 133
Astrolabe, 20, 28, 80, 82ff, 93, 133, 162ff, 169f, 175
Astrological houses, 9
Astrology, 131f, 192
Autolycus of Pitane, 2
Avempace, 5
Averroes, 7, 10, 134, 144, 158

Avignon, 1, 10, 13, 133
synagogue in, 178

Babylonians, 2
Baghdad, 10
Bagnols, 1; *see also* Leo
Bājja, Ibn, 5
Bar Ḥiyya, A., 9
Batecombe, 12
Battānī, al-, 9, 27, 94, 103, 110, 132, 134, 179, 187
Bayhaqī, al-, 8
Beaujouan, G., 14
Beaver, D., 158
Bellarmine, Cardinal, 6
Bessarion, Cardinal, 11
Bīrūnī, al-, 171, 183
Biṭrūjī, al-, 6f, 158
Black Death, 13
Bologna, observations in, 14f
Bonfils, Immanuel ben Jacob, 9, 12
Botarel, *see* Farissol Botarel
Bowen, A. C., 2
Brahe, T., 14, 143, 146, 162, 167
Brass, *see* Instrument
Britton, J. P., 170

Camera obscura, 2, 8, 48f, 140ff, 156, 188
combined with a Staff, 156f, 186, 192

Candle-light, 80; *see also* Instrument, illumination of
Cantera Burgos, F., 10
Carlebach, J., 72, 157
Carmody, F. J., 158
Cassini, G. D., 14
Cataract, 51, 144
Cavaillon, synagogue in, 178
Celestial bodies, movers of, 19
Chancellor, R., 164
Chords, 31ff, 139; *see also* Trigonometry
    table of, 135
Clement, VI, Pope, 1, 13
Clouds, 28; *see also* Refraction
Cohen, M. R., 156
Colson, N., 13
Comet, observation of, 107, 188
Commandino, F., 14, 142f, 146
Common sense, 52
Concentric model, 193f
Conjunction, planetary, *see* Planets, conjunctions of
Constable, G., 14
Copernicus, 6, 8
Cosmology, 2, 4, 15
Crabtree, W., 145
Creation, 1, 25, 131
Crescent-shaped, 8, 132
Cross Staff, *see* Jacob Staff
Crystalline lens, 51f, 54, 144, 146
Curtze, M., 14, 141, 154, 157

Danti, I., 14
Darkness of night, 69, 105f; *see also* Moonless night
Day-circle of the Sun, *see* Sun, day-circle of
Declinations, table of, 183
Dee, J., 146
de Santillana, G., 6
Digges, T., 14, 145f, 164
Dioptra, 146, 156; *see also* Vane
Drabkin, I. E., 156
Drake, S., 8
Dreyer, J. L. E., 2, 5, 14, 164

Duran, Profet, 9f

East–west line, 171, 177f
Eccentricity, 5, 114f, 157, 183, 192ff; *see also* Eye
Eclipse, 15, 93, 102, 142, 184, 187
    dated observation of, 155
Edelman, H., 157
Edges of the night, *see* Akronychos
Elongation
    instrument for measuring, 74ff, 159ff
    lunar, 158, 186
    with Venus as intermediary, 187
Ephod, *see* Duran
Epicyclic model, 116f
Equant, 3, 7, 192ff
    opposition to, 158
    skew, 192, 194
Equation of time, 10
Equinox, 178
Error, 27
    in arcsine, 138
    instrumental, 2, 80f
    observational, 20, 133
    random, 164
    systematic, 164
Espenshade, P., 135
Euclid, 31f, 34, 41, 44f, 85, 135ff, 140, 195
Eudoxus, 2, 7
Exodus, 72
Experiment, 2, 144, 146, 184f
Eye, *see also* Crystalline lens; Vision
    eccentricity of, 144ff, 154
    pupil of, 51

Faji, al-, 10
Farissol Botarel, M., 10
Feldman, S., 1
Finzi, M., 10, 12, 33, 136, 169
Florence
    astrolabe in, 164ff
    observations in, 14
Foscarini, 6

Friedländer, M., 5f

Galileo, 6, 8
Gandz, S., 169
Garin, E., 11f
Geminus, 4, 6
Gemma Frisius, 14, 143
Genesis, 72
Geographical latitude, 95, 106, 170f, 177, 180f, 183, 185
George of Trebizond, 10ff
Gibbs, S., 164
Gonzaga, 12
Grant, R., 164
Gunther, R. T., 164, 168

Haasbroek, N. D., 144
Ḥajjāj, al-, 191
Hamellius, P., 146
Harriot, T., 144
Hartner, W., 8, 132, 170
Haytham, Ibn al-, 4f, 7f, 131, 141f, 144, 158
Heath, T. L., 4, 31f, 41, 85, 135, 140
Heavens, sphericity of, 10
Heiberg, J. L., 135
Hipparchus, 3, 110, 156
Hirschfeld, H., 9
Hiyya, *see* Bar Ḥiyya
Homilius, 164
Homocentric model, 2, 7, 9
Horizon, artificial, 171
Horrox, J., 145
House for making observations, *see* Room
Ḥug ha-shamayim, 9
Humanists, 11

Ibn, *see* the next part of the name
Illumination, *see* Instrument, illumination of
Image
  inverted, 141f
  rounding of the corners of, 49, 141, 173
Inclination of the diameter, 132
Instruments, 20; *see also* Armillary sphere; Astrolabe; Camera obscura; Jacob Staff
  of brass, 154
  illumination of, 80, 149, 162
  observational, 27f
Interpolation, linear, 138f
Intromission theory, 144
ʿIrāq, Abū Naṣr Ibn, 183
Isaiah, 24, 71
Istanbul, 10
Iterative procedure, 134, 176f; *see also* Mechanical analogies

Jacob Staff, 2, 11, 13f, 52ff, 69ff, 134, 142, 158, 162
  and the center of vision, 144
  construction of, 55f, 146ff
  graduations on, 146ff
  modified, 74ff, 159ff
  mounted version, 67f, 154f
  and planetary diameters, 155f
Jerusalem, 10
Jervis, J. L., 14
Jupiter
  apogee of, 113
  apparent diameter of, 71, 106, 156, 188
  center for, 5
  daytime visibility of, 162
  observations of, 189
    dated, 133

Kaleb Afendopolo, *see* Afendopolo
Kellner, M. M., 1
Kennedy, E. S., 5, 134, 171
Kepler, J., 8
  on the center of vision, 143
  on pinhole images, 142f
Kinematics, 192ff
King, H. C., 164
Kremer, R. L., 158
Kunitzsch, P., 168

210 Index

Laban, 72
Langermann, Y. T., 9
Lay, J., 134
Leo, *see also* Levi ben Gerson
 de Balneolis, 11, 157
 Hebraeus, 11f
 Judeus, 11
Levi ben Gerson
 influence of, 9ff
 library of, 1
 life of, 1ff
 poems by, 71f, 157
 predecessors of, 2ff
 star list of, 9
Light
 behind observer, *see* Instrument, illumination of
 rectilinear propagation of, 140
Lindberg, D. C., 140ff, 144, 157, 173
Loewinger, D. S., 1

Maestlin, M., 143
Magnitudes, planetary, *see* Planets, sizes of
Mars
 apogee of, 113
 apparent diameter of, 105ff, 188
 brightness of, 7, 188
 correction for, 190
 elongation of, 191
Masha'allah, 168
Maimonides, M., 5ff, 132, 158
Manfredi, B., 13
Mantua, 10, 136
Manuscripts
 Cambridge, heb. Add. 1563, 13
 Florence, Laur., heb. Pl. 88/30, 9
 Istanbul, Aya Sofya, ar. 4832, 5
 Leyden, heb. Warner 43, 10
 London, British Library, heb. Add. 15,454, 10
 London, Montefiore Library, heb. 425, 9
 Lyon, lat. 326, 154

 Mantua, heb. 10, 9
 Oxford, Bodleian Library
  heb. Can. Misc. 334, 9
  heb. Can. Or. 64, 157
  heb. Lyell 96, 12, 136
  heb. Mich. 350, 10
  heb. Mich. 400, 170
  heb. Opp. Add. fol. 17, 134
  heb. Poc. 280b, 157
  heb. Regio 14, 10
 Naples, Biblioteca Nazionale, heb. IH F.9 (MS N), 10, 135, 169
 Paris, Bibiothèque Nationale
  heb. 724 (MS P), 8, 10, 132ff, 169, 183f, 192
  heb. 725 (MS Q), 10, 169
  heb. 1026, 10
 Rome, Casanatense, heb. 204, 9
 Vatican
  heb. 368, 9
  lat. 3380, 14
Mathematician, 22f
Measuring devices, 133; *see also* Instruments
Mechanical analogies, 30, 41, 134; *see also* Iterative procedures
Medicine, 51f, 144
Meer, M., 10
Mercury
 greatest elongation of, 188
 model for, 197
 observations of, 189
Meridian, 14f, 20, 67, 86, 95, 155, 170ff, 177ff, 185
Merton Astrolabe, 164
Minnaert, M., 162, 188
Mizrahi, Simon ben Jonah, 10
Models, *see* Moon, Planets, Sun
Monfasani, J., 10f
Moon
 diameter of, 20, 102, 157, 178, 186f
 distances of, 72, 157, 186
 elongation of, 158, 186
 inclination of the orb of, 103, 187
 model for, 159, 178, 192

errors in, 93, 102, 105
observations of, 103
opacity of, 25, 132
shading of, 117
Moonless night, observations on, 188; *see also* Darkness
Moulinas, R., 178
Mural Quadrant, 164

Naḥmias, Joseph ibn, 9
Nallino, C. A., 132, 179, 184
Natural philosophy, *see* Physics
Neugebauer, O., 2f, 183, 191, 193f, 197
Nolhac, P. de, 14
North, J., 12
Novak, B. C., 12

Obliquity, 95, 97, 171, 179, 182
Observatoire de Paris, *see* Paris
Occultation, planetary, *see* Planets, occultation of
On Animals, 113
Opacity, *see* Moon, opacity of
Optics, 7f, 140ff; *see also* Rays; Refraction; and Vision
Orange, 1
  synagogue in, 178
Orsini, F., 14
Ortive amplitude, 91, 99f, 171, 177, 184f
Oxford, Tables of, 12

Parallax, 93, 102f, 157, 159, 178, 186
Paris, Observatoire de, 15
Pedersen, O., 132, 135
Pegs, 56, 76
Penumbra, 157
Petrus de Alexandria, 13
Petz, H., 11
Peurbach, G., 143, 164
Pharaoh, 72
Phases, planetary, *see* Planets, phases of
Physics, 4, 6ff, 15, 19, 22ff, 72, 113, 131, 158, 192f
Pico della Mirandola, G., 11
Pines, S., 4, 8
Pinhole camera, *see* Camera obscura
Planets
  colors of, 25
  conjunctions of, 27f, 133
  eccentricities of, 5
  models for, 21, 108, 114ff, 189, 192ff
  observations of, 189
  occultation of, 133
  parallax of, 157
  phases of, 8
  positions relative to the Sun, 158
  self-luminosity of, 8, 132
  sizes of, 7f, 19, 21, 155ff, 187
Planetary Hypotheses, 5, 8, 188
Pliny, 191
Plumb line, 67, 82, 100, 155, 168, 177
Poel, Jacob ben Yomtov, 9
Poems, *see* Levi ben Gerson, poems by
Political philosophy, 131
Posidonius, 4
Practice, 134
Precession, 132, 134, 179
Problemata, 140
Proclus, 156
Profet Duran, *see* Duran
Prognostication, 13; *see also* Astrology
Psalms, 25, 71
Ptolemy, 1, 21, 24, 27, 29f, 72, 93f, 102f, 110, 155, 178f, 189, 193, 197; *see also Almagest, Planetary Hypotheses*
  lunar model of, 194
  lunar parallax according to, 186
  star list of, 187
  system of, 4f

Qabīṣī, al-, 5
Quadrant, 28; *see also* Mural

Raeder, H., 14, 158, 162
Ramus, P., 14, 146, 155

Rays, parallel, 71, 156
Refraction, 28, 80f, 107, 133, 162, 171, 188
Regiomontanus, 11, 13, 164
Regulus, 158
Reinhold, E., 143
Renan, E., 1, 9, 11
Retrograde motion, 3, 111
Revealer of Profundities, 71, 157
Ricius, A., 14
Righini Bonelli, M. L., 14
Roche, J., 13f, 142, 144ff, 154f
Rome, A., 133
Room for making observations, 86ff, 95ff, 170ff, 179ff
  size of, 171, 178
Rose, P. L., 11f, 14
Rosen, E., 8
Roth, C., 12
Rounded corners, *see* Image
Ruler, 88, 97

Sabra, A. I., 4, 7
Sachau, C. E., 183
Sagitta, 31; *see also* Versine
Salamanca, 10
Samuel, 72
Saturn
  apogee of, 113
  apparent diameter of, 106, 188
  observations of, 189
Saving the phenomena, 5
Sawyer, F. W., III, 179
Scale, *see* Jacob Staff; Transversal; and Vernier
Scheindlin, R., 157
Screen, 52
Scultetus, 164
Settle, T., 14
Sezgin, F., 5
Shadow radius during eclipses, 104
Shāṭir, Ibn al-, 5, 187
Shatzmiller, J., 1
Shehaby, N., 4

Sight, *see* Vane, Vision
Simon de Covino, 13
Simplicity, 8
Simplicius, 2, 4, 6
Sine, 20 31; *see also* Trigonometry
  of $\frac{1}{4}°$, 138
  table of, 40ff, 83, 139
Solomon, the brother of Levi ben Gerson, 1, 13
Solstice, 96f, 170f, 181ff
  dated observations of, 183, 192
Solutions, approximate, 152ff; *see also* Iterative procedure
Sosigenes, 2
Spain, 5f
Span, 55, 85, 92, 95, 98, 144, 146f, 171, 179f, 184f
Spherical geometry, 151; *see also* Trigonometry
Staff (*maqel*), 52; *see also* Jacob Staff
Stars
  list of, 9, 187
  positions of, 21, 56ff, 73f, 93, 102, 151ff, 158, 186f
  determined with Venus, 158
  sizes of, 69ff
  diminished near the horizon, 188
Stationary points, 3
Staub, J. J., 1, 131f
Steinschneider, M., 9f, 12, 136
Straker, S., 141ff
Sun, *see also* Solstice
  apogee of, 94, 98f, 113, 179, 182f, 185
  day-circle of, 174f
  eccentricity of, 49, 98f, 183f
  equation of, 93, 95, 179, 183, 185
  model for, 108f
  observations of, dated, 183f, 192
  orb of, 95
  size of, 20, 141, 157, 173, 180, 182ff, 192
Suryasiddhanta, 179
Swerdlow, N., 14
Symon de Phares, 13
Synodic month, 186

Synagogue, 14, 178

Tarascon, 9
Taton, R., 15
Taylor, E. G. R., 164
Thickness of vapors, *see* Refraction
Thoren, V., 189
Thorndike, L., 11, 13
Time of day or night, 133
Toledan Tables, 179
Toomer, G. J., 3, 131, 157, 178, 189
Toscanelli, P., 13f
Touati, C., 1, 9, 13, 131
Transit, 133
Transversal scale, 2, 55f, 83ff, 146ff, 163ff, 168ff, 181
Trial and reexamination, 134
Triangles, solutions of, 139f; *see also* Trigonometry
Trigonometric functions, capitalized, 134f
Trigonometry, 31ff, 134ff
Tuckerman, B., 133
Ṭūsī, al-, 5, 8
Twersky, I., 132

Uniform circular motion, 3ff, 114

Vane, 29, 67, 76, 82, 154, 164
Vapors, thickness of, *see* Refraction
Velocity, variation in, 192
Venus, *see also* Elongation
  brightness of, 187
  correction for, 190
  daytime visibility of, 162
  determination of stellar positions with, 158
  greatest elongation of, 190f
  observations of, dated, 133
  phases of, 8, 187
  self-luminosity of, 132
  size of, 71, 105, 156, 187f
Vernier scale, 164
Versine, 31, 136, 139; *see also* Trigonometry
Vision, 113, 192
  center of, 2, 20, 51ff, 56, 143ff, 156

Walls, non-perpendicular, 100f, 185f
Walther, B., 11, 13, 158, 187
Wars of the Lord, 1, 19ff; *see also* Levi ben Gerson
Weil, G. E., 1
Weinryb, B. D., 1
Westman, R., 8
Wickersheimer, E., 13
Wiedemann, E., 141
Wilson, C. A., 189
Window, polygonal, 48f; *see also* Room
Winter, H. J. J., 8, 132

Zachariah, 71
Zakkut, A., 10
Zarqāl, al-, 179
Zenith, 164, 167ff

Hebrew Text

הערות לפרק ד:33

נוסף בשולי נ (16ב): אמ"פ. וכבר יתבאר הנה ג"כ שאם היו הנכחו' והחץ ידועי' כי הנה הקוטר יהיה ידוע. וזה כי מפני שמדיעת שניהם נדע המיתר הנה נדע מרובע המיתר ולפי שהוא שוה להכאת החץ בקוטר הנה כאשר חלקנוהו על החץ יהיה העולה הוא האלכסון ר"ל הקוטר. ואם אין זה מכונת הספר לפי שהוא משים לעולם הקוטר בחזקה ידוע הנה יאות בחקירה.

נוסף בשולי נ (17א): אמ"פ. מתמונה האחרונה מזה הדבור יתבאר בדרך נקל מאד שאם החץ יהיה ידוע כי הנה הנכחו' ידוע. וזה כי נדע ממנו מרחקו מהמרכז ר"ל את אשר יחסר מס' או יעדיף עליו וכאשר גרענו מרובע זה החסרון או ההעדף ממרובע ס' ולקחנו שרש הנשאר הוא המבוקש.

151 ואולם כאשר תונח הנקדה אשר סביבה תהיה
תנועה השוה בלתי מונחת על הקוטר ההולך על מרכז
הארץ ומרכז הגלגל הנה תיוחד זאת התכונה מהתכונות
הקודמות שלא ימצא בה תכלית המהירות והמתינות
152 במרחק הרחוק והקרוב. אבל יהיה נמצא בקצוות הקוטר
העובר על מרכז הארץ ומרכז התנועה השוה וזה כלו
153 מבואר עם מה שקדם מהדברים. וידמה לפי מה שזכר
בטלמיוס מעניין תנועת כוכב שחאות זאת התכונה בו
וזה באר לפי מה שיראה מדבריו שתכלית המהירות
והמתינות בו אינו עובר על המרחק הרחוק והקרוב.
154 ולזאת הנקדה גם כן שסביבה תהיה התנועה השוה מצבים
מתחלפים כמו העניין בנקדה המונחת על הקוטר העובר על
המרחק הרחוק והקרוב ולכל אחד מהם סגולות ייוחד בהם
מסוג הסגולות הקודמות במה שידמה להם כשהיתה הנקדה
מונחת על הקוטר העובר על המרחק הרחוק והקרוב | ולא
הארכנו בהם להיותם כמבוארות בעצמם עם מה שקדם
מהדברים.
155 הנה אלו הם חלקי הסותר אשר אפשר שיצוייר
בחלוף הנראה בתנועת האורך כשהונחה תנועת האורך
156 פשוטה. והנה זכרנו עם זה התכונה המפורסמת בגלגל
ההקפה ואם היא מורכבת משתי תנועות להראות רוב
סגולותיה לסגולת התכונה אשר תהיה בגלגל יוצא
המרכז ותהיה התנועה סביב מרכזו בעבור שלא נכפל
המאמר בזה.

70א

---

153 כותב] נ: כוכב.   153 והמתינות] נוסף בכ"י פ
(למעלה): האיחור.   156 להראות] נ, פ: להדמות.
156 המאמר בזה] נ: המאמר.

ק 23ב 140	דטז החז הם שורת הנה יהיה יתרון זוית דטז על זוית דבז יותר גדול מיתרון זוית החז על זוית הבז והוא מה שרצינו לבאר. וכבר תיוחד עוד זאת התכונה שכבר יהיה בה תקון יותר גדול ליותר מצ׳ מעלה מתנועת הכוכב הנראית
141	ממה שימצא ממנו לצ׳ מעלה מתנועת הכוכב הנראית וזה לא ימצא באחת מהתכונות הקודמות. המופת שנשאיר התמונה הקודמת בזאת התכונה על עניינה ונבאר שכבר
142	ימצא שם ליותר מצ׳ מעלה מתנועת הכוכב הנראית תקון יותר גדול מזוית דטז. המופת שאנחנו נחלק קו דט
נ 69א	לחציין על נקדת ל ונקרה סביב נקדת ל עגלה תקיף
143	במשלש דטז והיא עגלת דטנז. ונדביק קו הלנ הישר ויחתך עגלת אבג בנקדת ס וזה מחויב לפי שקו לס הוא היותר קצר שבקוים המגיעים למקיף עגלת אבג מנקדת ל
144	שאינה על המרכז. ולזה יחויב שיהיה קו לס יותר
145	קצר מקו לט השוה לקו לב. ונקוה קו דסע הישר ויחתך
146	עגלת דט בנקדת ע ונקרה קוי זס זע. והוא מבואר שזוית דסז היא זוית התקון בזה המקום והיא גדולה מזוית דעז השוה לזוית דטז והוא מה שרצינו לבאר.
147	ואולם אם הונחה התכונה בגלגל יוצא המרכז
פ 34א	והונח מרכז הארץ במה שבין מרכז הגלגל ובין הנקדה אשר סביבה תהיה התנועה השוה הנה תיוחד זאת התכונה מהתכונות הקודמות שהמהירות לתנועה ימצא בה בעת היות שעור הכוכב יותר קטן במבט והאחור בעת הראותו
148	יותר גדול וזה מבואר בנפשו. ותיוחד עוד כאשר הונחה התחלת זאת התנועה ממקום המרחק הרחוק מהקוטר העובר על אלו המרכזים הנה יהיה התקון היותר גדול לצ׳ מעלה מהתנועה השוה ואולם מהתנועה הנראית היא ליותר מצ׳ מעלה כשעור התקון היותר גדול וזה מבואר
149	במעט עיון עם מה שקדם מהדברים בזה. ויהיה התקון הזה להוסיף על התנועה השוה ובשאר התכונות הקודמות
150	הוא לגרוע. ואם שמנו התחלת התנועה מהקצה האחד מהקוטר והוא המרחק הקרוב הנה תיוחד התכונה מהשאר התכונות אשר בגלגל יוצא המרכז שהתקון היותר גדול יהיה לצ׳ מעלה מהתנועה השוה ולפחות מצ׳ מעלה מהתנועה הנראית כשעור התקון היותר גדול וזה כולו
נ 69ב	מבואר עם מה שקדם מהדברים.

140 התכונה] נ: התמונה.   140 ממנו] פ: לנו.
143 לס] נ: לט.   143 אבג] פ, ק: אבגד.

פ33ב   מתנועתו הנראית | וזרית הבז היא זרית החלוף אשר
           תמצא לכוכב בהיותו בצ׳ מעלה מתנועתו השוה.
125   ונבדיל מקו הב קו הט שוה לקו זח ונקוה קו זט.
126   והוא מבואר שזרית הטז שוה לזרית החז כי זרית
           טהז שוה לזרית חזה והקוים המקיפים בזרית טהז
           שוים אל הקוים המקיפים בזרית חזה כל אחד לגילו.
127   ולזה יהיה יתרון התקון לצ׳ מעלה מתנועתו הנראית
           על התקון לצ׳ מעלה מתנועתו השוה כמו זרית טזב
           שהיא יתרון זרית הטז על זרית הבז.
128   וגם כן נשים בתכונה הזאת אשר אנחנו בה עגלת
           הגלגל היוצא המרכז עגלת אבג ותהיה מרכזה נקדת ה
129   ומרכז הארץ נקדת ז. ותהיה נקדת ד במה שבין שתי
           נקדות ה וז ותהיה סביבה תנועת הכוכב השוה.
130   ונדביק קו אהדזג הישר ונוציא משתי נקדרות ד וז
           קוי דב זט עומדים על זרית נצבה על קו אהדזג.
131   ויכלו אל המקיף מצד אחד. ונקוה קוי דט זב
           ויתבאר באופן הקודם שזרית דטז היא זרית התקון
           לצ׳ מעלה מתנועת הכוכב הנראית וזרית דבז היא
132   זרית התקון לצ׳ מעלה מתנועת הכוכב השוה. ונשים
           זרית דטז בזאת התמונה שוה לזרית החז בתמונה
נ68ב   הקודמת כדי שיהיה החלוף היותר | גדול בשתי אלו
133   התכונות בשעור אחד. ונבדיל מקו דב קו שוה לקו
           דט וזה אפשר כי קו דב הוא יותר ארוך מקו דט לפי
           שקו דט הוא יותר קרוב לקו דג שהוא שלמות הקוטר
134   מקו דב. ויהיה הקו השוה לקו דט קו דמ ונקוה קו
           מז ואומר שזרית דבז בזאת התמונה היא יותר קטנה
           מזרית הבז בתמונה הקודמת.
135   המורפת כי מפני שזרירות דטז החז שורת וזריות
           דזט הזח הם גם כן שורת הנה משלש דטז הזח הם
136   מתדמים. ולזה יהיה יחס קו דט אל קו דז כיחס קו
           הח אל קו הז ולפי שקו דמ שוה לקו דט וקו הב שוה
           לקו הח הנה יהיה יחס קו דמ אל קו דז כיחס קו הב
137   אל קו הז. ולפי שהזרית אשר יקיפו בה קוי דמ דז
           שוה לזרית אשר יקיפו בה קוי הב הז כי הם נצבות
           הנה משולשי דמז הבז מתדמים ולזה תהיה זרית דמז
138   שוה לזרית הבז. ולפי שזרית דבז קטנה מזרית דמז
139   הנה תהיה זרית דבז קטנה מזרית חבז. ולפי שזריות

124 החלוף...בצ׳[2] ב: התקון לצ׳.   127 הטז]
ק: הבז.   127[2] הבז] ק: הטז.

| נ 67ב
| השוה סביב מרכז הגלגל שענינו | הוא גם כן בזה
114 23א ק | החאר. שבזאת התכונה לא יהיה יתרון התקון שבין צ'
| מעלה מתנועתו הנראית ובין צ' מעלה מתנועתו השוה
| גדול כיתרון התקון שבין אלו המקומות בתכונה ההיא.
115 | וכבר תיוחד עוד זאת התכונה מהמחכונה ההיא כשהונח
| התקון היותר גדול שוה בשתי אלו התכונות שהתכונה
| ההיא תשים גרם הכוכב יותר מתחלף במבט כשהובט בו
| בגובה ובשפל ממה שיתחלף גרמו במבט לפי זאת התכונה
| כי חלוף מרחקו מהארץ יהיה יותר גדול בתכונה ההיא
| וזה מבואר מאד למעיין בזה הספר עם מה שקדם
| מהדברים בזה.
116 | ונניח עתה שיהיה מרכז התנועה השוה במה שבין
117 | מרכז הגלגל ובין מרכז הארץ. ואומר שכבר תיוחד
| זאת התכונה מהתבונות הקודמות בהנחת הגלגל יוצא
| המרכז כאשר יונח שעור התקון היותר גדול בזאת
| התכונה שוה לתקון היותר גדול בתכונות ההם הנה
| יהיה יתרון שעור הכוכב הנראה בהיותו במרחק הקרוב
| על שעורו הנראה בהיותו במרחק הרחוק יותר ממה שהיה
| באחת מהתכונות הקודמות כי היה יתרון המרחק הרחוק
| על המרחק הקרוב יותר רב בזאת התכונה וזה מבואר
118 | בנפשו למעיין בזה הספר. ואומר שזאת התכונה תשים
| יתרון התקון לצ' מעלה מהתנועה הנראית על התקון לצ'
| מעלה מהתנועה השוה יותר גדול ממה שהיה היתרון בזה
119 | באחת מהתכונות הקודמות. וזה שאשר ימצא בה זה
| היתרון יותר גדול בתכונות הקודמות היא התכונה אשר
| תהיה בגלגל יוצא המרכז ותהיה תנועתו השוה על
120 | מרכזו. ונבאר שזאת התכונה תשים זה היתרון יותר
נ 68א | גדול מהתכונה ההיא כשהיה התקון היותר גדול | בשתי
| התכונות האלו בשעור אחד.
121 | המופת כי בתכונה אשר בגלגל יוצא המרכז
122 | שהתנועה השוה היא על מרכזו. נשים עגלת הגלגל
| היוצא המרכז אשר עליה תהיה התנועה עגלת אבגד
123 | ותהיה מרכזה נקדת ה ומרכז הארץ נקדת ז. ונקוה
| קו אהזג הישר ונוציא משתי נקדות ה וז שני קוי הב
| זח עומדים על קוטר אג על זוית נצבת ויכלו אל מקיף
124 | העגלה בצד אחד ונקוה קוי הח זב. הנה זוית החז
| היא זוית החלוף אשר תמצא לכוכב בהיותו בצ' מעלה

112 הגלגל] חסר בכ"י נ.   122 אבגד] נ: אבג.

כשהרונה המרחק בין המרכזים בתכונה ההיא כמו מרחק מה
שבין נקדת ז ו ה בזאת הצורה הנה יהיה יתרון הקו שהוא
חצי הקוטר על השני בתכונה ההיא יותר גדול הרבה

102 מיתרון אחד מאלו הקוים על השני בזאת התכונה. ומזה
יתבאר מה שאנחנו רוצים לבארו בכמו האופן הקודם ר"ל
שנבדיל מקו חצי הקוטר קו שוה לקו השני.

103 והמשל שירונח בזאת התמונה בעינה נקדת ז הנקדה
אשר סביבה תהיה התנועה השוה ונקדת ה מרכז הארץ

104 ומרכז הגלגל תהיה התנועה השוה. ותהיה זוית זבה
זוית התקון בהירות הכוכב בצ׳ מעלה מתנועתו השוה
וזוית זטה תהיה זוית התקון בהירות הכוכב בצ׳ מעלה

105 מתנועתו הנראית. ונניח עגלת אבג בשעור בעינו

106 נ 67א שהונחה בתמונה הקודמת. ולפי שקו הט | הוא בזאת
התמונה חצי הקוטר הנה יהיה בהכרח יותר גדול משעור
קו הט בתמונה הקודמת כי הוא היה רחוק ממרכז העגלה
שעור קו הד ושעור קו זב הוא יותר קטן משעור קו

107 פ 33א זב | בתמונה הקודמת. ולזה תהיה זוית זבה בזאת
התמונה השנית יותר גדולת מזוית זבה בתמונה הקודמת.

108 וזה יתבאר בשנוציא קו זב אל ח ויהיה קו זח שוה לקו

109 זב בתמונה האחרת ונוציא קו הח. הנה אם כן זוית זחה
היא שוה לזוית זבה בתמונה הקודמת והיא קטנה מזוית
זבה בזאת התמונה.

110 וכזה יתבאר שזוית זטה תהיה בזאת התמונה השנית
יותר קטנה מזוית זטה בתמונה הקודמת בשנבדיל מקו הט
קו הכ שוה לקו הט בתמונה הקודמת ונוציא קו זכ.

111 ותהיה זוית זבה שוה לזוית זטה בתמונה הקודמת והיא

112 יותר גדולה מזוית זבה בזאת התמונה השנית. ולזה
יהיה יתרון זוית זבה על זוית זטה בזאת התמונה
השנית יותר גדול הרבה מיתרון זוית זבה על זוית
זטה בתמונה הקודמת וזה ממה שתיוחד בו התכונה אשר
אנחנו בה מהתכונה ההיא.

113 ובזאת ההנהגה יתבאר שכבר תיוחד התכונה מאלו
התכונות שתשים יותר גדול החלוף לצ׳ מעלה מתנועתו
הנראית מהחלוף אשר לצ׳ מעלה מתנועתו השוה מהתכונה
אשר על גלגל יוצא ממרכז העולם ויתנועע תנועתו

---

101 הרבה] פ: מאד.   102 ר"ל] פ: רצוני לומר.
103 הגלגל] נוסף בכ"י נ (למעלה): אשר עליו.
108 ח] נ: זח.   109 זחה] פ: זחא.   112 השנית
יותר] נ: יתרון.   113 ממרכז העולם] נ, ק: מרכז
העולם.

86	ואומר שזורית זטה שוה לזורית זבה. וזה שקו הט שוה
87	לקו הט שוה לקו זב לפי שמרחקם ממרכז העגלה שוה
	וקו הז משותף וזורית בזה שוה לזורית זהט ׀ לפי שהם
88	נצבות. וישאר קו זט שוה לקו הב וזה ריהיה המשלש שוה
89	למשלש וזויותיו שוות לזויותיו כל אחת לגילה. ולזה
	תהיה זורית זבה שוה לזורית זטה והוא מה שרצינו לבאר.
90	ומזאת התמונה יתבאר במעט עיון שאם היה במשלנו
	זה קו דה יותר ארוך מקו דז הנה תהיה זורית זטה יותר
	גדולה מזורית זבה ר״ל שהתקון יהיה יותר גדול לצ׳
	מעלה מתנועת הנראית מהתקון אשר לצ׳ מעלה מהתנועה
	השוה והפך זה יהיה אם יהיה קו זד יותר ארוך מקו
91	דה. המופת שנניח התמונה הקודמת על ענינה ויהיה
	תחלה קו דה יותר ארוך ולזה יחוייב שיהיה קו הט
92	יותר קצר מקו זב. ונבדיל מקו זב קו זל שוה הט
93	ונדביק קו לה. ויתבאר בכמו הבאור הקודם שזורית
94	זלה שוה לזורית זטה. והנה זורית זלה היא יותר
	גדולה מזורית זבה ומזה התבאר שזורית זטה היא יותר
95	גדולה מזורית זבה. ובכמו זאת ההנהגה יתבאר שאם
	היה קו זד יותר ארוך מקו דה שכבר תהיה זורית זבה
	יותר גדולה מזורית זטה והוא מה שרצינו לבאר.
96	וכבר תיוחד זאת התכונה מהתכונה הקודמת אשר
	בגלגל יוצא המרכז שתנועתו השוה היא על מרכזו כי
	כבר יונח התקון היותר גדול בזאת התכונה שוה ׀
	לתקון היותר גדול בתכונה ההיא ולא ימצא חלוף שעור
	הכוכב במבט כי אם פחות מהחלוף הנראה לו בתכונה
97	הקודמת. וזה כי המרחק בין מרכז הארץ ובין מרכז
	הגלגל הוא יותר קטן בזאת התכונה ממה שהיה בתכונה
98	הקודמת וזה מבואר למעיין בזה הספר. ולזה
	יהיה חלוף מרחקי הכוכב ממנו יותר מעטי בזאת התכונה
	מחלוף מרחקי הכוכב ממנו בתכונה ההיא.
99	וכבר תיוחד התכונה מאלו התכונות שתשים החלוף
	יותר גדול לצ׳ מעלה מתנועתו השוה מהחלוף אשר יהיה
	לצ׳ מעלה מתנועתו הנראית מהתכונה אשר על גלגל
	מרכזו מרכז העולם ויתנועע תנועתו השוה לא סביב
100	מרכז העולם. שהענין בו הוא גם כן בזה התואר שבזאת
	התכונה לא יהיה יתרון התקון שבין צ׳ מעלה מהתנועה
	השוה ובין צ׳ מעלה מהתנועה הנראית גדול כיתרון
101	התקון שבין אלו המקומות בתכונה ההיא. וזה כי

[88] 91 התמונה] פ: התנועה.

72 המגיע ממרכז הארץ למרכז הגלגל. ואם הונחה על יושר
הקו ההוא הנה אם שתהיה התנועה סביב נקדת יהיה מרכז
הגלגל בינה ובין מרכז הארץ או שתהיה הנקדה ההיא
בין מרכז הגלגל ובין מרכז הארץ או יהיה מרכז הארץ
73 בינה ובין מרכז הגלגל. וזאת חלוקה הכרחית אי אפשר
שיהיה לזאת הנקדה עם מרכז הגלגל ומרכז הארץ כשהיו
על קו אחד מצב רביעי.
74 והנה תיוחד התכונה מאלו התכונות אשר תהיה
התנועה השוה בה סביב מרכז הארץ שתנועתה תראה שוה
במבט ואמנם גודל הכוכב יראה בה מתחלף ר"ל שבהיותו
במרחק הרחוק יראה שעורו קטן מהשעור שיראה לו
75 65ב נ בהיותו | במרחק הקרוב וזה מבואר בנפשו. ובזה
האופן הניח בטלמיוס תנועת האורך בירח אלא שכבר
התבאר מדברינו שזה בלתי נאות למה שיראה משעור
קוטר גרם הירח מתחלף במקום מקום מתנועת האורך.
76 וזה כי לא יראה שעור קוטר גרם הירח מתחלף במקום
77 מתנועת האורך על האופן שיתחייב מזאת התכונה. וזה
כי לפי זאת התכונה היה ראוי שיראה קוטר הירח
בנגודים יותר קטן מהשעור ראוי שיראה לו ברבועים כמו
78 השליש. ואנחנו מצאנו הירח במבט ברבועים בלתי גדול
מהשעור שנראה לו בנגודים כי אם מעט כמו שזכרנו.
79 ואולם התכונה שיונח בה מרכז הגלגל בין מרכז
התנועה השוה ובין מרכז הארץ תיוחד משאר התכונות
האלו שכבר יקרה בה שיהיה התקון לצי' מעלה מתנועתו
80 השוה ולצ' מעלה מתנועתו הנראית שוה. וזה אמנם
יקרה כשיהיה מרכז הגלגל חולק המרחק שבין מרכז
התנועה השוה ובין מרכז הארץ לחציין.
81 המורפת שאנחנו נניח שתהיה עגלת אבג יוצאת
ממרכז העולם ומרכזה נקדת ד ומרכז הארץ נקדת ה
ונניח שתהיה הנקדה שיתנועע סביבה תנועתו השוה
82 נקדת ז. ונדביק קו אזדהג הישר ונדביק קוי זב הב.
83 ונניח שתהיה זוית אזב נצבת והיא זוית התנועה |
84 32ב פ השוה ויהיה התקון זוית זבה. עוד נשים זוית אהט
85 נצבה והיא זוית התנועה הנראית. ונוציא קו זט
נ 66א ויתבאר שזוית זטה | היא זוית התקון בזה המקום.

71 המגיע$_2$ ... בגלגל] חסר בכ"י נ, ק.
75 מתחלף] חסר בכ"י נ, פ.   76 וזה ... האורך]
חסר בכ"י ק.   76 וזה ... התכונה] חסר בכ"י פ.
79 השוה$_2$ ... מתנועתו$_2$] חסר בכ"י נ.

59 היא כמו זוית הקשת הראויה לנכחות ז'ל. ויתבאר
שזורית אזב שהיא זוית התנועה השוה בשתי התכונות
היא מוסיפה על צ' מעלה כשעור זוית לבז שהיא כשעור
זוית הקשת הראויה לנכחות חצי קו זה.
60 ואומר שהתקון ימצא יותר גדול בתכונת עגלת
אבגד ממה שהוא בתכונת עגלת חבטד בצדי התחלת
תנועת האורך עד שתי נקדות ב וד וההפך הוא בנשאר
61 מהעגלה. המופף שנניח בקשת אב נקדה | איך שתפול
והיא נקדת מ ונקוה קו זמ הישר עד שיחתך קשת חב
62 בנקדת נ ונקביק קוי המ הנ. הנה זוית זמה היא זוית
התקון בתכונת עגלת אבגד בזה המקום וזוית זנה היא
63 זוית התקון בתכונה עגלת חבטד. והנה זוית זמה
היא יותר גדולה מזוית זנה וכן יתבאר זה בכל תקון
64 שיהיה בצדי התחלת תנועת האורך עד שתי נקדות ב וד.
65 ונאמר שכבר ימצא הפך זה בנשאר מהעגלה. וזה שכבר
נניח בקשת בג נקדה איך שתפול והיא נקדת ס ונדביק
66 קו זס הישר ויחתך קשת בט בנקדת ע. ונוציא קוי הע
הס הנה | זוית זסה היא זוית התקון בזה המקום בתכונה
עגלת אבגד והיא יותר קטנה מזוית זעה שהיא זוית
התקון בזה המקום בתכונת עגלת הבטד.
67 והנה נשאר מהחלוקה הקודמת שתהיה התנועה השוה
על גלגל מרכזו יוצא ממרכז העולם ותהיה התנועה לא
68 סביב מרכזו. וזה גם כן לא ימלט מחלוקה אם שתהיה
התנועה השוה סביב מרכז הארץ או שתהיה על נקדה
יהיה מרכז הגלגל בינה ובין מרכז הארץ או שתהיה על
נקדה מונחת בין מרכז הגלגל ובין מרכז הארץ או
שתהיה על נקדה יהיה מרכז הארץ בינה ובין מרכז
הגלגל או שתהיה הנקדה ההיא בלתי מונחת על יושר הקו
69 ההולך ממרכז הארץ למרכז הגלגל ההוא. ואולם איך
70 יתבאר שזאת החלוקה מחוייבת הנה לפי מה שאומר. וזה
שלא ימנע משתהיה התנועה אם סביב מרכז הארץ אם על
71 זולת מרכזה. ואם הונחה על זולת מרכזה הנה אם שתהיה
הנקדה ההיא על יושר הקו המגיע ממרכז הארץ אל מרכז
הגלגל או תהיה הנקדה ההיא על זולת יושר הקו ההוא

63 והנה] ק: והיא. 68 נקדה]$_1$ ...על]$_2$ חסר בכ"י ק.
69 הנה] פ: הוא. 70 משתהיה] נ: שתהיה. 71 ההוא] חסר בכ"י פ.

על מרכזו תמצא תמיד התנועה יותר מתונית כשהיה
הכוכב יותר רחוק מן הארץ.

44 ואומר שאם הונח מרכז עגלת אבג במשלינו זה
מרכז הארץ והיתה תנועתו השוה סביב נקדה יוצאת
ממרכז הארץ הנה יהיה התקון היותר גדול לצ׳ מעלה
45 מתנועתו השוה מהתחלת תנועת האורך. המופת שנניח
התמונה הקודמת על ענינה ותהיה נקדת ה מרכז הארץ
46 ומרכז עגלת אבג. ותהיה התנועה השוה סביב נקדת ד
47 ונקוה קוי הב דב ונקוה קו אדהג הישר. ותהיה נקדת
א התחלת התנועה השוה ויתבאר באופן הקודם שהתקון
היותר גדול יהיה בזאת התכונה | כשתהיה זוית הדב
נצבת וזה אמנם יהיה לצ׳ מעלות קודם נקדת א ואחריה.
48 וזה כי שעור קו הב הוא בזאת התכונה ס׳ מעלה כי הוא
חצי קוטר עגלת אבג ולזה יחוייב מהבאור הקודם שיהיה
התקון היותר גדול בזאת התכונה כשתהיה זוית הדב
נצבת.
49 וכבר תיוחד עוד זאת התכונה מהתכונות הקודמות
שכבר יראה בה תמיד גודל קוטר הכוכב בשעור אחד בלתי
50 מתחלף כי הוא יראה תמיד ממרחק אחד. וכבר תיוחד
עוד זאת התכונה מהתכונות הקודמות כשהונח שעור קו
דה במשלינו זה שוה בשתי התכונות וזה עד צ׳ מעלה
מהתחלת תנועת האורך לפניה ולאחריה מהתנועה השוה
ויותר מזה כמו הקשת אשר נכחותה כמו חצי שעור קו
דה יהיה התקון בזאת התכונה יותר גדול ממה שיהיה
51 התקון בשאר התכונות הקודמות. ובמה שנשאר מהעגלה
ימצא הענין בהפך.
52 המופת שאנחנו נשים עגלת אבגד העגלה אשר תהיה
בה התנועה בזאת התכונה השנית ויהיה מרכזה ומרכז
53 הארץ נקדת ה. ותהיה תנועתה השוה סביב נקדת ז
54 ונדביק קו אזהג הישר. ונקוה סביב מרכז ז עגלת
חבטד שוה לעגולת אבגד תחתך עגלת אבגד על שתי
55 נקדות ב רד ויהיה קו חאזהטג ישר. ותהיה עגלת
חבטד עגלת הגלגל היוצא המרכז ומרכזה נקדת ז וסביב
נקדת ז תהיה תנועתה השוה כמו | הענין בתכונה
הקודמת באופן שתהיינה שתי עגלות אבגד חבטד משל |
56 לשתי אלו התכונות. ונקוה קוי זב הב הנה משלש זבה
57 הוא שוה השוקים. ונחלק קו זה לחציין על נקדת ל
58 ונדביק קו בל. ויתבאר שזוית זלב נצבת וזוית לבז

ג 64א

ג 64ב

ק 22א

55 ז וסביב] ק: ה וסביב.

33 יהיה בהיותו בנקדת השפל והחלוף אשר יראה בגודל קטרו בשני אלו המקומות הוא לפי מה שיתחייב מיחס שעור קו אה אל שעור קו הג בזה המשל. וזה כי הכוכב כשהיה בנקדת A יראה ממרחק קו אה וכשהיה בנקדת ג יראה ממרחק קו הג שהוא יותר קצר מקו אה כמו שעור כפל קו דה.

34 והנה תצוייר זאת התנועה בגלגל ההקפה כשיונח תקוע בגלגל אחד מקיף הארץ ויתנועע הגלגל המקיף הארץ בתנועת הכוכב השוה ויתנועע עוד בגלגל ההקפה תנועה שוה לתנועת הגלגל המקיף הארץ ר"ל שתשלם תנועת גלגל

35 ההקפה ותנועת האורך בזמן אחד. ויראה ממהלך הכוכב בגלגל ההקפה תוספת פעם וחסרון פעם מתנועת הכוכב

36 השוה על האופן שקדם. וכבר חיוחד התכונה בגלגל

37 ההקפה משאר התכונות אשר בגלגל מקיף הארץ. כי בתכונה בגלגל מקיף הארץ יראה לנו תמיד חלק אחד בעינו מהשטח המקיף בגרם הכוכב וזה מבואר בנפשו

38 חיובו. ואולם בתכונה בגלגל ההקפה לא יראה לנו תמיד חלק אחד בעינו מהשטח המקיף בגרם הכוכב אבל יראה לנו מהכוכב בהיותו באמתת הגובה שטח גרם הכוכב אשר לצד פנימי גלגל ההקפה ובהיותו באמתת השפל יראה לנו שטח גרם הכוכב אשר לצד | חיצוני גלגל ההקפה ובהיותו במרחקים האמצעיים יראה לנו קצת מהשטח החיצוני וקצת

39 מהשטח הפנימי. והנה מזה המקום נתבאר במה שאין ספק בו שאין הירח קבוע בגלגל ההקפה כמו שיראה בטלמיוס.

40 וזה שהצל הנראה בגוף הירח שהוא בהכרח אמתת ענין לא ראיה כמו שהתבאר בלמודיות ויתבאר בג"ה מדברינו במה שיבא שהוא נראה תמיד בשטח הירח הנראה לנו בצד אחד

41 מוגבל. וזה מופת שאנחנו נראה תמיד משטח גרם הירח המקיף בו חלק אחד | מוגבל וזה בלתי אפשר אם הונח הירח בגלגל ההקפה.

42 וכבר חיוחד עוד התכונה אשר בגלגל ההקפה מתכונה אשר בגלגל מקיף הארץ יוצא המרכז ותהיה התנועה על מרכזו כי בזאת התכונה יתכן שתהיה תנועת הכוכב יותר מהירה כשהיה יותר רחוק מן הארץ וזה אמנם יהיה כשהיתה תנועת הכוכב בגובה גלגל ההקפה לצד תנועת הארץ כמו שהניח בטלמיוס בחמשת כוכבי המבוכה.

43 ואולם בגלגל יוצא ממרכז הארץ המקיף אותה ומתנועע

ב63 ב

פ31 ב

39 נתבאר] ב: נבאר. 40 שיבא שהוא] נ, ק: שיבא הוא.

נ 62ב 21 אדהג והנה נקדת A ממנו היא נקדת הגובה ונקדת G
היא נקדת השפל. ונאמר שההחלוף היותר גדול שיראה
בזאת התנועה הוא ליותר מצ' מעלה מנקדת A | לפניו
22 ולאחריו כשעור התקון היותר גדול. וזה שאנחנו
נדביק קוי הב דב ותהיה תחלה זוית אהב והיא זוית
התנועה הנראית צ' מעלה ולזה תהיה זוית אהב נצבת.
23 והנה זוית אדב היא זוית התנועה השוה והיא גדולה
מזוית אהב כשעור זוית דבה ולזה תהיה זוית דבה
היא זוית התקון בזה המקום ואומר זוית דבה היא
פ 31א 24 זוית התקון בזה המקום. ואומר זוית דבה היא זוית
התקון | היותר גדול שאפשר שיתחדש מזאת התנועה
וזה שקו דה הוא נכחות הזורית הזאת לפי מה שקדם לפי
25 שזורית דהב נצבת ושעור קו דב הוא ס' מעלה. ואם
תהיה זוית אהב חדה או נרחבת הנה יהיה נכחות זוית
26 דבה שהיא זוית התקון יותר קטן מקו דה. וזה שכבר
יפול העמוד המגיע מנקדת D על קו בה המונח בלי
תכלית שהוא נכחות זוית דבה אם במה שבין שתי
27 נקודות ה ו ב או יפול לחוץ. ואין שהיה הנה יהיה קו
דה יותר ארוך הרבה מהעמוד ההוא כי הוא כחיי על
מרובע העמוד ההוא ולא מרובע מרחק נפילת העמוד
28 מנקדת ה. ולפי שהנכחות ההוא הוא יותר קטן מקו דה
והיה מבואר שכל מה שהיה נכחות הזורית יותר קטן מקו
שהזורית הוא יותר קטנה כשהיתה הזורית קטנה מצ' מעלה
29 כמו הענין בזאת התמונה. הנה הוא מבואר שזורית התקון
היותר גדול תמצא בזאת התכונה כשהיתה זוית דהב נצבת
ושהיא תהיה לקשת AB מהתנועה השוה שהוא מוסיף על צ'
מעלה כשעור זוית דבה.
30 והוא מבואר ממה שבאר בטלמיוס בספר המגסטי שבזה
ימשך אם הונח הענין בגלגל ההקפה והונח חצי קטרו שוה
נ 63א 31 לקו דה ותנועתו שוה לתנועת הגלגל היוצא המרכז המקיף
הארץ. ובכאן | התבאר שהתקון היותר גדול שימצא באלו
32 מתנועתו הנראית. והנה מה שישתתפו בו עוד שתי אלו
התכונות הוא שכבר יראה גודל קוטר הכוכב מתחלף
ק 21ב במקומות מתחלפים מזאת התנועה | ויהיה שעורו היותר
קטן לפי המבט כשהיה הכוכב בנקדת הגובה והיותר גדול

24 וזה] נ: והוא. 25 נרחבת] ק: נרחות; נ: נרוחת.
27 הנה יהיה] חסר בכ"י נ.

9 בצמחים שיש להם שנים ידרכו בהם אל החזק ושנים אחריהם ידרכו בהם אל החולשה. ובכלל הנה זה השורש מבואר מאד למעיין בזה הספר עד שהאריכות בבאורו הוא מותר.

10 ואחר שהתישב זה נאמר שהחלוף הנראה במבט בתנועת
11 האורך השוה יצוייר תחלה באחד משני פנים. אם בשיונח מרכז הגלגל המתנועע בה מרכז הארץ ותהיה התנועה סביב נקודה אינה על מרכזו אם בשיונח יוצא
12 ממרכז הארץ וזה יובן על שני פנים. אם שיהיה הגלגל היוצא המרכז מקיף הארץ או שיהיה בלתי מקיף הארץ ויהיה תקוע בגלגל מקיף הארץ והוא אשר יקראהו
13 בטלמיוס גלגל ההקפה. וכבר באר בטלמיוס והתבאר מדבריו שהמתחייב משתי אלו התכונות הוא שוה הן בשעור התנועה הנראית לכוכב ההוא הן במרחק הכוכב מהארץ כשהיה שעור יציאת המרכז שוה לחצי קוטר הגלגל הבלתי מקיף בשעור שבו יהיה חצי קוטר הגלגל המקיף
14 ס' חלק. וכבר תסבול החלוקה כשהונח מרכז הגלגל יוצא ממרכז הארץ שתונח תנועת הגלגל אם על מרכזו ואם על זולת מרכזו ואנחנו נבאר בג"ה סגולות על תכונה מאלו התכונות במה שתיוחד בו.

15 ונאמר שכבר תיוחד התנועה שתהיה סביב גלגל יוצא מרכז ממרכז העולם ותהיה התנועה על מרכזו היוצא היה הגלגל ההוא מקיף העולם או בלתי מקיף שהתקון היותר גדול יהיה בהם ליותר מצ' מעלה מתנועתו השוה מהתחלת זאת התנועה והוא הגובה בשעור התקון היותר גדול.
16 ר"ל שאם היה התקון היותר גדול לק"א י"א מעלות הנה ימצא התקון היותר גדול לק"א מעלות מהגובה או לרנ"ט מעלות
17 ממנו. ואמנם התנועה שתהיה על גלגל מרכזו מרכז העולם סביב נקודה יוצאת ממרכז העולם הנה יהיה התקון היותר גדול לצ' מעלה מהגובה ולר"ע מעלות ממנו מתנועתו השוה.
18 המופת שאנחנו נניח העגלה היוצאת ממרכז העולם אשר בה תהיה זאת התנועה עגלת אבג ונניח זאת העגלה מקפת הארץ כי מה שיתבאר בה מזה הוא מסכים למה שימשך מזה אם היתה התנועה על גלגל בלתי מקיף הארץ כמו
19 שזכרנו. ויהיה מרכז זאת העגלה נקדת ד ומרכז הארץ
20 נקודת ה. ויהיה הקוטר העובר על אלו המרכזים קו

---

11 $_1$ מרכז ... בשיונח$_2$ [חסר בכ"י נ. 13 בארו] פ,
ק: באר.

תצטרך באופן מה אל החוש כדי שנעמד ממנו על אמתת 45 תכונת הגלגלים והיא בזה דומה באופן מה לחכמה הטבעית שילקח בה ראיה מהמאוחר על הקודם. אלא שעם כל זה יש לאומר שיאמר שהענין בזאת החכמה הולך מהלך שאר 46 החכמות הלמודיות שנביא בהם מהקודם אל המאוחר. וזה שכבר יראה לפי מה שיתבאר במה שיבא שתמונת הגלגלים והכוכבים הם בעבור התנועות שימשכו מהם ר"ל שהם בזה 47 האופן כדי שיאותו להיות מהם כמו אלו התנועות. ולזה תהיינה תנועותיהם קודמות אצל הטבע לתמונותיהם כמו שנאמר שהכח הרואה קודם אצל הטבע לתמונה בעבור העין כי העין היא בעבור הראות לא הראות בעבור העין וזה מבואר בנפשו וכבר זכרו עוד הפילוסוף בספר בעלי 48 חיים. ואיך שהיה הנה זה בלתי מבואר בזה המקום.

הפרק הכ׳

1 ואחר שהצענו מה שראוי להציעו ממה שהושג בחוש מתנועות כוכבי לכת וחלופי שעוריהם לפי המבט ראוי שנזכור כל חלקי הסותר שאפשר שירונחו בתכונות גלגל 2 כוכבי לכת. ונבאר הסגולות אשר תיוחד בהם כל אחת מאלו התכונות כדי שיהיה לנו התחלת מופת על 3 תכונותיהם מה שעמדנו עליו בחוש בהם. וקודם זה ראוי 4 שנזכיר שני שרשים יכריח העיון אל ההודאה בהם. האחד הוא שהכוכבים הם תקועים בגלגלים והם מתנועעים 5 בתנועת הגלגלים תנועת החלק בכל. והשני הוא שתנועת הגרמים השמימיים יחוייב שתהיה שוה מצד עצמה ומה שנראה אותה מתחלפת הוא | ביחס אל מה שנראהו מזה לא 6 שתהיה תנועתם בעצמם מתחלפת. כי הגרמים השמימיים לא יתכן שיהיו פעם חלשים ופעם חזקים כמו שהתבאר בספר 7 השמים והעולם. ועוד כי אנחנו מצאנו סדור ישר ושוה בתנועותיהם בזה הזמן הארוך אשר הועתקו אלינו קורותיהם לא נתבלבל הסדור רק מה שאפשר שייוחס אל 8 הקירוב אשר במבט. ואם היה הענין בהם כמו הענין בצמחים שיש להם בחרות וזקנה בכל שנה ושנה והם דורכים מהבחרות אל | הזקנה ומהזקנה אל הבחרות היה ראוי שיתבלבל זה הסדור בהם באורך הזמן כמו הענין

נ 61ב

פ 30ב

ק 21א

45 בהם] פ, ק: מהם.   47 הרואה] נ: הנראה.
48 בעלי חיים] נ: ב"ח.   5 מזה לא] נ: מזה אל.

נ 60ב
34 זריחת השמש. ובכמו זה האופן יורגש זה כשבתאי
   וצדק ואם לא יתחלפו | בהם הזמנים חלוף רב.
35 וככה יורגש בירח קצת הרגש וידמה ממה שהבטנו
36 בכותב שהענין בו הולך על זה הסדר גם כן. וזה
   גם כן מוסכם כלו מהקודמים אין מחלוקת ביניהם
   בזה כלל.

פ 30א
37 וראוי עוד שנציע בזה המקים מה שנתבאר לנו
   ממבטינו שיהיה לנו ממנה עזר בזאת החקירה והנה
   קצתו הושב לקודמים ואם השיגורהו על מה שהוא עליו.
38 ונאמר שכבר מצאנו בשמש במבטינו בשעור ניצוצו
   הנראה מחלון הכלי שהמצאנו לזאת החקירה שכבר נראה
   במבט קטרו בהירותו סביב ראש סרטן בזמנינו זה כמו
   כ"ז דקים רנ' שניים ובהירותו סביב ראש גדי נראה
   קטרו יותר גדול מזה עד שכבר נראה אם שערו כמו
39 ל' דקים. ואולם במרחקים האמצעיים מצאנו שעור
   קטרו ממוצע בין אלו השעורים לפי יחס הקורבה מאלו
   המקומות וזה ממה שיתבאר ממנו במה שאין ספק בו
   שהשמש הוא קבוע בגלגל יוצא ממרכז העולם ־שגובה
   השמש הוא סביב ראש סרטן בזמנינו זה ושפלותו הוא
40 סביב ראש גדי. ונציע עוד בכאן על צד השרש המונח
   מה שיתבאר מדברינו אחר זה במה שאין ספק בו באמתתו
   והוא שתנועת הגובה תהיה בשמש מעלה אחת במ"ג שנה
   מצרית שהיא שס"ה יום וברל"ב יום ור' שעות וחצי שעה
   בקירוב מעט וזה הוא ממה שנתבאר ממבטינו ומבטי
41 בטלמיוס ואלבתאני ואם לא שער אחד בזה. וכן ידמה
42 שיהיה ענין בנגה. ואולם בשבתאי תהיה תנועת הגובה
   מעלה בכמו מ"ד שנה משס"ה יום ורביע בקירוב ואולם
   בצדק תהיה תנועת הגובה מעלה אחת בס' שנה משס"ה
   יום ירביע בקירוב ואולם במאדים תהיה תנועת הגובה
   מעלה אחת בששים ושש שנה בקירוב מעט.

נ 61א
43 וזה כלו בלתי מבואר בזה המקום אך | הצענוהו
   על צד השרש המונח עד שיתבאר מדברינו במה שיבא
   בג"ה כדי שיהיה לנו ממנו קצת עזר בזאת החקירה
44 בזה המקום. הנה זה מה שראינו להקדימו ממה
   שהושג בחוש לנו או לזולתנו כדי שנקח ממנו
   התחלת מופת על מה שנרצה לבארו כי זאת המלאכה

⚫

36 כלו] נ: לו. 38 מחלון] ק: בחלון. 38 לזאת]
פ: בזאת. 38 כ"ז] נ: כ"ו. 38 מזה] חסר בכ"י נ.
40 השרש] ק: ההקש. 42 בס'... מעלה אחת] חסר בכ"י
נ. 42 מעט] חסר בכ"י פ. 44 זה מה] ק: זה ממה.

23 אם למי שיכחוש המוחש כי זה יראה ברב אלו הכוכבים בזמן קצר.
וכבר עמדנו בזה מן החוש באלו הכוכבים מלבד כוכב
24 חמה שלא נזדמן לנו בו עדיין הכפל המבט. ובשבתי וצדק שלא ראינו אותם כי אם מהצד האחד ממקום התקון הקטן מצאנו בחוש הסכמת זה העניין בצד ההוא והוא עדות על
25 הצד השני שלא ראינו כמו שנבאר במה שיבא בג"ה. והנה יורגש זה בקלות במאדים ליחרון החלוף הנמצא בו בין תקון הקדימה ובין תקון האיחור במקום ממה שזכרנו.
26 וכבר עדמנו עוד במבטינו בנגה ומאדים כי מקום התקון היותר גדול הוא ביותר מצ׳ מעלות מתנועת החלוף

נ 60א ובפחות מר"ע מעלה | ממנה בכמו שעור קשת התקון אשר מפני תנועת החלוף בקירוב ר"ל שכבר מצאנו בנגה התקון היותר גדול בקירוב לקל"ב מעלה מתנועת החלוף ולרכ"ח
27 מעלה ממנה. וזה אמנם מצאנוהו בכלי המבט אשר המצאנו לזה שראינו בהיותו סביב זה המקום שלא נראה ממהלכו ביום אחד כי אם מה שראוי שייוחס לתנועת הארוך.
28 וכבר השגנו זה באופן אחר יותר קל והוא שכבר מצאנו מהרושם מתקן תנועת החלוף בנגה מעת היותו דורך אל השקיעה הערבית אל עת היותו דורך אל השקיעה הבקרית כמו מה שנמצא מהרושם מתקן תנועת החלוף מעת היותו דורך אל השקיעה הבקרית עד עת היותו דורך
29 אל השקיעה הערבית. וכבר מצאנו הזמן אשר מהתחלת השקיעה הערבית עד התחלת השקיעה הבקרית יותר קצר מאד מהזמן אשר מהתחלת השקיעה הבקרית עד התחלת
30 השקיעה הערבית. והנה הזמן הקצר לא מצאנוהו מגיע לשתי חמימיות הזמן השני וכבר תוכל לעמוד על זה בקלות.
31 והנה במאדים מצאנו העניין גם כן באופן מתיחס לזה בקירוב כשהבטנו בו קודם היותר בעניין שקראו בטלמיוס
32 קצות הלילה ואחריו. וזה ממה שלא יקשה לעמוד עליו באחד מאלו הצדדים וזה שכבר יראה במבט כי הזמן אשר מעת היות מאדים אחר השמש צ׳ מעלות עד הזמן אשר הוא בו קודם השמש צ׳ מעלות הוא יותר ארוך יותר מהכפל

ק 20ב מהזמן אשר בין היות מאדים צ׳ מעלות | לפני השמש
33 עד היותו צ׳ מעלות אחר השמש. ורצוני באמרי לפני השמש שיהיה מאדים מזרחי לשמש כי אז יזרח קודם

27 מה] נ: ממה. 28 בנגה] נ: כנגד.

12 הקירוב יותר מעטי ויקריבנו יותר אל הנכונה בזמן
הסבוב האחד מזה הכוכב וזה מבואר מאד. ולזה הוא
מבואר שכבר תצטרך זאת המלאכה לעדות וראוי שילקח
העדות ממי שיתבאר לנו שהוא ראוי לסמוך עליו במבטיו
כמו אברכס ובטלמיוס ואלבתאני ושאר הקודמים המדקדקים
בזה.
13 ונאמר כי הוא ממה שראוי שנציעהו בזה המקום שכבר
ימצאו לכוכבי לכת זולתי השמש שני מינים מהתנועות.
14 האחד הוא תנועת הארך וזאת התנועה תראה מתחלפת
במקומות מתחלפים מגלגל המזלות בשמש וביתר כוכבי
15 לכת זולתי הירח. וזה שכבר יראה שם מקום תהיה התנועה
במה שקרב אליו יותר מהירה. ומקום מקביל לו תהיה
16 התנועה במה שקרב אליו יותר מתונית. והשני הוא
תנועת החלוף וזה המין מן מתנועה בלתי נמצא לשמש
והנה יראה בעבורה לכוכבים קדימה ואיחור עד שכבר
יהיו נזורים בחלק מזאת התנועה זולתי הירח לבדו.
17 וזאת התנועה תראה בכל מיניה בכל מקום ממקומות גלגל
המזלות וזה דבר תוכל לעמוד עליו בחוש בכל סבוב וסבוב
מזאת מתנועה אשר ישלם בזמן בלתי ארוך. אלא שבקצת
18 המקומות מגלגל המזלות יראה זה התקון יותר גדול
ובקצתם | יראה יותר קטן ומה שקרב אל מקום התקון
היותר קטן ישים תקון הקדימה והאיחור מקרבצים יותר
19 קטן ממה שרחק. וגם כן | הנה כל מה שקרב אל המקום
האמצעי בין מקום התקון הקטן ובין מקום התקון הגדול
ישים החלוף בין תקון הקדימה ובין תקון האיחור יותר
20 גדול. וזה ימצא בחמשת הכוכבים הנבוכים באופן שאומר
והוא כי במרחקים האמצעיים אשר אחר מקום התקון הקטן
ימצא אצל התחלת תנועת החלוף תקון הקדימה יותר גדול
מתקון האיחור הדומה לו ר"ל שהתקון לל' מעלות מתנועת
החלוף שם הוא יותר גדול מהתקון לש"ל מעלות מתנועת
21 החלוף. ואולם אצל ק"פ מעלות מתנועת החלוף ימצא
הענין שם בהפך כי תקון האיחור יהיה שם יותר גדול
מתקון הקדימה הדומה לו כאלו תאמר שהתקון לק"נ מעלות
מתנועת החלוף הוא שם יותר קטן מהתקון לר"י מעלות
22 מתנועת החלוף. והפך זה ימצא במרחקים האמצעיים אשר
קודם מקום התקון הקטן וזה כלו ממה שאין ספק בו כי

נ 59ב
פ 29ב

12 שילקח] נ: שיחלק. 19 מקום התקון הקטן] נ: התקון○
19 החלוף] פ: החלוף והתמורה. 19 גדול] פ: רב.
20 הנבוכים] חסר בכ"י נ.

נ 58ב 13 באלו הכוכבים צודקת מזה הצד. ולקיים שאין הטעות הנמצא בהם במבט מפני סבה מהסבות האלו יצטרך למאמר
14 ארוך ולמבטים רבי המספר. ולזה ראוי שנמתין זאת החקירה עד מאמרנו ברכוכב כוכב לפי התכונה שיתבאר היותה צודקת בהם.

הפרק הי"ט

פ 29א

1 וראוי שנציע בזה המקום על דרך השרש וההתחלה מה שיראה במבט מסדר תנועות אלו הכוכבים ממה שאין ספק באמתתו ואחר כן נחקור על התכונה אשר ימשך ממנה זה
2 הסדר. ואם לא תמצא כי אם תכונה אחת ימשך ממנה זה הסדר יתאמת במה שאין ספק בו שהתכונה ההיא צודקת.
3 ואם תמצא יותר מתכונה אחת ימשך ממנה זה הסדר עם הסכמת מה שיראה בחוש מחלוף שעור קוטר הכוכב הנה מה שיתאמת לנו מהחקירה הזאת הוא שאחת מהתכונות ההם
4 היא הצודקת לא זולתה מהתכונות. ולזה ראוי שנחקר על כל התכונות אשר אפשר שימשכו מהם סוגי אלו התנועות הנראות לכוכבים ונחקור על סגולות תכונה
5 תכונה מהם והם יהיו חלקי הסותר בזאת החקירה. והנה התכונה שידבקו בה סגולות כל מה שיראה בחוש מתנועות הכוכב באורך וברוחב וחלוף שעור קטרו הנראה במבט היא אשר אפשר שתונח בכוכב ההוא ואשר לא ידבקו בה אלו הסגולות היא אשר אי אפשר שתונח בכוכב ההוא.
6 והנה לזאת הסבה נציע בזה המקום מה שהשגנו
7 במבטינו והושג לקודמים שהסתכלו בזה. ונציע גם כן מה שהושג לקודמים ראע"פ שלא הושג לנו עדיין אחר שלא נמצא בו מחלוקת ביניהם לבעבור יהיה לנו מזה קצת עזר
8 אל השלמת זאת החקירה. וזה כי זאת המלאכה לא יספיק
9 נ 59א בה מה שאפשר להשיגו בחוש בזמן החיים האנושיים. אבל תצטרך לזמן יותר רחב מזה הרבה ובפרט בתנועות שלא ישלמו כי אם בזמן ארוך כמו תנועת הכוכבים הקיימים
10 ותנועת שבתאי וצדק. והנה לא נעמד באמתות על שעור תנועת כוכב מהכוכבים השוה אם לא נעזר בזה במבטי
11 הקודמים לנו זמן רב. וזה כי הקירוב אשר במבט כל ק 20א מה שיקיף מה | שבין שני המבטים בזמן יותר ארוך יהיה

1 תנועות] נ, ק: תנועת. 4 כל] נ: כן. 6 שהשגננו] ק: שהשגנוהו. 6 נמצא בו] פ: נמצא.

2 שזכרנו. הנה היה ראוי שנעיין עוד בזה מצד מה שיראה
3 מתנועותיהם. וזה שאם היה הענין במה שיראה
מתנועותיהם באורך וברוחב בלתי הולך על הסדר המתחייב
לפי תכונת בטלמיוס יתאמת לנו גם כן מזה הצד שאין
4 הענין בתכונותיהם על האופן שיניחהו בטלמיוס. עם
שזה גם כן יהיה לנו פתח ושער לעמד על תכונה אשר
לכוכב כוכב מאלו כי ראוי שתונח בצד יאות אל שימשך
ממנה מה שיראה מהתנועה לכוכב כוכב.
5 58א ב אלא שאנחנו רואים שאם עשינו החקירה | הזאת בזה
המקום יקרה בדברינו הכפל כי המבטים בעינם שהיינו
זוכרים בזה המקום הם המבטים אשר נקחם להתחלת מופת
6 לקיום התכונה אשר נקימה בכוכב כוכב. עם שכבר תקשה מאד
החקירה הזאת בזה המקום ונצטרך בבאור הנעשה בה למאמר
7 ארוך מאד. וזה כי לא יתכן שנקיים מהמבטים המושגים
לנו בחמשת כוכבי המבוכה בטול התכונה שהניח בהם
בטלמיוס כשיראו המבטים בלתי מסכימים למה שיתחיב
מתנועתו אלא אם התבאר לנו תחלה מקום השמש האמצעי.
8 כי יש לאומר שיאמר שהוא אפשר שתהיה הסבה בזה הטעות
חשבנו מקום השמש האמצעי בזולת המקום שהוא בו. כי זה
יביא לחשב מקום נגה ורוחב באורך על זולת מה שהוא
כי מקומם האמצעי באורך הוא מקום השמש האמצעי באורך
וזה גם כן יביא לחשוב מקום שבתאי וצדק ומאדים
מתנועת החלוף על זולת מה שהוא לפי שכבר התבאר
שהמקום האמצעי באורך לכל אחד מאלו הכוכבים השלשה
כשחובר עם מקומו מתנועת החלוף יהיה שוה למקום השמש
האמצעי באורך.
9 ואפשר גם כן שייוחס הטעות הזה בנגה ורוחב שיהיה
מקומם מתנועת החלוף על זולת מה שנחשבהו או שיהיה
10 מקום הגובה בהם על זולת מה שנחשבהו. וכן הענין
בשבתאי וצדק ומאדים ר"ל שאפשר שייוחס הטעות הנמצא
בהם במבט אל שיהיה מקום הגובה בהם על זולת מה
שנחשבהו או שיהיה מקומם מתנועת האורך על זולת מה
11 שנחשבהו. ואפשר עוד שייוחס הטעות הנמצא בהם בכלל
במבט אל שיהיה מקום הכוכבים הקיימים אשר הובטו עמהם
12 על זולת המקום שנחשבהו. כי בכל אחד מאלו הפנים

---

3 המתחייב] נ: המחויב.   4 מה שיראה]נוסף בכ"י
פ (למעלה)(:) לחוש.   5 לקיום] נ: לקיים.
7 שיתחייב] נ: שיחויב.   8 וכותב] ק: וכוכב.
10 ר"ל] חסר בכ"י פ.   11 או שיהיה ... שנחשבהו]
חסר בכ"י נ.

23 החלוף. והיה ראוי גם כן לפי תכונת בטלמיוס שיראה
קוטר גרם צדק יותר גדול בהיותו אצל ק"פ מעלה
מתנועת החלוף מהשעור שיראה בו בהיותו אצל התחלת
תנועת החלוף קרוב לשלישת השעור וזה דבר בלתי נמצא
כלל כשהיה נראה צדק בחשך באופן שלא ישלוט בו אור

24 השמש כלל. כי כשהיה הענין כן לא ימצא חלוף מורגש
בין השעור שיראה בו בהיותו בזה המקום ובין השעור
שיראה בו בהיותו בק"פ מעלה מתנועת החלוף.

25 וראוי שתדע כי מה שמצאנוהו ראשונה במאדים
מגודל שעורו אצל הנזורות בהיותו באריה יחסנוהו

26 תחלה אל עננים דקים נראה אז באמצעותם. ואמנם
עשינו זה מפני מה שמצאנוהו בשעורו בהיותו נזור
בעקרב שלא היה נוסף שם שעורו ומפני שלא ראינו זה
התוספת על הסדר שהיה ראוי שיהיה אלו היה זה מצד

נ 57ב 27 קרבתו | ממנו. כאלו תאמר שאחר זמן מועט שב להראות
בקוטן שעור יותר מהראוי הרבה לפי היחס הראוי עד
ששבנו לראות שעורו בהיותו נזור בגדי ומצאנוהו נוסף

28 בדמיון מה שמצאנוהו באריה ויותר מזה מעט. והנה
שבנו ליחס מה שמצאנוהו מהעדר התוספת בשעור מאדים
בהיותו בעקרב אל עובי האידים אשר היה נראה אז

29 באמצעותם. וכבר הורה שכן היה מה שנראה אז מהכוכב
פ 28ב המזונב | אשר התמיד הראותו יותר מג' חדשים שהיה
מוצא האיד ההוא מתחת עקרב ונמשך משם עד למטה
מהקוטב הצפוני שעור מה ומשם התחיל ההתלהבות וכלה
בעקרב.

30 וזה ממה שיחזק מה שקימנו תחלה שהיה מיוחס
התוספת בשעור הנראה למאדים אל עננים דקים נראה אז
באמצעותם. והנה נצטרך עוד בזה אל השנות בחוש אלא
שאיך שהיה הוא מבואר ממה שזכרנו שזה בלתי הולך על
הסדר המתחייב מתכונת בטלמיוס והנה נרחיב עוד המאמר
בזה בג"ה במה שיבא.

הפרק הי"ח

ק 19ב 1 ולפי שהענין בתכונת קצת הכוכבים התבאר שאינו
על האופן שהניח בטלמיוס מצד מה שיראה משעוריהם כמו

24 בהיותו בזה] נ: בזה.   25 כי מה] נ: מה.
25 ראשונה] פ: תחלה.   26 בשעורו] ק: משעורו.
26 זה מצד] פ: מצד.   29 הורה] נ: הותר.

שקדם וזה אמנם יקרה כשלא יהל אורו אז הירח והיה
באשון חשך.

11 והתבאר לנו עוד בחוש ממאדים שאין הענין הולך על
הסדר המחוייב מתכונת בטלמיוס וזה שלא יראה מאדים
יותר | גדול אצל ק"פ מעלה מתנועת החלוף מהשעור
שיראה בו אצל התחלת תנועת החלוף על האופן שהיה ראוי
12 שיראה לפי תכונת בטלמיוס. וזה שלא ימצא קטרו יותר
גדול בהיותו אצל ק"פ מעלה מתנועת החלוף מהשעור
שיראה בו אצל התחלת תנועת החלוף מה שיגיע לכפל שעורו
13 הנראה שם. ואולם לפי תכונת בטלמיוס היה ראוי שיראה
מאדים יותר גדול בעת מה בהיותו אצל ק"פ מעלה מתנועת
החלוף ממה שהוא נראה בו בעת מה בהיותו אצל התחלת
14 תנועת החלוף יותר משישה כפלים. וראוי שתדע שכבר
מצאנו בזה הענין במאדים קצת מבוכה וזה כי בהיותו
נזור בארית מצאנו שעורו יותר גדול משעור שבתחי שעור
מרחש וכן הענין בהיותו נזור בגדי אלא ששם נראה שעורו
15 יותר גדול ממה שנראה בארית. ואולם בהיותו נזור
בעקרב לא נוסף שעורו לעין ולא נראה יותר גדול אז
16 משבתחי. וזה כלו ממה שיתן מבוכה בעניננו אלא שעל כל
פנים לא נמצא התוספת בשעורו מגיע אל הכפל כמו שזכרנו.
17 ואולם ביתר כוכבי לכת לא נתבאר לנו עדין איך
18 הענין בהם מזה הצד. וזה בכוכב חמה אי אפשר שיתבאר
זה כי הוא לא יראה באקלימנו כי אם אצל גדול מרחקו
19 מהשמש ימים מעטי | המספר. ויתכן שייוחס הקוטן הנראה
בשעורו כשיקרב יותר אל השמש ממשלת אור השמש בו אז
שישים שעורו יותר קטן כי הוא לא יראה אז כי אם קרוב
לעליית השמש ויחזק האור באופק וישים שעור כל הכוכבים
20 הנראים אז יותר קטן. ואולם בשבתחי וצדק לא עמדנו
עדין על אמתת זה למיעוט החלוף הנמצא במבט בשעוריהם
אלא שעל כל פנים יתבאר בשבתחי וצדק מזה הצד שאין
הענין הולך בהם על הסדר המתחייב מחשבון בטלמיוס.
21 וזה כי לפי תכונתו היה ראוי שיראה קוטר שבתחי יותר
גדול בהיותו אצל ק"פ מעלה מתנועת החלוף מהשעור
שיראה בו בהיותו אצל התחלת תנועת החלוף קרוב לחמשית
השעור וזה דבר בלתי נמצא כלל כשהיה נראה שבתי בחשך
22 באופן שלא ישלוט בו אור השמש. כי כשהיה הענין כן
לא ימצא חלוף מורגש בין השעור שיראה בו בהיותו בזה
המקום ובין השעור שיראה בו בהיותו בק"פ מעלה מתנועת

22 בו בהיותו בזה] פ בזה.

הפרק הי"ז

1 וכבר זכרנו בירח דברים יתבאר מהם במופת שאין ספק בו שאין העניין בתכונתו על האופן שהניח בטלמיוס.
2 וזה כי לפי תכונתו היה ראוי שיראה קוטר הירח יותר גדול ברבועים בעת ממה שהוא נראה בנגודים בעת מה קרוב מהכפל ר"ל כשהיה בנגודים בהתחלת תנועת החלוף
3 והיה ברבועים בק"פ מעלות ממנה. והנה לא מצאנו אנחנו רושם מורגש מאד בחלוף שעור קוטר הירח הנראה לא מפאת הגובה ולא מפאת תנועת החלוף ונכפל לנו המבט בזה פעמים רבות בזה הכלי אשר אי אפשר שיקרה בו טעות כמו שבארנו.
4 והנה התבאר לנו עוד בחוש ששעור קוטר נוגה הוא יותר גדול במבט כשהיה בגדול מרחקו מהשמש ממה שהוא בהיותו אצל התחלת תנועת החלוף או אצל ק"פ מעלה ממנו ולא מצאנוהו גדול במבט בהיותו אצל ק"פ מעלה מתנועת החלוף ממה שהוא נראה בו בהיותו אצל התחלת תנועת החלוף.
5 וזה כלו הוא חלוף מה שיחייב מתכונת בתלמיוס כי לתכונתו היה ראוי שיראה קוטר נגה יותר גדול בהיותו אצל ק"פ מעלות מתנועת החלוף ממה שיהיה נראה בה בהתחלת תנועת החלוף יותר משהה כפלים.
6 ואמנם עמדנו על אמתת מה שזכרנו מחלוף שעור נגה כשהתבוננו בחוש בעת עת ביחסו אל גודל הכוכבים הקימים הנראים בעת הראותו שהם מהערך הראשון או מהערך השני וזה דבר כולל לעמד בו על חלופי השיעור הנראים בעת עת לאחד אחד מהכוכבים הנבוכים.
7 וכבר יובן לנו בנגה לעמד על זה מפני הראותו ביום עם השמש וזה כי כשהיה נגה אצל גדול מרחקו מהשמש תמצאהו נראה גם עם אור השמש בצהרים. וכאשר קרב אל השמש כמו עשרים מעלה או פחות מזה לא יוכל
8
9 להראות עם אור השמש. וכבר יתאמת | לך זה גם כן כשתעיין | בגודל קוטר נגה בשני אלו המקומות
10 וכשתעיין בגודל קוטר נגה בהיותו אצל התחלת תנועת החלוף ואצל ק"פ מעלה ממנו. וכבר יתאמת לך זה גם כן בנגה מצד ניצוצו הנכנס בחלון הכלי שזכרנו במה

---

4 או אצל ק"פ ... התחלת] נ: ממה שהיא נראה בו בהיות אצל התחלת.   5 מה שיחייב מתכונת] נ: שיחייב מתנועת.
7 בנגה] נ: בנגד.   7 גדול] פ: גודל.

20 ממה שהוא בנגודים כי אם מעט. וכבר נכפל לנו המבט
בזה פעמים רבות כמו שנזכר בג"ה אצל המאמר בירח.
21 וכאשר נתבאר לנו זה שבנו לחקור על המופת שחייב
ממנו בטלמיוס שיהיה התחלפות ההבטה ברבועים על
22 השעור ההוא שהניחו. ומצאנו שהמופת ההוא הוא בנוי
על שתכלית מרחב הירח לצפון ולדרום מגלגל המזלות
הוא חמש מעלות והנה אין זה צודק כמו שנבאר במה
23 שאין ספק בו אצל המאמר בירח. וזה שם יתבאר שרוחב
הירח הוא על תכליתו ארבע מעלות וחצי כמו שהסכים
עליו אלבתאני ועם זה יהיה | מה שהושג לחוש לבטלמיוס
בלתי חולק על מה שהתבאר בכאן שאין ההחלפות ההבטה
יותר גדול ברבועים ממה שהוא בנגודים כמו שנבאר
24 בג"ה במה שיבא. וזה מסגלת האמת שאי אפשר שימצא
שיחלק לעצמו אבל הוא מסכים לעצמו כמל צד ומכל פנה.
25 וכבר אפשר שנעמד באופן אחר יותר קל על מקום
26 הכוכבים הקיימים. והוא שנביט בירח בעת הלקות הירחי
ונביט מקיף עגלת הצל אשר בו עם כוכב אחד מהכוכבים
27 הקימים. ונעמד על שעת המבט ונדע המרחק אשר בארך
מגלגל המזלות בין הכוכב הקיים ומקיף עגלת הצל לפי
28 הנראה במבט. והנה חצי קוטר עגלת הצל ידוע השעור
בקירוב והוא מ"ב דקים כמו שנבאר אצל המאמר בירח
ושעור התחלפות ההבטה לעגלת הצל הנראה בירח ידוע
29 בקירוב נפלא. וכאשר תקננו מקום עגלת הצל בלוח
התחלפות ההבטה נתאמת לנו המרחק האמתי בין מקיף
30 עגלת הצל ובין זה הכוכב הקים בזה העת. ומזה נעמד
בקלות על המרחק האמתי בין מרכז עגלת הצל ובין זה
הכוכב הקיים בעת המבט לפי שחצי קוטר עגלת הצל ידוע
ולפי שמקום השמש האמתי בעת המבט ידוע היה מקום
מרכז עגלת הצל ידוע כי הוא רחוק ממרכז גרם השמש
31 ק"פ מעלות. ולזה ישאר מקום הכוכב הקיים המובט
עמו ידוע והנה הורה לנו מקום הכוכב הקיים ההוא
מקומות שאר הכוכבים הקיימים החקוקים בלוחות בארך
כשנחשב ענינם לפי מה שמצא בטלמיוס מקשת המרחק
32 ביניהם. וזה מספיק בזה המקום להיות פתח אל שנשתמש
במבטים במקומות כוכבי לכת בזה האופן שהישרנו אליו
עד שיתבאר לנו בזה אם הענין בזה הולך על הסדר
המתחייב מתכונת בטלמיוס אם לא.

22 כמו] נ: במה.   32 בהם הולך] פ: הולך בהם.

10 ברבועים ממה שמצאנוהו בנגודים כי אם שעור מועט כמו
שנזכר אצל המאמר בירח. והנה חקרנו על גודל קוטר
הירח בזה הכלי בכל הצדדים אשר זכרנו ובכלם מצאנו
שאין קוטר הירח נראה יותר גדול ברבועים מהשעור
11 שהוא נראה בו בנגודים כי אם מעט. וזה חמצאהו כן
12 בלי ספק אם תחקר בו על האופן שהישרנו אליו. וזה
ממה שיתבאר ממנו בלי ספק שאין הירח יותר קרוב
13 אלינו ברבועים ממה שהוא בנגודים כי אם מעט. ועם
כל זה לפי שכבר יחשב שבטלמיוס באר שעור התחלפות
ההבטה ברבועים במופת אמתי חקרנו עוד על זה בזה
האופן.
14 והוא שכבר הבטנו הירח אצל היותה ברבועים בקירוב
עם כוכב מהכוכבים הקיימים שהם אצל עגלת המזלות
שיאות בו יותר זה המבט לפי מה שקדם ועמדנו על
15 מרחקה ממנו באורך ועל שעת המבט. ואחר זה ארבע או
חמש שעות שבנו להביט הירח עם הכוכב ההוא ועמדנו על
מרחקו ממנו באורך ועל שעת המבט וכאשר ידענו זה
הנה מצאנו מהלך ירח הנראה בזה הזמן שבין שני המבטים.
16 וכאשר גרענוהו ממהלכו הראוי להיות בזה הזמן לפי
תכונת בטלמיוס כי לא יזיק בזה המקום אם נסמוך על
החשבון אשר לפי התכונה ההיא כי לא ימצא בו טעות
מורגש בכמו זה הזמן הקטן ובפרט ברבועים כמו
17 שיתבאר במה שיבא. הנה יהיה הנשאר שעור מה שהוסיף
התחלפות ההבטה במבט השני על התחלפות ההבטה במבט
הראשון אם היו שניהם לגרוע ממקום הירח האמתי או
שעור שני התחלפויות ההבטה אם היה הראשון להוסיף
והשני לגרע ממקום הירח האמתי או שעור מה שהוסיף
התחלפות ההבטה במבט הראשון על התחלפות ההבטה במבט
השני אם היו שניהם להוסיף על מקום ירח האמתי.
18 ומזה המקום אפשר לעמד על שעור התחלפות ההבטה
כשנחשב זה לפי לוחות בטלמיוס ונראה אם היה ראוי
שיהיה הנשאר שהוא מיוחס להתחלפות ההבטה על האופן
19 שמצאנוהו. וכאשר חקרנו על זה בזה האופן מצאנו
שאין ראוי שיונח שעור התחלפות ההבטה ברבועים יותר

---

13 ועם כל] ק: ועל כל.   13 שבטלמיוס] נ: בטלמיוס.
16 להיות] חסר בכ"י ק.   17 נ: והשני לגרוע ממקום
הירח אמתי] פ, ק: והשני לגרוע.   18 שמצאנוהו]
נ: שהמצאנוהו.

75 נקדות למ. והנה נדע זוית גבל באופן שידענו זוית
76 בגה. ונניח העמוד הנופל מנקדת ל לקו גב המונח
   בלי תכלית קו לנ וידוע קו זנ באופן שנודע קו זט.
77 ולזה נודע קו טנ והוא אורך הכותל הצפוני הראוי לפי
   זה הדרוש ובזה האופן יודע קו נל והוא מרחק נקדת ל
78 הראוי מהכותל הצפוני. ובזה האופן בעינו תפול
   מנקדת מ עמוד מס על קו גב המונח בלי תכלית וידוע
   קו טס והוא אורך הכותל הצפוני הראוי לפי זה הדרוש
   וידוע קו סמ והוא מרחק נקדת מ הראוי מהכותל הצפוני.
79 ובזה האופן יתישר לך זה הבית לזה המעשה כאלו היה
   נצב הזוריות והוא מה שרצינו לבאר.

הפרק הי"ו

1 ולפי שאי אפשר לנו לקחת מבט מהמבטים לכוכבי לכת
   אם לא ידענו תחלה מקומות הכוכבים הקיימים אשר יובטו
2 עמהם. והיה במקומות הכוכבים הקיימים ספק עצום מצד
   רבוי הדעות אשר מצאנום לקודמים בתנועת גלגל השמיני.
3 הוא מבואר שהוא ראוי שנאמת תחלה מקומות הכוכבים
   הקיימים קודם שנדע מהם מקומות כוכבי לכת הנראים
4 עמהם. וזה אמנם יתכן לנו בכלי שיובט בו יחד בשמש
   ובירח כי השמש הוא ממה שידענו מקומו בעת המבט בלי
5 ספק. וכאשר הודיענו זה הכלי המרחק אשר בינו ובין
6 הירח היה מקום הירח הנראה ידוע לנו בלי ספק. ומזה
   נדע מקום הכוכבים הקיימים הנראים עם הירח סמוך לעת
   זה המבט כמו שקדם אלא שכבר ישאר בכאן ספק מפני
   סכלותינו בשעור התחלפות ההבטה.

7 נ 54ב   והנה על צד | הקירוב נניח בכאן שיהיה התחלפות
   ההבטה במולדות ובניגודים מסכים למה שיתחייב מלוחות
   בטלמיוס כי זה ימצא מסכים בקירוב נפלא בלקריות
8 השמשיים. ואולם בשאר ימי החדש נגלה לנו במה שאין
   ספק בו שאין שעור ההתחלפות הזה יותר גדול ממה שהוא
9 בנגודים כי אם שעור מועט. וזה התבאר לנו ממה שאמר
   והוא שכבר הבטנו גודל קוטר הירח מצד ניצוצו הנכנס
פ 27א   בתלון | הכלי שזכרנו ולא מצאנוהו יותר גדול

79 הזוריות] נ: הזוית.   7 צד] נ: זה.   9 ממה
שאומר] ק, נ: מהצד שאומר.

59 הזמן אשר נהיה בו כל אחד מאלו המבטים ידוע והיה
ידוע עם זה הזמן שנכנס בו השמש בראש סרטן. הנה
60 נעמד מזה באופן שזכרנו בתמונת קו חצי היום על שעור
תקון השמש ועל מקום גובה גלגל השמש. וכבר נבאר עוד
הדרך בזה בג"ה המאמר בגלגל השמש.
61 ואם תרצה לעמוד בקלות על אורך הקוים המגיעים
מהחלון אל רשמי הכותל ראוי שיהיה בבית ישר
הכותלים נצב הזויות כמו שהתנינו ותמוד ארכו ושעור
מרחק ראש החלון מהכותל הצפוני ושעור מרחק הרשמים
62 מהכותל הצפוני ואז תדע בקלות אורך אלו הקוים. ואם
היה הבית בלתי נצב הזויות יהיה בזה קושי מה אלא
63 בנ 53 שהדקדוק ימצא בו יותר לרוחב מרחב הזריחה שם. וכזה
קרה לנו בבית אחד הכינוהו אותו לאלו המבטים
64 פ 26א ומצאנוהו | בלתי נצב הזויות. והנה ראוי שנישירך
איך תעשה זה.
65 והמשל שיהיה רושם הכותל המערבי בו בקרקע הבית
קו בא ורושם הכותל הצפוני בו קו בג ורושם הכותל
66 המזרחי אשר בו החלון קו גד. והמקום שיפול בו אבן
העופרת בקרקע הבית ממקום ראש החלון שיבא ממנו
67 הניצוץ הנרשם בכותל המערבי הוא נקדת ה. והנה מדדנו
68 קו בג בקנה המדה וכן קו גה וכן קו אב. ונרשם תחלה
בקו בג נקדה איך שתפול והיא נקדת ז ונקרה קו זה
69 ונדע שעורי קוי גז זה בקנה המדה. הנה קוי משלש
70 גזה ידועים ולזה תודע זוית הזג. ונוציא מנקדת ה
עמוד על קו גז המונח בלי תכלית והוא עמוד הט
71 ונדביק קו הגט הישר. הנה מפני שזוית טזה ידועה
הנה תודע זוית זהט כי היא שלמות צ' מעלה מזוית טזה
ולפי שקו זה ידוע השעור הנה קו זט שהוא מיתר זוית
72 זהט הידועה ידוע ולזה ידוע גם כן קו טה. ולפי שקו
ק 18א זט ידוע | וקו זב ידוע בקנה המדה הנה קו טב ידוע
והוא אורך הכותל הצפוני אם הונחה הזוית המזרחית
צפונית נצבת וזה אמנם יצדק לו הונחה הזוית הצפונית
מערבית נצבת.
73 ואם לא היתה נצבת הנה יצטרך לתקן אורך הכותל
הצפוני מהצד ההוא גם כן לפי המקום שיהיה בו הרושם.
74 נ 54א ונניח שתהיינה | מקומות שתי נקדות הרשמים בכותל
המערבי כשהפלנו מהם אבן העופרת בקרקע הבית שתי

58 והיה ידוע] חסר בכ"י נ.   62 מרחב] נ: מרחק.

כאלו תאמר הצד הצפוני ונרשם בקו השפל במקום ההוא
קו יעמד עליו על זוית נצבה עד שיגיע הקו אל הקו
העליון ויחתכהו ושם יהיה רושם זה הניצוץ בעת במבט.
47 והנה תוכר לך הנטיה בכל יום היכר נפלא להיות מעלות
הנטיה בזה האופן גדולות מאד מצד גודל קו הניצוץ
ומצד שהם שורות בקירוב למרחב הנטיה בעת הזריחה בזה
האופק כמו שיתבאר במעט עיון ממה שנאמר במה שקדם עד
שתהיה כל מעלה ממעלות הנטיה יותר מזרת בזה הקו אשר
48 בו יורשמו אלו הרשמים. ואם נהגת באלו הרשמים
המנהג שזכרנו ברשמים אשר בקו חצי היום יתבאר לך דרך
משל מקום הרושם בעת היות השמש בראש סרטן ומתי נכנס
השמש בראש סרטן וזה בדקדוק נפלא.
49 ונאמר שכבר נעמד מזה על שעור תקון השמש ועל
50 מקום הגובה לו. וזה שכבר בארנו במה שקדם איך נדע
מאלו הרשמים מקום הזריחה האמתית לשמש בהיותו במקום
51 שהוא בו מגלגל המזלות בעת המבט. ולזה נעמד מזה על
מקום הזריחה האמיתית בהיות השמש בראש סרטן וכן נעמד
מזה על מקום הזריחה האמתית בהיות השמש באי זה שיהיה
52 מאלו הרשמים. וכאשר ידענו שני אלו המקומות מדדנו
מה שביניהם בקנה המדה ושעור | הקו אשר מראש החלון 53א נ
מהצד אשר נרשם בו מהניצוץ אל כל אחד מאלו המקומות
53 ויהיה בידינו משלש ידוע הקוים. ולזה יודעו כל
זויותיו כמו שקדם ומפני זה תודע הזוית אשר אצל ראש
החלון והוא מרחב הזריחה באופק ההוא מהמעלה שהשמש בה
54 בעת היותו בזה הרושם עד ראש סרטן. ולפי שהנטיה
ידועה לנו וגובה הקוטב בזה האופק ידוע לנו כמו שקדם
הנה מרחב זריחת ראש סרטן בזה האופק ידוע לנו באופן
55 שיתבאר במה שיבא. וכאשר נגרע זאת הזרית שאצל ראש
החלון ממרחב זריחת ראש סרטן היה הנשאר שוה לזרית
56 מרחב זריחת המעלה שהשמש בה. וכבר נעמד מזה בקלות
על המעלה שהשמש בה בעת זה המבט שיאות לה זה המרחב
בזה האופק וזה בלתי צריך ביאור למי שיש לו בחכמה
הזאת חלק ראוי להשתדל בהשלמת זאת החקירה.
57 וכן יודע מזה מקום השמש האמתי בעת שובו אל זה
58 הרושם או אל רושם אחר. וכאשר ישלם לנו זה ויהיה

47 למרחב הנטיה] נ: למרחק הנטיה. 54 שהנטיה] ק:
שהשנה. 55 נגרע] ק: נגיע. 57 אחר] נ: אחד.
58 ויהיה ידוע] חסר בכ"י נ.

37 ואחר שידעת זה הנה תוכל לדעת בזה הקו על זה האופן
38 גובה השמש בכל יום שתרצה. וכאשר ידעת גובה השמש
נ 52א ושעור הנטיה בכללה | הנה תעמד מזה בקלות בלוחות
הנטיה על המעלה שהשמש בה שתאות לה זאת הנטיה ועל
החלק מהמעלה ותדע מזה מקום השמש האמתי בכל יום
39 שתרצה. ולו תחקור בזה האופן על מקום השמש האמתי
מגלגל המזלות כמו ארבעים או חמשים יום קודם היות
השמש בראש סרטן ועל מקומו האמתי כמו ארבעים או
חמשים יום אחר היות השמש בראש סרטן הנה יתבאר לך
מזה מקום גובה השמש ושעור התקון על תכליתו בכמו
האופן שבאר זה בטלמיוס בכוכבים וכבר נישירך על
הביאור בזה בג"ה אצל המאמר בגלגל השמש.
40 ואולם אם יש לשמש גלגל יוצא ממרכז העולם אם לא
הנה זה יתבאר לך אחר שידעת מקום הגובה שתחקר בהיות
השמש אצל המקום ההוא על גודל קוטר השמש מצד הניצוץ
41 הנכנס בחלון המקל. עוד תחקר על גדלו בהיותו אצל
השפל ואם מצאת הגודל שוה ידעת כי אין גלגל השמש על
מרכז יוצא ממרכז העולם באופן שהניחו בטלמיוס ואם
מצאת הגודל מתחלף ידעת כי הוא על מרכז יוצא ממרכז
42 העולם. ובכלל הנה איך שיהיה שיראה גודל קוטר השמש
במקרומות מתחלפים מהגלגל מתחלף השעור ויהיה זה הולך
על סדר ידענו כי הוא על מרכז יוצא ממרכז העולם.
43 וכבר תוכל לדעת מקום השמש האמתי בכל יום בזולת קו
חצי היום ותעמד על עת היות השמש בראש סרטן ובראש
גדי ועל מקום הגובה לגלגל השמש ועל שעור התקון
התכליתי לשמש אחר שידעת גובה הקוטב באופן שאתה בו
ושעור נטיית גלגל המזלות מגלגל משוה היום באצטורלב
שקדם זכרו.
44 וזה אמנם יהיה בבית יהיה בו חלון ישר הקוים
לפאת מזרח או לפאת מערב ויהיה ארכו כמו מ' זרתות
נ 52ב או יותר | כי כל מה שהיה הבית יותר ארוך יהיה
ק 17ב 45 הדקדוק נמצא בו יותר. ויהיה | הבית נצב הזוית
פ 26א ישר הכותלים ונרשם בכותל הנכחי לחלון | קו יהיה
נכחי לקו קרקע הבית יהיה שפל מראש החלון כמו ששה
זרתות או יותר ותחתיו כמו זרת או יותר נרשם קו שני
46 נכחי לראשון. וכשיגיע ראש הניצוץ בקו העליון נראה
באי זה מקום יחתוך הקו השפל מאחד מצדדי הניצוץ

---

40 בחלון] נוסף בכ"י נ (למעלה):עם.   46 פ, נ:
בקו] לקו ק.

22 אחריו עד עשר מעלות או יותר מעט.
ונניח גם כן שלא ישוב לנקודת ג בעינה אבל ישוב
23 קרוב לה והיא נקדת ד. ויהיה הרושם שהיה בו השמש
ביום הנלוה ליום שהיה בו בנקדת ג רושם ה ונדע יחס
24 קו גד אל קו גה וכיחס ההוא נקח מיום אחד. והמשל
שאם היה קו גד ששית קו גה ידענו כי בד׳ שעות שהוא
ששית היום ילך השמש זה המהלך אשר בין שתי נקודות
25 ג וד. ואם היתה נקודת ד במה שבין שתי נקודות ב וג
נוסיף אלו הד׳ שעות על הזמן שבין רושם ג ובין רושם
ד ואם היתה נקדת ד בצד האחר נגרע מהזמן אלו הד׳
26 שעות. ומה שיהיה מהזמן אחר התוספת או אחר הגרעון
אם חלקנו אותו לחציין יגיע לנו הזמן שבו היה השמש
27 בראש סרטן בקירוב | נפלא. ובזה האופן נרשם בהיות
השמש אצל ראש גדי ונדע רושם ראש גדי בזה האופן
28 והעת שנכנס בו | השמש בראש גדי. ואם היה הזמן
אשר מעת היות השמש בראש סרטן אל עת היותו בראש גדי
שוה אל הזמן אשר מעת היות השמש בראש גדי אל עת
היותו בראש סרטן ידענו כי הגובה הוא בראש סרטן.
29 וזה אמנם נעמד עליו אם מצאנו הזמן שמעת היות השמש
בראש סרטן אל עת היותו בראש גדי שוה לחצי אורך
30 השנה. ואם לא היה שוה הנה נודע שהגובה הוא בחלק
31 אשר יתחכהו השמש בזמן יותר ארוך. וכשיהיה הענין
כן הנה נעמד על מקום הגובה באופן שיתבאר אחר זה.
32 והנה כאשר היה אצלנו רושם השמש בחצי היום
בהיותו בראש סרטן ורושם השמש בהיותו בראש גדי הנה
נדע מזה שעור נטיית אופן המזלות מאופן המישור ושעור
33 גובה הקוטב במקום ההוא. וזה שאנחנו נניח בתמונה
הקודמת שתהיה נקדת ב רושם ראש סרטן ונקדת ז רושם
34 ראש גדי. ונדביק קו אבגז הישר ויהיה ראש החלון אשר
יגיע ממנו זה הניצוץ נקדת ט כמו שקדם ונוציא קוי
35 טב טז. ונעמד על שעורי קוי טב טז בז בקנה המדה
ותודע מפני זה זוית בטז והיא זוית כפל הנטיה וזוית
36 טזב והיא גובה ראש גדי. וראוי לפי מה שקדם שתוסיף
על הגובה ראש גדי כמו זוית חצי קוטר השמש ואם תחבר
אליו עוד חצי זוית בטז יגיע בידך גובה ראש טלה באופק
ההוא ושאריתו מצ׳ מעלות הוא גובה הקוטב באופק ההוא.

25 במה שבין שתי] פ: בין.   30 נודע] ק: נדע.
35 והיא זוית] נ: והיה זוית.

10 מראש החלון ונרשם זה בכל יום שנרצה.   ותמצא שכל
מה שיתקרב השמש יותר לראש סרטן יתקרב יותר ראש
הניצוץ אל כותל החלון עד שכבר תכיר בזה הקו הנטיה
לשמש במהלכו ליום אחד גם בהיותו סמוך לראש סרטן.
11 ולו תחקר בזה תמצא שאם היה גובה ראש החלון כמו
כ"ד זרתות יהיה אורך המעלה מהנטיה אצל היות השמש
בראש סרטן בקו חצי היום הנרשם בזה הבית הנזכר קרוב
12 לחצי זרת באופק שיגבה הקוטב בו מ"ד מעלות.   והוא
מבואר שזה השיעור יחלק לדקים מבוארי השעור יתכן
התחלקם לשנים בקירוב בכמו האופן שחלקנו מעלות
המקל לדקים.
13 והנה אצל היות השמש בראש סרטן כמו עשרה ימים
או יותר מעט הבטנו בכל יום ניצוץ השמש באופן הקודם
וכתבנו במקום הרושם יום המבט ההוא וכן עשינו זה עד
שהיה הרושם היותר קרוב שאפשר מן החלון ורשמנו שם.
14 וכאשר הבטנו בזה אחר זה בעת ששב השמש לרדת לפאת
דרום קרוב למקום הרושם שרשמנו כמו עשרה ימים קודם
היות השמש בראש סרטן ראינו אם שב למקום הרושם
15 בעצמו או שב קרוב אליו.   ואיך שהיה הנה נוכל לעמוד
מזה בקירורב נפלא על רגע הכנס השמש בראש סרטן.
16 והמשל שיהיה סימן לראש החלון נקודת ט ויהיה הקו
17 המערבי מהחלון קו טח.   ונדביק אותו ביושר עד קרקע
הבית והוא קו טחא הישר ויהיה | קו טחא נצב על שטח
קרקע הבית ויהיה קו חצי היום בקרקע הבית קו אבג.
18 ונניח תחלה שיהיה שב השמש אל מקום הרושם ביענו
שיהיה בו כמו עשרה ימים קודם היותו בראש סרטן.
19 ויהיה הרושם בהיותו בראש סרטן נקדת ב והרושם שהיה
בו כמו עשרה ימים קודם היות השמש בראש סרטן נקדת ג.
20 ונניח שיהיה שב ברושם ההוא בי"ט יום אחר הזמן
21 הראשון שהיה בו ברושם ההוא.   ונאמר שכבר נתבאר לנו
כי בתשעה ימים וי"ב שעות שוות אחר המבט הראשון היה
השמש בראש סרטן כי הוא ראוי שיהיה השמש מהלך בחצי
הראשון מזה הזמן שוה בקירוב נפלא למהלכו האמתי
בחצי השני כי שעור התקון לא יתחלף דבר מורגש בזה
הזמן | המועט ראע"פ שיהיה הגובה לפני ראש סרטן או

---

10 ותמצא] פ: ונמצא.   12 מבוארי] פ: מבואר.
14 הרושם שרשמנו] נ: הדרום שרשמנו.   19 ויהיה
הרושם בהיותו בראש סרטן] חסר בכ"י נ.   21 נתבאר]
נ: התבאר.

הקירוב מצדדים רבים כמו שקדם והטעות המעטי בגובה
12 השמש ירביל לטעות גדול בידיעת | מקומו באורך. ועוד
כי בתקון תנועת השמש האמצעית ספק וזה שבטלמיוס יראה
שתכלית התקון הוא כמו ב׳ מעלות וכ"ג דקים והבאים
13 אחריו יראה שאינו מגיע לב׳ מעלות שלמות. ועוד כי
במקום גובה גלגל השמש ספק כי בטלמיוס יראה שהוא
תמיד בה׳ מעלות וחצי מתאומים וקצת מהבאים אחריו
יראו שהוא בי"ז מעלות וחצי מתאומים כמו י"ב מאות
14 שנה אחר מבט בטלמיוס. ולאלבתאני יהיה בכמו כ"ט
15 מעלות וט"ו דקים מתאומים בזה הזמן שזכרנו. וזה
כלו ממה שיוסיף קושי בידיעת מקום השמש האמתי.

הפרק הט"ו

1 ולפי שכבר התבאר שאין דרך לידיעת מקום הכוכבים
הקימים באורך וברוחב אם לא נדע תחלה מקום השמש
2 האמתי. והיתה ידיעת מקום השמש האמתי ממה שתקשה מאד
מפני הספק אשר ימצא במקומו האמצעי באורך ובשעור
נטית גלגל המזלות ממשוה היום ובשעור התקון בו
3 ובמקום הגובה. השתדלנו אנחנו למצא דרך ירבילנו
לידיעת מקום השמש האמתי בכל יום ולדעת שעור התקון
התכליתי לתנועת השמש האמצעית ולדעת שעור נטית קוטב
גלגל המזלות מקוטב גלגל משוה היום ולדעת מקום גובה
גלגל השמש.

4 ואופן המעשה בזה הוא שכבר עשינו בבית אחד חלון
5 גבוה לפאת דרום נצב הזויות ישר הקוים. והיה קו
ראשו נכחי לקו קרקע הבית והקו המערבי נצב על שטח
6 קרקע הבית שהוא נכחיי לשטח עגלת האופק. והפלנו
אבן העופרת כנגד הצד המערבי מהחלון בישר בקרקע
הבית ושמנו קרקע הבית ישר באופן שיעמוד עליו כותל
7 החלון על זוית נצבה. ורשמנו שם בקרקע במקום שיפול
בו אבן | העופרת וממקום הרושם הוצאנו קו בקרקע
הבית | יהיה קו חצי היום באופן שבארנו במה שקדם.
8 ולפאת מערב מזה הקו הוצאנו קו בקרקע נכחי לו רחוק
מהקו הראשון בכמו שעור חצי זרת או פחות מעט.
9 וכאשר יגיע קו הניצוץ המערבי בזה הקו השני נרשם
בקו הראשון באי זה מקום יחתכהו ראש הניצוץ הבא

---

13 פ: מהבאים] נ, ק: הבאים.   13 בי"ז] פ: בז׳;
ק: בי"ח.   14 כ"ט] נ: נ"ט.   3 לידיעת] ק: לדעת.

83 באחד מצדדיו אל הקרקע ותרשם שם נקודה. וממקום
הרושם הוציא קו יעמד על זוית נצבה על הקו שהיה
משל לקו בהג והוא יהיה קו חצי היום ר"ל שכאשר
יעבר ניצוץ מרכז השמש מזה החלון בקו ההוא אז יהיה
84 חצי היום. והנה אם היה זה הבית ארוך כמו ארבעים
או חמשים זרתות הנה ילקח בזה תועלת נפלא בהודעת
רוב הדרושים אשר ידיעתם נכספת מענין השמש כמו
שיתבאר במה שיבא.

הפרק הי"ד

1 וראוי שלא יעלם ממנו מה שיש מהקושי בעמידה על
האמת במקומות הכוכבים הקיימים באורך וברוחב כדי
שיהיה לנו פתח נכנס בו | לידיעת מקומות כוכבי
2 המבוכה במבט. וזה שכבר יחשב שלא יהיה בזה קושי עם
מה שהישרנו אליו מהמבטים לפי שתנועת הירח נודעה
בלי ספק לפי מה שיחשב אלא שכאשר עיננו בזה התבאר
לנו כי לא תספיק הידיעת אשר לנו במקום הירח להודיע
3 מקומות הכוכבים הקיימים הנראים עמה. וזה שהירח לא
ימצא אפי' בנגודים ובמולדות במקום המתחייב מהלוחות
4 הנעשות בזה לפי תכונת בטלמיוס. וזה דבר תעמד עליו
מעתות הלקיות הירחיים שלפעמים יהיה זה קודם העת
אשר לפי תכונת בטלמיוס ולפעמים אחריו וזה אמנם
5 ימצא הטעות בו מורגש. וכל שכן בשאר ימי החדש שכבר
מצאנו בזה מהטעות פעם להוסיף פעם לגרוע יותר ממעלה
6 אחת וחצי. ועוד כי בשעור התחלפות ההבטה לירח מן
7 המבוכה וההספק לפי מה שהתבאר לנו ממבטים רבים. וזה
הוא אינו באופן המתחייב מחשבון בטלמיוס כמו שנבאר
8 בג"ה במה שיבא. וזה ממה שיפיל ספק עצום במקומות
הכוכבים הנראים עמה.
9 ועוד שאע"פ שיתאמת מהלקיות שמרחק הירח מהשמש
הוא לפי מה שיתחייב מתכונת בטלמיוס הנה לא נוכל
לדעת מקום הירח האמתי אם לא ידענו מקום השמש האמתי.
10 והנה בידיעת מקום השמש האמתי מהקושי העצום עד שכבר
כתב בטלמיוס שהוא לא ידע רגע הכנס השמש בראש טלה
ולא ידעוהו אשר לפניו ולא ידעוהו הבאים אחריו.
11 והסבה שהביאתהו לזה המאמר היא כי כלי המבט שהיו
עשויים | אז ללקיחת הגובה כמו האסטורלב היה בהם

1 לידיעת מקומות] פ: לידיעת.   7 המתחייב] נ:
המחויב.   9 מתכונת] נ: בתכונת.   10 כתב] חסר
בכ"י ק.

69 קו עפ על קו עש אלא שלא נצטרך לכל זה הדקדוק בזה המקום כי גובה הכוכבים בהיותם אצל קו חצי היום לא יתחלף דבר מורגש מפני הקירוב שאפשר שימצא בזה בקו חצי היום.   ואם תרצה תוכל לדקדק יותר מזה ותשיג
70 המבוקש בקירוב נפלא.   והוא שתראה אם הקו המגיע מחלון אל מקום הזריחה האמתית בכותל עומד על זוית
71 נצבה על קו סע בתמונה הראשונה אם לא.   ואם לא יעמד עליו על זוית נצבה תדע זויתו אם היתה חדה או שאריחה
72 מק״פ מעלה אם היתה נרחבת.   וכיחס המיתר המחוצה לזוית ההיא אל ס׳ מעלה כן יחס קו עש בזאת התמונה אל
73 קו עפ.   ולזה תוסיף על קו סע לפי יחס יתרון קו עפ
74 על קו עש ותשיג המבוקש בקירוב נפלא.   ואם תשוב לדקדק בזה האופן אחר שהוספת על קו סע זה היחס הנזכר הנה תשיג האמת בזה בקירוב נפלא.
75 וכאשר תשיג ידיעת נקדת הזריחה האמתית בכותל המערבי תשים האנך על הנקדה ההיא ותרשם נקדה בקרקע
76 הבית במקום שיפול שם האנך.   וכן תשים האנך על ראש החלון מהצד | שהגבלת ניצוצו ותרשם נקדה בקרקע הבית
77 במקום שיפול שם האנך.   ותקוה אחר זה הקו ישר בקרקע מהרושם האחד אל השני רזה הקו הוא דומה לקו הס
78 בתמונה השנית הקודמת.   ואחר זה תדע בלוחות מרחב זריחת המעלה ההיא שהשמש בה באופק שאתה בו והוא קשת
79 בז בזאת התמונה השנית שקדם זכרה.   ותעשה בקרקע קו יחתך הקו האחר אשר בקרקע באופן שיקיף עמו בזרית
80 שוה לזוית בהז בזאת התמונה שהיא מרחב זריחת המעלה שהשמש בה.   ותשים נטיית זה הקו השני אשר בקרקע מהקו הראשון בצד המערבי לפאת צפון כשהיה השמש במזלות הצפוניים ולפאת דרום כשהיה השמש במזלות הדרומיים והפך יהיה אם היה החלון בכותל המערבי
81 רזה כלו מבואר הסבה למעיין בזה הספר.   והנה זה הקו השני הוא משל לקו בהג בזאת התמונה והנה ראשו אשר לפאת מזרח הוא מקום זריחת ראש טלה וראש מאזנים
82 וראשו אשר לפאת מערב הוא מקום עריבתם.   ואם היה לכותל הדרומי מהבית חלון תפיל האנך מראש החלון

71 נרחבת] נ: נרוחת.   73 בקירוב נפלא] חסר בכ״י פ.   76 ניצוצו] נ: ניצוץ.   77 בתמונה] נ, פ: בתמונת הקו.   79 ותעשה] נ: ועשה.   79 האחר] נ: האחד.

54 הקטרים קצתם אל קצת ידוע. ואם תוציא הקשת המלא
לזה המיתר יהיה בידך הקשת המגיע מז אל ל מעגלה
55 גדולה. ויהיה מיתרה המחוצה והוא קו למ בזאת
התמונה ידוע וכבר היה שעור קו לב ידוע וישאר שעור
קו מנ ידוע כי קו למ מוסיף על קו לב בכח כמו מרובע
56 קו מנ. ולפי ששלשת קוי משלש למנ ידועים הנה
זריותיו ידועות ולזה תודע זוית למנ שהיא שוה לזוית
הנטייה הנראית לעלית השמש זאת הקשת המונחת שהיא קשת
זל.

57 ונוציא בתמונה הראשונה מנקדת נ בכותל המערבי
לפאת דרום זוית סנע שוה לזוית מלנ בתמונה הזאת
השנית ויהיה קו סע נכחי לקרקע הבית שהוא נכחי לעגלת
58 האופק. ויתבאר שהזריחה האמתית לניצוץ מרכז השמש
האמתי היא בנקדת ע לפי שזוית סענ היא שוה לזוית למנ
שהיא שוה לזוית הנטיה הנראית לעלית השמש כמו שהתבאר.
59 וראוי שתדע שזה יהיה צודק אם היה הכותל המערבי נצב
על קו עה בזאת התמונה וזה כי קו מנ הוא נצב על קו
60 זס בתמונה השנית. ואם לא היתה נטיית הכותל המערבי
באופן שיהיה נצב על קו עה הנה תהיה זאת הנטיה יותר
גדולה מעט לפאת דרום בכותל וזה יתבאר ממה שאומר.

61 נשים קו הכותל המערבי הנכחי לקו קרקע הבית אשר
בו מקום הזריחה האמתית בלתי עומד על קו זס בזאת
התמונה השנית אל זוית נצבה והוא קו עפ בזאת התמונה
62 השנית. ותהיה נקדת פ מקום הזריחה האמתית כשהיה השמש
בנקדת | ז ותהיה נקדת ה למשל לנקדת ראש החלון ונקדת
63 ע משל לנקודת ס בתמונה החלון. ויהיה קו רע עמוד על
קו עפ והוא נרשם בכותל מלמעלה למטה כמו העניין בקו
סנ בתמונה הקודמת ויהיה קו ער שוה לקו נס בהמונה
64 הקודמת. ונוציא מנקדת ע עמוד על קו סה והוא קו עש
65 ונקוה קו רש. והוא מבואר ממה שקדם שזוית עשר שוה
לזוית למנ הידועה וזוית רעש נצבת ותשאר זוית ערש
66 ידועה. ולפי שקו ער ידוע הנה ידוע שאר הקוים ידועים
67 ולזה יהיה קו עש ידוע. והוא מבואר שקו עפ מוסיף
68 עליו בכח כמו | מרובע קו פש. ולזה תהיה הנטיה
לזריחת השמש האמתית יותר גדולה ממה שהונחה כיתרון

נ 48ב

פ 24א

ק 16א

57 סנע] ק: סענ. 57 מלנ] נוסף בכ"י פ (למעלה):
המונחת. 58 למנ ... לזוית] חסר בכ"י נ. 58 לזוית
הנטיה] ק: לזאת הנטייה. 58 הנראית] חסר בכ"י נ.
62 ע משל לנקודת] חסר בכ"י נ. 65 עשר] חסר בכ"י נ.

ק 15ב   שנבאר שעור זאת הזרית אשר יעלה בה השמש זאת העליה
המונחת שהיא מהזריחה האמתית עד הגיעו בנקדת ז
במשלינו לפי מה שיראה במבט כי היא זולת הזרית אשר
תעשה עגלתו הנכחית עם האופק.
40   ונשים שטח עגלת האופק עגלת אהדג ומרכזה נקדת
ה ונקוה קו בהג הישר ויהיה קו בהג סימן לקוטר עגלת
משוה היום ויחתכהו קו אד בנקדת ה על זוית נצבה.
41   ויהיה סימן לחתוך המשותף לעגלת האופק והעגלה
פ 23ב   הנכחית אשר ילך בה השמש בהירותו בראש גדי | קו זח
42   ויחתך קו אד על זוית נצבה בנקדת ט. ונניח שיהיה
השמש עולה לפאת מזרח מהעגלה הנכוחית קשת זל ויהיה
השמש בנקדת ל ויהיה קו לכ עמוד על קו חז ויחתך זה
43   הקו בנקדת כ. ונוציא מנקדת ל עמוד על שטח עגלת
44   אבגד והוא קו לנ ונקוה קו כנ בשטח עגלת אבגד. הנה
זוית לכנ היא שוה לנטיית עליית כל העגלות ונכהיות
על האופק והוא הנשאר מתשעים מעלה כשחוסר מהם גובה
הקוטב ותשאר זוית כלנ שוה לזרית קשת גובה הקוטב.
נ 47ב   45   וכבר תדע שעור קו לנ אם תעמד על | גובה השמש בעת
ההיא באסטורלב שקדם זכרו כי המיתר המחוצה לגובה
46   ההוא היא שוה לקו לנ. והוא מבואר שאם היה החלון
בנקדת ט היתה זאת הנטייה אשר לעגלות הנכוחיות על
47   האופק היא הנטייה הנראית לעליית השמש בו. ואמנם
48   מקום החלון הוא בנקדת ה בקרוב. ולזה כאשר נרצה
לידע שעור הנטייה הנראית לעליית השמש מנקדת ז
לנקדת ל הנה ראוי שנגיע מנקדת ז אל נקדת ה קו זהס
49   הישר. ונדמה שיצא קשת מעגלה גדולת ז אל נקדת ל
ונוציא מנקדת ל מיתר זל המחוצה יפול בקוטר העגלה
50   שהוא קו זהס על נקדת מ והוא קו למ. ונגיע קו מנ
והוא מבואר שזוית למנ היא הנטייה הנראית לשמש בזאת
העלייה והיא זולת זוית לכנ.
51   והנה מגובה השמש תדע הקשת העולה מהעגלה הנכחית
והיא קשת זל באופן שהתבאר בחשבון המהלכות ויתבאר
52   במה שיבא בג"ה. וכבר תדע בלוח הקשתות והמיתרים |
נ 48א   שיעור מיתרה המלא והוא קו זל הישר בזאת הצורה
בשעור שבו יהיה חצי קוטר העגלה הנכחית ההיא ששים
53   חלק. ולזה תדע שעור קו זל הישר בשעור שבו יהיה
חצי קוטר עגלת משוה היום ששים חלק כי יחס אלו

42 לכ] פ, נ: לב.   42 כ] נ: נ.   43 לג ... קו]
חסר בכ"י נ.   44 עלייה] פ: עליות.   44 ותשאר
זוית] נ: ותשאר.

23 בכלי שהמצאנו לזה כמו שבארנו במה שקדם. והיא אצל ראש
גדי ט"ו דקים בקירוב מעט לפי מה שעמדנו עליו במבט.
24 והנה נקדת ל היא במה שבין שתי נקדות ז ו ד בקו גד
ונדביק קו הד ונדע שעורו בקנה המדה וכזה נדע שעור
25 קו זד. הנה משלש זהד ידוע הקוים ולזה הוא מבואר
26 שהוא ידוע הזוירות ותודע מפני זוית הזל. וכבר נודעה
זוית זהל ותשאר זוית זלה הנשארת מהמשולש ידועה.
27 וכבר היה שעור קו זה ידוע ולזה יודע שעור קו זל כמו
28 שהתבאר במה שקדם. ואחר שנודע שעור קו זל | הנה
נוציא בכותל מנקדת ל קו לנ נכחי לקו קרקע הבית
29 ותהיה נקדת נ דרום מנקדת ל. ונניח שתהיה זוית להנ
שוה לזוית חצי קוטר השמש וידוע שעור קו לנ בכמו
30 האופן הקודם. הנה נקדת נ היא מקום הזריחה האמתית
למרכז השמש בזה המשל ונשים קו נס בכותל המערבי
עומד על זוית נצבה על קו לנ באופן שיהיה גבה נקדת
31 ס שוה לגובה נקדת ה. ונמדוד בקנה המדה שעור קו נס
והוא מבואר שגובה ניצוץ השמש האמתי בעת הזריחה
האמתית היא כגובה נקדת ס אלא שהוא נוטה מנקדת ס
בכותל המערבי לפאת דרום.
32 וראוי שנעמד על מקום הזריחה האמתית בכותל
33 המערבי. ונציע לזה הצעה מבוארת והיא כשעור
נטיית שטח עגלת משוה היום על האופק כן שעור נטיית
34 שאר העגלות הנכחיות עליו. והשמש כשיעלה על אלו
35 העגלות יעלה על זה השטח הנוטה. ועוד נציע על שהוא
מבואר למעיין בזה הספר שניצוץ השמש המגיע לנו
בהירותו בעגלות הנכחיות לעגלת משוה היום אינו בשטח
עגלתו הנכחית כי הוא עובר על מרכז עגלת משוה היום
36 לא על מרכז עגולתו הנכחית. וזה שהארץ היא על מרכז
עגלת משוה היום כמו שבארו הקודמים ויתבאר עוד
מדברנו במה שיבא בג"ה והיא כלה כנקודה בקירוב ביחס
37 אל גלגל השמש. ולפי שזה הניצוץ יעבור ביושר במרכז
עגלת משוה היום הוא מבואר שהוא יכלה אל העגלה
38 הנכחית אשר בצד השני שוה לעגלה שהשמש בה. ונשים
החלון בזה המקום משל למרכז הארץ כי זה לא יזיק
בזה המקום ולזאת הסבה שזכרנו תמצא שכל מה שהיה
השמש יותר נוטה לפאת צפון יהיה ניצוצו בכותל המערבי
39 יותר נוטה לפאת דרום. ולפי שהשמש יעלה באפקים
בעגלה נוטה לא בעגלה נצבת על האופק הנה ראוי |

25 הזל] נ: קול.

10 ראש סרטן מאופן המישור. ובזה האופן מהחקירה
מצאנו ששעור זאת הנטיה הוא כ"ג מעלות ול"ג דקים
בקירוב נפלא.

11 וכאשר נתאמת לנו זה הנה בהיות השמש בראש סרטן
או בראש גדי או סמוך לזה כדי שלא יקרה במבטינו טעות
מורגש מפני מה שנסכל ממקום השמש האמתי נביט
בניצוצו בבית אחד נכחיי הכותלים ויהיה קרקעיתו

12 ישר נכחיי לשטח האופק. ויהיה בו חלון לפאת
המזרח או לפאת מערב בקירוב ויהיה נצב הזויות ישר
הקוים ויהיה עליון החלון נכחיי לקו קרקע הבית

13 וקוי ימינו ושמאלו נצבים על שטח קרקע הבית. ואם
היה החלון מזרחי בקרוב נרשם בכותל המערבי הנכחי
לו קו נכחי לאופק שפל מגובה ראש החלון שעור מה כאלו

14 תאמר חמשה זרתות או יותר. ותחת הקו ההוא נרשם
בכותל ההוא כמו מרחק זרת או יותר קו נכחי לקו

15 הראשון. והנה בבקר כשיגיע עליון הניצוץ היוצא לקו
העליון נרשם בקו השפל ממנו באי זה מקום יחתוך
הניצוץ הקו השפל ויגיע הרושם ביושר עד הקו העליון

16 ובאשר יחתכהו נניח שיהיה רושם ראש הניצוץ. ומזה
נעמד על מקום ניצוץ השמש בכותל | ההוא בעת הזריחה
האמתית.

17 המשל שאנחנו נשים בכותל המזרחי קו אב עמוד על
קרקע הבית ויהיה זה הקו ממשש הצד הצפוני מהחלון
ונשים נקדת ה בו משל לנקדת ראש הצד הצפוני מהחלון.

18 ונרשם בכותל המערבי קו גד נכחי לקו אב ויהיה זה
הקו ממשש הניצוץ הנכנס מהצד הצפוני מהחלון המזרחי.

19 ותהיה נקדת ז בו משל לנקדת ראש זה הניצוץ מזה הצד
ותהיה נקדת ט משל למרכז השמש ויהיה חצי קטרו השפל

20 הנכחיי לקו גד קו טח. וכבר נתבאר ממה שקדם שהניצוץ
האמתי אשר יגיע ממרכז השמש הוא למטה מנקדת ז בשעור
זוית חצי קוטר השמש ולפאת דרום מנקדת ז | בשעור

21 זוית חצי קוטר השמש. ונבאר איך נדע הנקדה שהיא
למטה מנקדת ז כמו זוית חצי קוטר השמש ומזה נתישר
לדעת הנקדה שהיא לפאת דרום ממנה בשעור זוית חצי

22 קוטר השמש. ונדביק קו זה הישר ויכלה לנקדת ח ונשים
זוית זהל שוה לזוית חצי קוטר השמש שהיא ידועה

ק 15א
נ 46א
פ 23א

11 בניצוצו] נ: בניצוץ.   15 היוצא] חסר בכ"י ק.
15 הרושם] נ: הרשום.   17 משל] נ: המשל.   17 הצד]
חסר בכ"י נ.   18 ממשש] נ: הממשש.   22 ויכלה]
נ: יכלה.

חלוקת כל המעלות האלכסרוניות לט"ו זוירות שוות.
51 ואולם חלוקת המעלות בזה האופן לא תקשה באופן
שהתבאר מאקלידס הדרך לחלק כל קו על יחס חלוקת קו
52 מונח. וזה שכבר נחלק קו ארוך בכמו זאת החלוקה
ויתחלק אחר כן זה הקו שהוא חצי קוטר עגלת הכלי על
53 יחסו באופן שהתבאר שם. וראוי שתדע שכל מה שיהיה
כלי האסטורלב יותר גדול יהיה הדקדוק נמצא בו יותר
ורואי שלא יהיה חצי הקוטר בו פחות מזרת ואז תוכל
54 פ 22ב לדקדק בו כפי מה | שתרצה. ואולם השלמת
המעשה ביתר הדברים שיכללם האסטורלב כבר דבר בו
זולתינו מה שיש בו די.

הפרק הי"ג

1 ואחר שהיישרנו אל השלמת המעשה באסטרלב באופן
לא יקרה טעות בלקיחת הגובה ונוכל לדקדק בו לדקים
ולחלקים מהם הוא מבואר שבכמו זה הכלי יקל לדעת קו
2 חצי היום בקירוב נפלא. והוא שתעיין בו גובה השמש
סמוך לחצי היום ותראה שיתוסף הגובה מעט עד שלא
3 יתוסף עוד גובהו ואז הוא חצי היום בקירוב. ולו
תקוה בקרקע הבית על יושר קו ניצוץ השמש הנכנס שם
אז מן החלון הנה יהיה זה הקו קו חצי היום בקירוב.
4 וכבר נעמד על קו חצי היום באופן אחר בתכלית הדקדוק.
5 והוא שנחקר תחלה על גובה הקוטב במקומנו באצטורלב
הנזכר כשנעיין בהיות השמש בראש סרטן ובראש גדי או
6 בסמוך לזה מעט. כי הטעות שיקרה שם מצד הקירוב אשר
בידיעת מקום השמש בארך לא יהיה לו רושם בגובה
נ 45ב השמש | דבר מורגש למיעוט חילוף הנטיה שתקרה מפני
7 מעלה אחת שם. והנה כאשר נחבר אל גובה השמש בהיותו
בראש גדי חצי תוספת גובה ראש סרטן על גובה ראש
גדי היה המחובר שוה לגובה ראש טלה באופק ההוא.
8 וכאשר חוסר גובה ראש טלה מהשעים מעלה היה הנשאר
9 שוה לגובה הקוטב באופק ההוא. והנה חצי תוספת
גובה ראש סרטן על גובה ראש גדי הוא שוה לנטיית

51 מאקלידס] נוסף בשולי נ: בי"ג מששי. 1 לא
יקרה] פ: שלא יקרה. 2 מעט] ק: מעט מעט.
6 למיעוט] נ: מיעוט.

32 ול"א שניים. ולזה אם נקוה עגלה סביב מרכז הכלי יהיה
חצי קטרה נ"ט מעלות י"ב דקים ול"א שניים בשעור שבו
יהיה חצי קוטר עגלת מעלות האצטורלב ס' מעלות.
33 הנה תחלק כל אלו המעלות האלכסוניות בזה האופן
34 ר"ל שתהיה זוית החלק השפל מהם ד' דקים. ותהיה עתה
זורית זAג ח' דקים וכבר התבאר שזורית זגA היא ד' מעלות
ונ"ט דקים ותשאר זוית גזA הנשארת מהמשלש שלמות ק"פ
35 מעלות והוא ק"ע"ד מעלות ונ"ג דקים. ולזה הוא מבואר
כי בשעור שבו יהיה שעור קו גA ה' מעלות כ"א דקים וג'
שניים בו יהיה שעור קו זA ה' מעלות י"ב דקים ומ"ג
36 שניים. ובשעור שבו יהיה שעור קו גA ס' מעלות בו
יהיה שעור קו זA נ"ח מעלות כ"ו דקים ומ"ה שניים.
37 ונשים חצי קטר העגלה השלישית בזה השעור. ובזה
38 האופן יתבאר שחצי קטר העגלה הרביעית הוא נ"ז מעלות
39 ומ"א דקים ומ"א שניים. ושחצי קוטר העגלה החמשית
40 הוא נ"ו מעלות נ"ז דקים ונ' שניים. ושחצי קוטר
העגלה הששית הוא נ"ו מעלות ו"ד דקים ונ"ט שניים.
41 ושחצי קוטר העגלה השביעית הוא נ"ה מעלות ל"ו דקים
42 וכ' שניים. ושחצי קוטר העגלה השמינית הוא נ"ד מעלות
43 נ"ב דקים נ"ג שניים. ושחצי קוטר העגלה התשיעית הוא
44 נ"ד מעלות י"ג דקים י"ד שניים. ושחצי קוטר העגלה
45 העשירית הוא נ"ג מעלות ל"ד דקים כ"ב שניים. ושחצי
קוטר העגלה האחת עשרה הוא נ"ב מעלות נ"ו דקים כ"ו
46 שניים. ושחצי קוטר עגלת השתים עשרה הוא נ"ב מעלות
47 י"ט דקים ול"א שניים. ושחצי קוטר העגלה השלש עשרה
48 הוא נ"א מעלות מ"ג דקים כ"ח שניים. ושחצי קוטר
העגלה הארבע עשרה הוא נ"א מעלות ח' דקים י"ג שניים.
49 ושחצי קוטר העגלה החמש עשרה הוא נ' מעלות ל"ג דקים
50 מ"ו שניים. ואולם חצי קוטר השש עשרה הוא כפי זה
החשבון נ' מעלות וככה הונח | שעורו ובזה האופן נשלמה

35 כ"א ... וג' שניים] חסר בכ"י פ.   35 וג'] נ:
רנ"ג; בשולי נ: נ"ל ג' שניים.   36 כ"ו] נ בטקסט
ט'; נוסף בשולי נ: נ' כ"ו דקי' ל"ג שניי'.   39 נ"ו
מעלות] נ: נ"ז מעלות.   40 רנ"ט] נ: רט"ו.
43 י"ג] נ: נ"ג.   43 י"ד] ק: ד'; נ: נ"ד.
44 ל"ד] ק: ל"ב.   44 כ"ב] נ: נ"ב.   46 נ"ב] נ:
נ"ט.   46 ול"א שניים] חסר בכ"י נ.   47 כ"ח] נ: נ"ח.
49 נ'] חסר בכ"י נ.   49 ל"ג] נ: ל"ז.

מהעגלה הפנימית מכל אחד מהצדדים וננהג בשלשית לה
מכל אחד מהצדדים המנהג שנהגנו בה וכן נעשה עד
שישלם זה המעשה בכל החצי השפל.

19 ואחר שישלם זה תמצא בכל מעלה ממעלות הכלי מעלה
אלכסונית ארכה כמו אורך עשר מעלות ממנה בקירוב.

20 והנה נבאר לך איך תחלק כל מעשה מהמעלות האלכסוניות
לט"ו זריות שוות באופן שיהיה שעור כל חלק מהם ד'

21 דקים. נניח שיהיה נקדת א' מרכז הכלי ותהיה מעלה

22 אחת ממעלות העגלה החיצונה קשת בג. ונקוה קוי אב אג
הישרים ונוציא מנקדת ג עמוד על קו אב והוא קו גה

23 ויהיה שעור קו בד שעית קו אב ונקוה קו דג. והוא
מבואר מלוח הקשתות והמיתרים כי בשעור שבו יהיה
שעור קו אג ס' מעלות בו יהיה שעור קו הג מעלה אחת
וב' דקים ונ' שניים בקירוב ובו קו הב ל"ג שנים
בקרוב וישאר שעור קו הד ט' מעלות ונ"ט דקים וכ"ז

24 שנים בקירוב. ולפי שקו גד מוסיף עליו בכח כמו
מרובע קו גה הנה יהיה שעור קו גד י' מעלות | ב'

25 דקים מ"ז שנים בקרוב. ונרשם בקו גד נקדת ז ונקוה

26 קו אז ונניח תחלה שתהיה זרית גאז ד' דקים. ונרצה
לעמד מזה על שעור קו אז והוא מבואר כי מפני שזרית
גאה היא מעלה אחת שזרית הגא היא שלמות הזרית הנצבה

27 שהוא פ"ט מעלות. וכאשר חסרנו ממנה זרית הגד תשאר
זרית דגא ידועה והנה זרית הגד היא ידועה לפי שקוי

28 משלש גהד ידועים. הנה בשעור שבו יהיה שעור קו גד
עשר מעלות וב' דקים ומ"ז שנים בו יהיה שעור קו גה
מעלה אחת וב' דקים ונ' שנים ובשעור שבו יהיה שעור
קו גד ספ מעלות בו יהיה שעור קו גה ו' מעלות וט"ו

29 דקים וי"ו שנים. ולזה תהיה זרית גדה ה' מעלות
ונ"ט דקים ותשאר זרית הגד שלמות הזרית הנצבה והוא
פ"ד מעלות ודק אחד ותשאר זרית דגא ד' מעלות ונ"ט

30 דקי'. וכבר היתה זרית גאז ד' דקים ותשאר זרית
גזא הנשארת מהמשלש שלמות ק"פ מעלה והוא קע"ד מעלות

31 ונ"ז דקים. ולזה הוא מבואר כי בשעור שבו יהיה שעור
קו גא ה' מעלות וי"ו דקים ונ"ג שנים בו יהיה שעור
קו אז ה' מעלות י"ב דקי ומ"ג שנים ובשיעור שבו יהיה
קו גא ס' מעלות בו יהיה קו אז נ"ט מעלות י"ב דקים

23 בשולי פ: ובו קו הב] בשולי נ: וקו בה; חסר בכ"י
ק. 24 מ"ז] בטאקסט נ: מ"ו; בשולי נ: נ' מ"ד.
28 ומ"ז] נוסף בשולי נ: נ' מ"ד. 28 וי"ו] פ: וי'

21ב פ | | יפול בו בלוח בעת במבט.
3 | | ראינו להישירך לעשות זה הכלי באופן שלא יקרה
4 | | ממנו טעות מאלו הצדדים. ואופן המעשה יהיה שתעשה
5 | | הלוח בעל השני דפין הנקובים קודם שתקוה קוטר הכלי
 | | וקודם שתחלקהו למעלות ורשם בקצהו האחד. ואחר זה
6 | | תחלה הכלי בהיות השמש בקו חצי היום בקירוב ויהיה
 | | בשפל הכלי טבעת רחבו כעובי הכלי. ותחלה בו אבן
7 | | העופרת באופן שיהיה מקום שפל הכלי תמיד בנקדה
 | | אחת ולא יתבלבל לסבה מהסבות. ותתקן הלוח באופן
8 | | שיכנס ניצוץ השמש בנקב האחד ויפול בתוך הנקב השני
 | | ותרשום שם במקום נפילת הלוח בצד ההוא. עוד תהפך
9 | | הלוח בצד השני וישאר תחתון הנקב שהונח תחלה תחתון
 | | ותרשם שם במקום נפילת הלוח כשיכנס ניצוץ השמש בנקב
10 | | האחד ויפול בתוך הנקב השני. ותקח זה המבט תכף אחר
 | | שלקחת המבט הראשון כדי שלא ישתנה גובה השמש בדבר
11 | | מורגש. ואחר כן תחלק מה שבין שני הרשמים בתחתית
 | | הכלי לחצאין ותרשם שם במקום החלוקה. והוא יהיה
12 | | תכלית הגובה משני הצדדים ר"ל ששם יפול הלוח בהיות
 | | המובט מבין שני הנקבים נכח הראש. ותקוה על מקום
13 | | החלוקה ועל מרכז הכלי וזה הקו יהיה קוטר הכלי לפי
 | | זה הלוח ולפי מצבו שרשמת בו אלו הרשמים. ואחר כן
 | | תחלק העגלה ממקום החלוקה לד' רבעים שוים בתכלית מה
 | | שאפשר מהדקדוק ותחלק כל רובע לתשעים מעלות שוות
 | | בתכלית מה שאפשר מהדקדוק וזה יעמידך על מעלות
43ב נ | | הגובה | בזולת טעות.
14 | | ואולם העמידה על החלק מהמעלה תהיה כשנחלק חצי
15 | | קוטר עגולת המזלות לששה חלקים שוים. ונקוה סביב
 | | מרכז הכלי במרחק חמשה ששיות חצי קוטר העגולה עגלה.
16 | | ונחלקה באופן חלוקת העגולה החיצונה כשנשים הלוח
 | | אשר הוא תקוע במרכז הכלי על מקום חלוקת מעלות
 | | העגולה החיצונה ובמקום שיחתך הלוח העגלה הפנימית
 | | נרשם שם רושם עד שתתחלק העגולה הפנימית בזה האופן
17 | | לש"ס מעלה ויחתך חציה השפל לק"פ מעלה. ואחר זה
 | | נשים קו המישור על רושם מעלת העגלה הפנימית ועל
 | | רושם המעלה אשר לפניה ולאחריה מהעגלה החיצונה ונקוה
18 | | אלו הקוים משני הצדדים. ואחר כן נדלג רושם אחד

5 ק, בשולי פ: כעובי] בטאקסט פ, נ: כרחב.   9 כדי]
נ: כי.   16–17 העגולה הפנימית בזה ... קו] חסר
בטאקסט נ; בשולי נ: כלה בזה האופן אחר כן תשים קו.

ממרכז גרם הכוכב האחד אל מרכז גרם הכוכב השני נכחי
לקוי הלוח אשר בו יהיה המבט בזולת נטיית דבר מורגש
4 וזה מבואר הסבה למעיין בזה הספר. וראוי שתדע שלא
יתן המבט האמת כשהיו עננים אצל הכוכב כי כבר יקרה
5 שיראהו הענן | בזולת מקומו. וזה דבר כולל לכל המבט
עד שכבר תראה ביום המעונן באצטורלב גובה השמש על
זולת מה שהוא ותראה שהענן כשיתנועע יתנועע גובה
6 השמש. ולזה גם כן אין ראוי שתביט לכוכב קרוב מאד
7 לאופק לעובי האיד שם. וראוי שתדע שכבר יראה גם כן
הענן גודל הכוכב על זולת מה שהוא והנה כשהיה הענן
חזק העובי יראה הכוכב קטן משעורו וכאשר היה הענן
8 דק יראהו יותר גדול משעורו. וכן יראה אור השמש
הכוכבים הקרובים לו יותר קטנים משעורם עד שכבר יהיו
באופן מהקורבה שיסתיר אותם בשלמות.
9 וראוי שתדע שכאשר נרצה לעמד מזה הכלי על מקום
כוכב מכוכב המבוכה מצד מרחקו מכוכב קיים אין ראוי
שיהיה המרחק ביניהם יותר מארבעים מעלה שעור רב וזה
כי במרחק הרב לא יחזק הראות לראות שני הכוכבים יחד.
10 ואם היה רחבם בלתי ידוע באמתות ראוי שיהיה המרחק
ביניהם בלתי פחות מעשרים מעלות כי במרחק המעטי יהיה
הטעות המעט אשר יקרה בידיעת רוחב הכוכבים מביא
לטעות גדול באורך הכוכב וכל שכן כשהיה המרחק רב
11 השעור. ולזה ראוי שיהיה המרחק בין הכוכבים המובטים
באופן שיהיה הטעות הרב אשר יקרה בידיעת רחב הכוכבים
12 מביא לטעות המועט. וזה כלו ממה שלא יעלם ממי שראה
דברינו בדרך ידיעת המרחק בין הכוכבים המובטים בזה
הכלי.

פרק י"ב

1 ולפי שהתועלת אשר יגיע מכלי האצטורלב נפלא
לקלות ההשתמשות בו ובפרט ללקיחת גובה הכוכב באי זה
2 מקום שיהיה. והיה הטעות נופל במעשהו מצדדים רבים |
אם תחלה מצד נטיית | קוטר הכלי ואם שנית מצד השבוש
הנופל במעשה הלוח בעל השני דפין הנקובים ואם שלישית
מצד הקושי אשר יהיה בעמידה על החלק מהמעלה אשר |

---

5 שהענן] חסר בכ"י פ.   6 האיד] ק: האויר.
11 המועט] ק: מועט.   12 בזה הכלי] חסר בכ"י נ.

54 האלכסוני. ונניח הצורה על ענינה ואומר שכבר יודע
55 גם כן מזה הכלי מרחק הירח מהשמש לפי הנראה. ונניח
שיהיה הניצוץ עובר מנקדת ד לנקודת ל במשל הקודם
56 ויהיה הירח נראה אז בנקודת מ. ואומר שמרחק הירח
מהשמש לפי הנראה הוא ידוע.
57 המופת שנוציא קוי דל אמ ונוציא מנקודת א קו אנ
58 נכחי לקו דל וקו אס נכחי לקו דמ. ונאמר שמרחק הירח
59 מהשמש לפי הנראה שוה לזוית נאמ. וזה כי מפני שקוי
דל נא נכחיים הנה בעת שיבא ניצוץ השמש מנקודת ד
60 לנקודת ל יבא מנקודת נ לנקודת א. וזה כי חצי קוטר
כדור הארץ היא כנקודה ביחס אל מרחק השמש מן הארץ
וכל שכן שיהיה כנקודה מרחק שני אלו הקוים ולזה יראה
61 השמש והירח מזוית נאמ. ואומר שזוית נאמ היא שוה
62 לזוית לדמ ושלמות ק״פ מעלות מזורית במא. וזה ששתי
זויות | במא מאס הם שוות לשתי זויות נצבות וזוית סאנ
שוה לזוית לדמ לפי שהקוים המקיפים בשתי הזויות הם
63 נכחיים כל אחד לגילו. הנה אם כן זוית נאמ שוה לזוית
לדמ ולשלמות ק״פ מעלה מזורית במא שהוא כמו זורית מאס.
64 ויהיה גם כן ניצוץ השמש עובר מנקודת ד לנקודת
65 ה בעת שיהיה הירח נראה בייושר קו אמ. ואומר שזורית
מרחק הירח מהשמש שוה לשלמות ק״פ מעלה מזורית במא.
66 וזה שאנחנו נניח הצורה על ענינה ויהיה קו אס נכחי
67 לקו דה. ויתבאר באופן הקודם שזורית סאמ היא זורית
מרחק הירח מהשמש והיא שלמות ק״פ מעלה מזורית במא.
68 ובכאן התבאר מה שרצינו לבארו מזה הכלי.

פרק י״א

1 וממה שראוי שתנהג בו בהשתמשותך בזה הכלי בדרך
2 שלא יקרה בו טעות. אם תחלה תשים אור הנר מאחוריך
כשתביט בו בלילה בדרך שיתפשט האור בשטח הלוח הנראה
לך אשר בו יהיה המבט כדי שיתבאר לך אם הכוכבים
ממששים במבט שני קצות הלוח רצוני שיראה הכוכב
האחד ממשש במבט לקצה הלוח האחד והכוכב האחר ימשש
3 במבט לקצה השני. וראוי גם כן שיהיה הקו המגיע

---

55 ד לנקודת] חסר בכ״י נ.   61 ואומר ... נאמ]
חסר בכ״י נ.   66 סאמ] נ: אמ.   3 שיהיה] נוסף
בכ״י פ (למעלה): שתדע.

פ 20ב 38   המשל שאנחנו נשים המקל קו אבג | ותהיה נקודת א
נ 41א   ממנו בישר מקום | מרכז העין באופן שקדם ותהיה נקודת
39   ב על יושר מקום חתוך הלוח המקל בעת המבט. ויהיה
המשלש נצב הזוית הנרשם בזה הלוח משלש דהז ותהיה זוית
40   ה נצבת וזויות ד ו ז חדות. ותהיה נקודת ד במקום הזוית
החדה אשר בקצה הלוח הראשון והנה בנקדות הלוח הראשון
כתבנו דחט ובמקום נקדרות הלוח השני כתבנו זכל.
41   ונניח תחלה שיהיה הניצוץ עובר מנקדת ד לנקדת ל ולזה
42   תהיה זוית הדל ט"ו מעלות. ויהיה עתה קוטר המשלש
בין הראות ובין הלוח הישר ונשים בקצה הלוח הראשון
שאין שם לוחות המשלש אשר נראה מרכז הירח ממשש לו
43   נקדת מ. ותהיה זוית אבמ היא נצבת ושעור קו אב
44   הוא ידוע ושעור קו במ הוא ידוע. ולזה תודע זוית באמ
באופן שקדם ותודע ממנה זוית במא כי היא שארית הזוית
45   הנצבה. ונניח זוית במא ע"ה מעלות וכבר היתה זוית
הדל ט"ו מעלות ואומר שמרחק הירח מהשמש בעת המבט
הוא שלמות ק"פ מעלות שהוא צ' מעלות.
46   המופת כי מפני שאצל קצות קו דמ היו קוי דל מא
בשטח אחד ושתי הזוירות אשר יחדשו עם קו דמ הם פחות
47   משתי זויות נצבות. הנה שני אלו הקוים יפגשו ובמקום
פגישתם תהיה הזוית שלמות שתי זויות נצבות. ולפי
48   שמקום פגישתם הוא כמו מרכז זוית המבט וזה כי
למיעוט מרחק הנקודה ההיא ממרכז הראות לא יהיה הבדל
מרחש במבט הזה בין שיובט מרחק הירח מהשמש מהנקודה
49   ההיא ובין שיובט ממקום מרכז הראות. הנה יהיה זה
הזוית מרחק הירח מהשמש לפי הנראה וכאשר תוקן
מקום הירח לפי התחלפות ההבטה הראוי לו בעת המבט
נ 41ב 50   עמדנו מזה על מקום הירח האמתי בעת | המבט. והוא
מבואר שאם הושם במבט מקיף ירח ממשש לנקודת מ יצטרך
להוסיף על מרחק הירח מהשמש כשעור חצי קוטר הירח
שהוא כמו ט"ו דקים בקירוב לפי מה שיתבאר בג"ה במה
שיבא.
ק 13ב 51   ונניח | עוד שיהיה הניצוץ עובר מנקדת ט לנקדת
ז בזה המשל הנה זוית הזט היא ט"ו מעלות ותשאר זוית
52   הטז ע"ה. ונניח בתמונה הקודמת שיהיה זוית
במא ע"ה מעלות ותשאר זוית מרחק הירח מהשמש הנראה
בעת המבט שלמות ק"פ מעלות והוא ל' מעלות והקש על
53   זה. ויהיה עתה הלוח הישר המהמשלש בין הראות והלוח

38 מקום מרכז] פ: במקום מרכז.   39 דהז] נ: דמז.
43 ותהיה] ק: והנה.   43 זוית] נוסף בכ"י פ: נקודת.
49 הראוי] פ: הראויה.

ק 13א
23 המשלש שאצל קצה הלוח הראשון קוים ישרים יעברו | על
אלו הנקדות שנרשמו על הלוח השני. וכן נעשה מהזוית
24 החדה שאצל קצה הלוח השני על הנקודות שנרשמו על הלוח
הראשון. ונעשה בנקודות הזויות החדות מזה המשלש שני
25 נקבים תהיינה אלו הנקדות מרכזיהם ויהיו עומדים על
שטח המשלש. ונעמיד שם שני יתדות בראשם דף נקוב נקב
קטן מתנועעות לאי זה צד שנרצה להניעם מימין אל שמאל
או משמאל אל ימין ועל מקומות הנקודות הנרשמות בשני
אלו הלוחות נעמיד יתדות בעלות דפים נקובים יהיו
26 עומדים על שטח המשלש. ויעמדו שרשי היתדות האלו על
זוית נצבה עם הקוים הנרשמים על אלו הנקודות.
27 ובמקום חתוך היתדות הקוים אשר בשרשם נעמיד קו
ביתדות יהיה עמוד על שטח המשלש ובקו ההוא יהיה מרכז
נ 40ב
נקב דף היתד ונקבי אלו | היתדות כלם יהיו עומדים על
28 שטח היתד. ונשים שוה גבה הנקבים כלם על שטח המשלש.
29 וכן נעשה בצד השני מאלו הלוחות משלש אחר בכמו
זה האופן כדי שנוכל לעיין בזה הכלי השמש והירח יחד
בהיות השמש מזרחי מהירח או מערבי.
30 והוא מבואר שזה הלוח יהיו לו שני מצבים המצב
האחד שיהיה הלוח האלכסוני בין הראות ובין הלוח הישר
הנצב על המקל והמצב השני שיהיה הלוח הישר בין הראות
31 ובין הלוח האלכסוני. והנה נביט בירח ממקום הראות
במקל באופן שימשש הירח קצת הלוח השני שאין בו המשלש
והשמש יכנס אז מנקב אחת הזויות החדות במשלש אל אחד
32 מנקבי הלוח שהוא מיתר הזוית הזאת. וכשהיה השמש
מזרחי לירח נשים קצה המשלש לפאת ימין וההפך בהיות
33 השמש מערבי. וכשהיה מרחק השמש מהירח עד צ׳ מעלה
בקירוב נשים הלוח האלכסוני בין הלוח הישר ובין הראות.
34 וכשהיה מרחק השמש מהירח מצ׳ מעלות עד ק״פ מעלות
נשים הלוח הישר בין הלוח האלכסוני ובין הראות.
35 ונאמר שכאשר היה ניצוץ השמש עובר מנקב אחת הזויות
החדות אל אחד מנקבי הלוח שהוא מיתר לה והובט אז הירח
ממקום הנחת הראות בזה המקל הנה יודע מרחק הירח מהשמש
36 לפי הנראה. ומזה נעמוד על מרחקו האמתי מהשמש כשידענו
37 שעור תקון התחלפות ההבטה. וכבד ניישיר לעמוד על
אמתת שעור התחלפות ההבטה במה שיבא בג״ה.

22–23 השני ... על הלוח] חסר בכ״י נ. 25 שמאל או
משמאל] נ : שמאל. 34 מהירח] חסר בכ״י נ.
36 ומזה] ק: וממנו.

קבענו בשטח רחב הלוח אחר יעמוד עליו על זוית נצבה
ויהיה שטחו חותך שטח הלוח הראשון על זוית נצבה.
10 ושמנו אורך זה הלוח על השטח ההוא שוה אל האורך
11 הנשאר מהלוח הראשון בצד היותר ארוך. ואחר זה קבענו
שתי אלו הלוחות בלוח שלישי דבק על קצוותיהם בצד אשר
12 יעשה בו משלש שוה השוקים. ואחר זה נקבנו זה הלוח
השלישי באופן שכשיכנס המקל בשני נקבים יעמד הלוח
13 הראשון על המקל על זוית נצבה. ואחר זה נרשם משלש
נצב הזוית באמצע עובי לוחות המשלש ונקח מקרי זה
המשלש משני צדדי הזוית הנצבה קו יהיה יחסו אל כל
הקו ההוא כיחס נכחות קשת ט"ו מעלות אל נכחות קשת
14 ע"ה מעלות. ונרשם שם שתי נקודות משני צדדי הזוית
15 הנצבה יחלקו הקוים ההם לפי היחס הנזכר. והוא מבואר
שכאשר הרונח הענין כן והוצאנו קוים מאלו הנקודות אל
קצות קוי המשלש נצב הזוית הנרשם באלו הלוחות שכבר
יתחדשו מזה שני משלשים נצבי הזוית תהיה זויתם החדה
האחת ט"ו מעלות וזויתם החדה השנית ע"ה מעלות.
16 והמשל שיהיה המשל הנרשם באלו הלוחות משלש דהז
פ 20א 17 ותהיה הזוית שאצל נקדת ה נצבת. ונקח מקו דה | קו
הט יהיה יחסו אל קו הז כיחס נכחות קשת ט"ו מעלות
אל נכחות קשת ע"ה מעלות ולזה תהיה זוית הטז ע"ה
מעלות וזוית הזט ט"ו מעלות ונקח מקו הז קו הל שוה
נ 40א 18 לקו הט. עוד נקח בזה האופן מקוי משלש דהז במשלנו
זה משני צדדי הזוית הנצבה קו יהיה יחסו אל כל הקו
ההוא אשר לוקח ממנו כיחס נכחות קשת ל' מעלות אל
נכחות קשת ס' מעלות ונרשם שם שתי נקודות באופן
19 הקודם בנקודות האחרות. והוא מבואר שאם הוצאו קוים
מאלו הנקודות אל שתי נקודות ד ו ז במשלנו זה שכבר
יתחדשו מזה שני משלשים נצבי הזוית תהיה זוית האחת
20 ל' מעלה וזויתם השנית ס' מעלה. והוא מבואר שזויות
משלש דהז החדות תהיה כל אחת מהם מ"ה מעלה לפי שכל
21 אחת מהן היא חצי הזוית הנצבה. ובזה האופן ימצאו
במשלש דהז הזויות עד צ' מעלות רשומות מט"ו מעלות
לט"ו מעלות ר"ל שכבר ימצאו בו זויות ט"ו מעלות ול'
מעלות ומ"ה מעלות וס' מעלות וע"ה מעלות וצ' מעלות.
22 ואחר שמצאנו אלו הנקודות נגיע מהזוית החדה מזה

11 שלישי] פ: השלישי. 12 נקבנו] נ: קבענו.
17 הז] נוסף בשולי נ: ' הד. 20 מ"ה] פ: מ'.

60 הכלי לעמוד על אורך הכוכב הנראה עמו. וכן אם ידענו
אורך כוכב מהכוכבים ועמדנו בזה הכלי על שיעור גבהו
בחצי שמים הנה נעמד מזה על שיעור נטיית הכוכב ההוא
מגלגל המזלות כשידענו גובה המעלה שהוא בה | בחצי
61 השמים במקום ההוא. וכבר נבאר אחר זה בג"ה איך נדע
זה באופק אופק אלא שלא נודע לנו באמת אורך כוכב
62 מהכוכבים ולא רחבו. וזה דבר עמדנו עליו במבטים
רבים כשסמנו מקום הכוכבים באורך ורחבם | לפי הלוחות
כי כבר נתבאר מהמבטים ההם במה שאין ספק בו שאין
63 מקומם באורך וברחב לפי מה שהונח בלוחות. וזה ממה
שיוסיף קושי בזה כי הוא ראוי שנחקור תחלה איך יודעו
64 מקומות הכוכבים הקיימים בארך וברחב. ואחר זה יתכן
שנישר מהם לדעת במבטים מקומות הכוכבים הנבוכים
באורך וברחב בזה הכלי.

הפרק העשירי

1 ולפי שכבר התבאר שהוא מחוייב שנחקר תחלה באי
זה מקום מגלגל המזלות בארך וברחב יהיו הכוכבים
הקיימים בעת מבטינו קודם שנתישר מהם לדעת מקום
2 הכוכבים הנבוכים הנראים עמהם. והיה הדרך בזה כמו
שיתבאר אחר זה שנשים התחלת החקירה ממקום השמש כי
3 הוא אשר נוכל לדקדק על מקומו האמתי בכל עת. והיה
מבואר מעניננו שאין כוכב נראה עמו בזולת קושי רב
4 זולת הירח. הנה היישרנו בזה הכלי לעמוד ממנו על
מקום הירח בארך כשנביט יחד בשמש ובירח בזה הכלי
5 באופן שנבאר אחר זה. וכאשר לקחנו זה המבט דרך
משל קודם השקיעה מעט ואחר זה נראה הירח בתחלת
הלילה או סמוך לו עם כוכב מהכוכבים הקיימים ונודע
6 המרחק אשר ביניהם באורך מגלגל המזלות. וידענו
מהזמן האמצעי בין שני המבטים מהלך הירח האמתי
באותו העת לפי המקום שהוא בו מתנועת החלוף ומהגובה
לפי חשבון תכונת בטלמיוס שאין בו מהקירוב מה שירגש
בכמו זה הזמן הקצר לפי מה שנתבאר לנו אחר החקירה
7 הארוכה. הנה הורה לנו אז | מקום השמש על מקום הכוכב
הקיים ההוא.
8 וזה תאר הלוח עשינו לוח ישר השטחים נקוב באמצעו
9 באופן מעשה הלוחות הקדומום. ואחר הנקב שעור מה

62 עליו] פ, נ: אליו.   5 הקיימים] חסר בכ"י נ.

[ב]

35 למה תקראוני חובלים	צעק מקל מר צורח
36 ושמי נועם יעידו בי	חיתו ארץ עוף פורח
37 יצא חוטר מבית דוד	בקצות אשר יתן ריח
38 שמץ בתהלותי אגיד	לכם בהם אשתבח
39 בי תוכחות מוסר יחכם	תלמיד תלמודו שוכח
40 בי משען אל גבר יחלש	בי בחשך אור זורח
41 כי ילך במים ובטיט	בי יבין דרכו אורח
42 עינים לעור אהיה	אף רגלים לפסח
43 מני לכלים ידים	בי ישוב אלם נובח
44 וירמיני איש בסאון	נגד עם רב ומנצח
45 כי אכרת הן עוד אחליף	מגזעי נצר צומח
46 מרוב חילי טעה בי איש	ישתחוה לי כי בוטח
47 לא יאמין כי לא אל אני	ימחה חטאו הסולח
48 בשאת ארון קדש הנני	ובנחלתו אסתפח
49 עם עולת תמיד שם אני	בעלותה על המזבח
50 אותות בי נעשו על פרעה	שמו איש רהב שוחח
51 לולי ידי הנה לבן	את יעקב ריקם שלח
52 איש הביט בי כוכבי מרום	דלתי שמים פתח
53 ידע תבנית גלגליהם	ונתיבות שמש ירח
54 כי ימדו הנמדדים	וירמיני שחק טפח.

נ 38ב

55 וראוי שתדע כי זה הכלי המצאנוהו תחלה לבאר ממנו אם יש שם גלגל יוצא מרכז כי בו יכולנו לאמת שעור קוטר הירח הנראה בהיותו בכל מרחקיו הארבעה לפי

56 תכונת בטלמיוס. והעירנו על זה מה שראינו ממחלוקת

57 הטבעיים על תכונת בטלמיוס. וכאשר נתאמת לנו בו שאין העניין בזה כמו שהניח בטלמיוס הוצרכנו לחקור על תכונה ימשך ממנה מה שיראה לחוש מהתנועות לגרמים השמימיים

58 באופן יאות חלוף המרחקים למה שנראה בחוש. ובכלל הנה זה הכלי הוא אשר היישירנו השם ית' בו לעמד על האמת בזאת המלאכה כמו שתראה.

59 והוא מבואר ממה שקדם שאם היה ידוע לנו אורך כוכב אחד ורחבו ורחב הכוכב השני הנראה עמו הנה יישירנו זה

39 יחכם] נוסח קארלעבאך: בי יחכם. 48 קדש] נוסח קארדלעבאך: הקודש. 53 ידע] נ: ידעת. 53 ירח] לפי קארלעבאך צ"ל: וירח. 57 לחקור] פ: לחקור בג"ה.

שוה לזוית אחב שהיא זוית קוטר הכוכב ביחס אל העגלה
הנזכרת.
25 והוא מבואר למי שעיין בזאת החכמה כי בשמש וצדק
ונגה לא יהיה הבדל מורגש מאד בין נקדת ח בין מרכז
26 הארץ. ואולם בירח יתבאר ההבדל לקורבתו מהארץ ולזה
יחויב כאשר נדע יחס קוטר הירח ביחס אל זאת העגולה
הנזכרת שנבאר יחס קטרו ביחס אל זאת העגולה שמרכזה
27 מרכז הארץ. וראוי שתדע שכל מה שיהיה המקל יותר
ארוך יעמידנו יותר על אמתת שעור קוטר הכוכב המאיר.
28 הנה זה הוא מה שנעזור בו בזאת החכמה בזה הכלי אשר
המצאנו ללקיחת המבטים והוא נפלא מאד כמו שתראה.
29 והנה קראנו שם זה הכלי מגלה עמוקות כי בו נתאמתו לנו
30 עמוקות רבות בזאת המלאכה בעזר השם ית'. ולהעיר
המעיינים על עוצם מדרגת זה הכלי ולמשוך לבבם אליו
כתבנו עליו שני שירים יעידו על מעלת זה הכלי בזאת
החכמה והם אלו.

[א]

נ 38א 31 להנחיל לאנוש יש   חננו צור שכלו
            בנעמו יחזה בו   יבקר בהיכלו.
       32 ולו הוכן כל כלי   להשכילו בינה
            לדעת כל סתום   בסוד היצור ופעלו.
פ 19א 33 יבוננהו סודו   בכל כוכבי שחק
            יבינהו דרכו   ומרחקו וגדלו.
       34 מלאכת שמים   הליכות מסלולו
            בעצו ישאל איש   ומקלו יגיד לו.

26 ולזה ... הארץ] חסר בכ"י נ.   26 פ: זאת₂ [זאת] חסר
בכ"י ק.   30 יעידו] ק: יעירו.   [א] בשיר²הזה
בנוסח קארלעבאך נמצאות שבע שורות. כאן נמצאות רק
ארבע מהן ולא כסדרן שם: 31-33 מקבילות ל-5-7 שם,
ו-34 מקבילה ל-4 שם.   33 יבינה] נוסח קארלעבאך:
הבינהו.

הנקב ויודע באופן הקודם שעור קוטר הכוכב בעגלה אשר
יסרב בה.   רזה כי שעור רחב הנקב הוא שעור לוח
המבט בזה המקום.
ז 18 פ

10

11   והצד השלישי והוא מיוחד בכוכבים המאירים כמו
12 השמש והירח.   וכבר נעזר בו בנגה וצדק ואם בו מן
הקושי להולשת אורם אבל יהיה זה באישון חשך כשיעדר
אור שאר הכוכבים רזה יהיה בחשך לילה שיכנס אורם מן
13 החלון ולא יתערב שם אור מאיר זולתם.   וזה יהיה
כשנשים אורך המקל כמו ששה עשר זרתות או יותר ונקים
בראשו האחד לוח עומד עליו על זוית נצבה ויהיה שטחו
14 חותך שטח אורך המקל על זוית נצבה.   ויהיה בו נקב
עגול ידוע הקוטר ביחס אל מעלות המקל כאלו תאמר
15 מעלה אחת או ב׳ מעלות.   ויהיה בקצה המקל השני לוח
אחר נכחי ללוח הראשון יראה בו ניצוץ המאיר הנכנס
16 בנקב הלוח הראשון.   וכאשר יודע שעור הניצוץ ההוא
הנה יודע יחס קוטר המאיר אל העגולה שיהיה מרכזה על
שטח הארץ שבו יהיה המבט ויהיה חצי קוטרה במרחק
17 המאיר מהשטח ההוא.   רזה שכאשר נדע יתרון רחב הניצוץ
על רחב הנקב כבר נדע הקשת אשר יגבילה המיתר ההוא
המלא כאשר יושם אורך המקל ששים מעלה והוא יהיה
שיעור קוטר המאיר ביחס אל מקיף העגולה הנזכרת.
18   המשל שיהיה אורך המקל בין שתי הלוחות מאה מעלות
ויהיה רחב הנקב ב׳ מעלות ויהיה רוחב הנצוץ ג׳
ג 37 נ
19 מעלות.   והנה יתרון הרחב הוא מעלה בשעור שבו יהיה
אורך המקל מאה מעלות ובשעור שבו יהיה אורך המקל ס׳
20 מעלות בו יהיה שעור היתרון הזה ל״ו דקים.   והנה
הקשת המלא להם הוא ל״ד דקים וכ״ב שניים ולזה יהיה
קוטר המאיר בשיעור הנזכר ל״ד דקים וכ״ב שניים.
21 המופת שיהיה קוטר המאיר קו אב וקוטר הנקב הנכחי
לו קו גד וקוטר הניצוץ הנכחי להם בלוח השני קו
ק 12א
הז ונוצייא קוי אדז בגה הישרים ויתחתכו על נקודת ח.
22 הנה זוית אחב היא זוית קוטר הכוכב ביחס אל העגולה
שיהיה מרכזה נקדת ח שהיא אצל שטח הארץ וזוית החז
23 שוה לה.   ונבדיל מקו הז קו הט שוה לקו גד ונוציא
24 קו סד הנה קוי הט גד נכחיים ושוים.   ולזה יהיה קו
סד נכחי לקו הח ולזה תהיה זוית סדז שוה לזוית החז
והיא

9 קוטר] ק: קטב.   10 בזה] חסר בכ״י נ。   16 הנה]
חסר בכ״י נ.   24 סדז שוה לזוית] בשולי נ: שוה
לזוי׳ סדז.

נ 36ב

10 על כמה מעלות נפל מהמקל וכמה דקים והוא המרחק.
ונדע גובה ראש זה הלוח על שטח המקל כאלו תאמר
חמישים מעלות או יותר או פחות לפי שיעור
הלוח שיהיה בו המבט והוא אשר נקראהו גובה בזה
הפרק.

11 וכאשר | ידענו הגובה והמרחק הנה נחבר מרובע
הגובה עם מרובע המרחק ונוציא שרש המחובר והוא חצי
12 הקוטר המתוקן. ונחלק עליו מעלות הגובה אחר
שכפלנום בס׳ מעלה והיוצא מהחלוקה הוא נכחות קשת
גובה הכוכב על הארץ וממנו תעמד על קשת גובהו על
13 הארץ והוא המבוקש. והנה ילקח לזה המבט הלוח היותר
נכון לקחת אותו בו לפי גובה הכוכב על הארץ ובכלל
כל מה שילקח המבט במרחק יותר רב יהיה יותר מדוקדק.
14 ובזה האופן נדע מזה הכלי מרחב הכוכב מאופן המישור
15 כשלקחנו גבהו בעברו על קו חצי היום. ובזה האופן
נדע גם כן מזה הכלי השעה שאנחנו בה מהיום או מהלילה
כשנעמוד ממנו על גובה השמש ביום או על גובה כוכב
מהכוכבים בלילה.

הפרק התשיעי

1 וכבר נעזר בזה הכלי לעמוד על שיעור קוטר הכוכב
ביחס אל העגולה אשר יסוב בה וזה אמנם יהיה מג׳
2 צדדין. הצד האחד הוא שנעיין תחלה במרחק זה הכוכב
מכוכב מה מזולת שיכנס בזה המרחק דבר מקוטר הכוכב
3 שנרצה לדעת שעור קטרו. אך נקח המרחק שבין הכוכב
4 והמקיף הסמוך לו מזה הכוכב. וכאשר ישלם לנו זה
נשוב לעיין במרחק זה הכוכב מהכוכב המונח ונכלל בזה
5 המרחק על קוטר הכוכב שנרצה לדעת שעור קטרו. וזה
יהיה בשנקח המרחק שבין הכוכב והמקיף הנכחי לו מזה
6 הכוכב. והנה יתרון זה המרחק על המרחק הראשון הוא
7 שעור קוטר הכוכב ביחס אל העגלה אשר יסוב בה.
והצד השני הוא שיהיה בלוח שיכנס בו המקל נקב
8 נצב הזוירות ידוע השעור ביחס אל מעלות המקל. ויהיה
גובה הנקב על שטח המקל מסכים | לגובה מרכז הראות
עליו ותקריבהו אל הראות עד שימלא קוטר הכוכב רחב
9 הנקב ההוא. ותדע מרחק הלוח בעת המבט ושעור רחב

נ 37א

13 היותר] חסר בכ״י פ.   14–15 מזה הכלי מרחב ...
נדע] חסר בכ״י נ.   1 לעמוד] חסר בכ״י נ.

תחקור תמצא שזה לא יסבב מהטעות בקו מ͏ח כי אם פחות משני אחד.

125 ואם היה קשת המרחק לʹ מעלות הנה לא יהיה בזה
126 גם כן רושם בקירוב המגיע מזה. וזה שכבר יתבאר באופן הקודם שקו זה איננו יותר גדול מקו ז͏ם כי אם כמו דק אחד וזה לא יוסיף על מרובע קו ז͏ם כי אם כמו נ״ג שניים ולו תחקור תמצא שזה לא יסבב מהטעות בקו מ͏ח כי אם פחות משני אחד הוא דבר בלתי
127 מזיק במבטים. ולזה יהיה הנאות להשתמש במבטים בזה
128 הדרך האחרון לקלותו. ובכאן נשלם הבאור באופן ההשתמשות בזה הכלי בזה האופן.

הפרק השמיני

1 ‎‏נ 36א וראוי שתדע שכבר נעמד מזה הכלי ‎| בדקדוק נפלא
   ‎‏פ 18א על גובה השמש והירח או אי זה כוכב שיהיה ‎| בעברו בקו חצי היום וזה מועיל מאד בחקירה הזאת אשר
2 אנחנו בדרכה כמו שיתבאר לך ממה שיבא בג״ה. וזה
3 אמנם יהיה כשנעשה למקל רגלים קיימות על הארץ שתים באמצעו ושתים בראשו אצל הראות. ונשים בראשו שאצל הראות דף נקוב נקב קטן סמוך לשטח המקל וממקום הנקב נתחיל לכתוב מספר מעלות המקל לא ממרכז הראות.
4 וירושם המקל הזה נכחי לאופק ביושר קו חצי היום והנה נבאר במה שיבא איך נמצא קו חצי היום.
5 ויהיו לנו לוחות רבות נקובות בקצה האחד מהם באופן שכאשר נכניסם במקל יעמדו עליו על זויות
6 נצבות בתכלית מה שאפשר מן הדקדוק. ונשים אורך הלוח האחד ששים מעלות או יותר ואורך הלוח השני מʹ מעלות ואורך הלוח השלישי לʹ מעלות ואורך הלוח הרביעי כʹ מעלות ואורך הלוח החמישי ט״ו מעלות ואורך הלוח הששי יʹ מעלות. והנה יתבאר לנו שהמקל
7 ‎‏ק 11ב הוא נכחי לשטח ‎| האופק כשהכנסנו המקל בלוח והושם האנך על אמצע ראש הלוח ויפול אבן העופרת על אמצע הלוח.
8 וכאשר ישלם לנו זה הנה נקריב הלוח אל הראות
9 ונרחיקהו עד שנראה הכוכב מהנקב כאלו הוא ממשש לראש הלוח. ונדקדק תמיד שיהיה הלוח נצב על המקל ונראה

127 הנאות] פ: נאות.   1 בעברו] נ: בעבור.
3 בראשו] נ: בראש.   8 הראות] ק: הראיה.

110 שוה לזורית טהמ. ולפי שמשלש הגח הוא שוה השוקים
111 הנה תהיינה זוריות טהח החל שורת. ולפי ששתי זוריות
    טהח טהמ הם שורת לשני נצבות הנה זוריות טמה החל
112 שורת להם כל אחת לגילה שורת לשתי נצבות. ולפי שכבר
    עמדו על קו מח שני קוי מט חל והיו הזוריות הפנימיות
    שחודשו שורת לשתי נצבות הנה שני קוי מט חל הם נכחיים.
113 ולפי ששני קוי מח טל אשר ביניהם הם גם כן נכחיים
    הנה הם גם כן שוים והוא מבואר שכאשר גרענו ממרובע
    קו זח מרובע קו זמ יהיה יסוד הנשאר שוה לקו מח.
114 ולזה יהיה יסוד הנשאר שוה לקו טל שהוא שוה לקו מח
115 כמו שבארנו. והנה יתבאר מזה שאם גרענו ממרובע קו
    זח מרובע קו זמ כמו שזכרנו היה יסוד הנשאר שוה לקו
116 טל. אלא שאנחנו לקחנו קו זה תמורת קו זמ ולזה
117 יהיה בזה קירוב מעט. אלא שכבר נבאר שזה הקירוב
    בלתי מזיק במבטים באופן שהתחנינו בענין ההשתמשות
    בזה הכלי במה שיבא והוא שיהיו הכוכבים בלתי רחוקים
    מאזור גלגל המזלות כי אם כמו שבע מעלות ושלא יהיה
    המרחק ביניהם יותר משלשים מעלה ולא פחות מעשר
    מעלות.
118   והנה יתבאר שאע"פ שיהיה מרחב האחד מהכוכבים
    ז' מעלות ולאחר אין מרחב כלל לא יזיק זה הקירוב
    במבט וזה שחץ ז' מעלות הוא פחות מכ"ז דקים
119 רנניח שיהיה המיתר המלא המתוקן עשר מעלות לבד ר"ל
    קו | אב בזה המשל ונשאיר הצורה על ענינה ונוציא
120 מנקדת ג עמוד על קו אב והוא קו גב. ולפי שחץ הצי
    קשת אב הוא פחות מי"ד דקים הנה יהיה קו גב יותר
121 מנ"ט מעלות ומ"ר דקים וקו אג הוא ס' מעלה. וגם כן
    הנה מפני שקוי חמ אב הם נכחיים ועמד ביניהם קו אה
122 הנה הוא ישים הזוריות המומרות שורת. ולזה תהיה
    זורית זהמ שוה לזורית גאן וזורית זמה הנצבת שוה
    לזורית גנא הנצבת ותשאר זורית מזה הנשארת ממשלש זהמ
123 שוה לזורית אגן הנשארת ממשלש גאן. ולזה הוא מבואר
    שמשולשי זהמ גאן הם מתדמים ולזה יהיה יחס קו זמ
124 אל קו זה כיחס קו גב אל קו גא. ומזה יתבאר שאין
    קו זה יותר ארוך מקו זמ כי אם פחות מז' שניים וזה
    לא יוסיף על מרובע קו זמ כי אם פחות מו' שניים ולו

---

117 גלגל] חסר בכ"י ק.   121 אה] נ: אג.
122 לזורית גנא] פ: לזורית גבא.   124 מז'] נוסף
בשולי נ: נ' מו' שניים וי"ח שלישיים.

94 לשניהם. והוא שכבר תדע המיתר השני באופן הקודם
   ואחר כן חגרע ממרובעו מרובע יתרון חץ המרחב הרב על
95 המעט. ותוציא יסוד הנשאר והוא המיתר השני המתוקן
96 כאלו היו הרחבים שוים. הוסף חצי יתרון חץ הרב על
   חץ המעט על שארית חץ הרב מס׳ מעלה אם היה לשניהם
97 נ 34ב מרחב והנשאר הוא שארית החץ המתוקן. או אם לא היה
   מרחב כי אם לאחד מהם הוסף חצי חץ בעל המרחב על
   שארית החץ מס׳ מעלה והעולה הוא שארית החץ המתוקן.
98 ערוך המיתר השני המתוקן על ס׳ מעלה וחלק על שארית
   החץ המתוקן והנה היוצא מהחלוקה הוא בקירוב המיתר
   המלא לקשת המרחק וממנו תעמד בקרוב על קשת המרחק.
99 מופת זה שאנחנו נשים קשת המרחק בזאת הצורה קשת
100 אב על מרכז ג ונוציא קוי אג גב. ונדע המיתר השני
    באופן הקודם ונבאר שזה הדרך שזכרנו יוביל אל האמת
101 בקרוב בלתי מזיק במבטים. וזה יתבאר אחר שנבאר שאם
    הוצאנו משתי קצות המיתר השני במשלש אגב קוים נכחיים
    למיתר קשת המרחק שהוא קו אב והוצאנו עמוד מהאחד אל
102 השני. הנה כשגרענו ממרובע המיתר השני מרובע העמוד
    והוצאנו יסוד הנשאר הנה הוא שוה אל הקו הנכחי
103 היוצא ממקום חצי יתרון החץ הגדול על הקטן. המופת
    שאנחנו נשים המיתר השני בזאת הצורה קו זח ונניח
104 שיהיה קו אז יותר קצר מקו בח. ונוציא משתי קצות
    קו זח במשלש אגב קוים נכוחיים לקו אב והם קוי זד
105 חה ויהיה קו זה הוא יתרון קו בח על קו אז. ונחלק
   פ 17ב קו זה | לחצָיין על נקדת ט ונוציא מנקודת ט במשלש
106 אגב קו טל נכחי לקו אב. ונוציא מנקודת ז עמוד על
    קו חה המונח בלי תכלית והוא קו זמ ונדביק קו חהמ
107 ק 11א הישר. ונוציא קו טמ ויתבאר שקו טמ שוה לקו טה לפי
   נ 35א שמשלש זמה הוא נצב הזוית וקו ז הוא קוטר | הזרית
108 הנצבה. וחולק לחצָיין בנקודת ט הנה אם כן תהיה
109 נקדת ט היא מרכז העגלה המקפת במשלש זמה. ולזה
    יתבאר שקו טמ שוה לקו טה ולזה תהיה זוית טמה

96 והנשאר] נוסף בכ"י נ (למעלה): נ׳ והעולה.
97 וחלק] נוסף בשולי נ: העולה.   99 גב] פ: גד.
100 אל] נ: כל.   105 נקדת ט] ק: נקדת ד.
106 חהמ] ק: זהמ.

הקודם שזוית דטה היא נצבת ולזה הוא מבואר שכאשר
חוסר מרובע קו דה שהוא המיתר המתוקן מרובע קו טד
היה יסוד הנשאר שוה לקו טה שהוא שוה לקו זח שהוא
82 המיתר השני. ומזה המקום התבאר שאם היו הרחבים
שוים והם לצד אחד שהמיתר המתוקן הוא שוה למיתר
פ 17א 83 השני. וכאשר | התישב שהמיתר השני והוא קו זח ידוע
ולכבר התבאר ששני קוי חג אג הם ידועים הנה תודענה
כל זויות משלש חגז ותודע מזה זוית חגב שהיא מגבלת
קשת המרחק שבין שני אלו הכוכבים.
84 וכבר יודע זה באופן אחר והוא כי מפני שכל קוי
משלש חגז הם ידועים הנה שעור העמוד הנופל מנקדת ח
אל קו אג המונח בלי תכלית הוא ידוע והוא קו חל.
85 ונוציא מנקדת ב אל קו אג המונח בלי תכלית קו במ
נכחי לקו חל והוא מבואר שקו במ הוא נכחות קשת
86 המרחק בין אלו הכוכבים. והוא מבואר כי יחס קו גח
נ 34א אל חל הוא כיחס | קו גב אל קו במ לפי ששני משולשי
87 חגל בגמ הם מתדמים. ולזה הוא מבואר ששטח השני
88 בשלישי הוא שוה לשטח הראשון ברביעי. ולזה הוא
מבואר שאם הכינו קו חל בקו בג וחלקנו אותו על קו
חג יהיה היוצא מהחלוקה שוה לקו במ שהוא נכחות קשת
המרחק בין אלו הכוכבים.
89 וכאשר היו הרחבים שוים והיו לצד אחד הנה יתבאר
מזה הביאור שהמיתר המתוקן הוא שוה למיתר השני ולזה
90 יודע קו חז שהוא המיתר השני. ונוציא קו אב ולפי
שקו אז שוה לקו חב וקו אג שוה לקו גב הוא מבואר
שכבר חולקו שני קוי אג גב על יחס אחד בקו זח.
91 ולזה יהיו קוי אב זח נכחיים ויהיו משלשי זגח אגב
מתדמים ולזה יהיה יחס קו זג אל קו זח כיחס קו אג
92 אל קו אב. ולזה הוא מבואר בהנהגה הקודמת שאם
הכינו קו זח בקו אג וחלקנו העולה על קו זג היה
היוצא מהחלוקה שוה לקו אב שהוא המיתר המלא לקשת
המרחק שבין אלו הכוכבים מעגלת המזלות.
93 וכבר תוכל לעמד בקלות בקירוב בלתי מזיק
במבטים על המיתר המלא לקשת המרחק כשהיה לאחד מהם
או לשניהם מרחב והיו הרחבים בלתי שוים לצד אחד או

84 אג] נ : זג. 85 אג] נ : זג. 91 זח] נוסף בכ"י
נ, ופ (וסימנים בשניהם להשמיט את המלים האלה):
כיחס קו אג אל קו זח. 93 המרחק] חסר בכ"י ק.
93 אחד או לשניהם] ק: אחד או לשני צדדים.

לצד אחד והיו שוים הנה המיתר המתוקן הוא המיתר השני ותנהג בו באופן הקודם להוציא נכחות קשת המרחק שבין
65 שני הכוכבים. ואם תכפל המיתר השני על ששים ותחלק העולה על שארית חץ קשת המרחב ממשים הנה יהיה העולה הוא המיתר לקשת המרחק הוצא קשתו המלא והוא המבוקש.
66 והנה נביא לך משל על כל אחד מאלו הדרכים.
67 נניח שיהיה שטח עגולת המזלות עגלת אב על מרכז נקדת ג ויהיה קשת המרחק שבין שני אלו הכוכבים בארוך קשת
68 אב. ויהיה קשת מרחב הכוכב האחר קשת אד וקשת מרחב הכוכב השני קשת בה ותהיה קשת אד לפאת צפון וקשת בח לפאת דרום והנה קו דה הוא המיתר המתוקן.
69 ונוציא קוי אג גב ויהיה קו דז נכחות קשת אד ויהיה קו הח נכחות קשת בה והם ידועים לפי שקשתות המרחב
70 ידועות. ובזה הוא מבואר שחצי אלו הקשתות הם ידועים והם קוי אז בח וישארו קוי גז גח ידועים כי הם שלמות
71 ס' מעלה מכל אחד מהחצים הידועים. ונוציא קו זח והוא המיתר השני ונוציא קו דז לפאת דרום בשעור קו חה
72 והוא קו זט. הנה קו דזט נכחי לקו חה ולזה יהיה קו
73 זט נכחי ושוה לקו חה. ונוציא קו טה ויתבאר ששני קוי זח טה הם נכחיים ושוים כי הם בין שני קוי טז
74 חה שהם נכחיים ושוים. והנה זוית דזח היא נצבת כי קו דז הוא עומד על שטח עגלת אבגד ולזה תהיה זוית
75 דטה נצבת כי קו טה הוא נכחי לקו זח. ולזה הוא מבואר שכאשר חוסר מרובע קו דט ממרובע קו דה | שהוא המיתר המתוקן היה יסוד הנשאר שוה לקו טה ולזה יהיה שעור קו טה ידוע. ומפני זה יהיה שעור קו זח ידוע
76 שהוא המיתר השני כי הוא שוה לקו טה.
77 ואם היו הרחבים לצד אחד והיו בלתי שוים הנה יתבאר בכמו זה הבאור שכבר יובילנו הדרך שזכרנו אל
78 ידיעת המיתר השני. וזה שכבר נניח התמונה על עניינה אלא שאנחנו | נשים שני קוי זד חה לצד אחד ונוציא קו זח שהוא המיתר השני ונניח שיהיה קו זד יותר
79 ארוך מקו חה. ונבדיל ממנו קו זט שוה לקו חה ונוציא
80 קו טה. ולפי ששני קוי זט חה הם נכחיים ושוים הנה
81 שני קוי זח טה הם נכחיים ושוים. ויתבאר באופן

ג 33ב

ק 10ב

65 והוא המבוקש] פ, נ: והנה המבוקש. 71 נוסף בכ"י פ (למעלה), ובשולי נ: דרום] פ, ק, נ: צפון. 75 לקו טה] פ, ק: למרובע טה.

50 גם כן ידוע וישאר שעור קו חה ידוע כי הוא שלמות חצי
הקוטר. ונוציא קו אז והוא המיתר המתוקן למרחק אשר
51 בין אלו הכוכבים ונוציא קו אח. והוא מבואר ממה
שקדם שזוית אחז היא נצבת כי קו זח הוא עמוד על שטח
עגלת אבגד ולזה יהיה קו אז כחיי על שני מרובעי
52 קוי אח זח. וכשגרענו ממרובע קו אז הידוע מרובע קו
זח הידוע היה הנשאר שוה למרובע קו אח ולזה יהיה קו
53 אח ידוע והוא המיתר השני. ולפי שכל קוי משלש אחה
ידועים הנה העמוד | הנופל מנקדת א אל קו חה המונח
בלי תכלית ידוע לפי מה שקדם ויהיה העמוד ההוא קו
54 אט. הנה אם כן שעור קו אט הוא ידוע והוא נכחות
קשת המרחק שבין שני הכוכבים וממנו תעמוד על קשת
55 המרחק והוא המבוקש. וכבר יתבאר זה באופן אחר והוא
שזויות משלש אחה הם ידועות לפי שכל קויו הם ידועים.
56 ולזה תודע זוית אחה המגבלת קשת המרחק בין שני אלו
הכוכבים.
57 ואם היה לשני הכוכבים מרחב הנה אם היה לשני
צדדים הדרך בזה הוא שתקבץ נכחויות קשתות המרחב
והמקובץ שמור ואם היו הרחבים לצד אחד תקח ההבדל
שבין שתי הנכחויות ושמור. הוצא מרובע השמור איך
58 שהיה משני אלו האופנים ממרובע המיתר המתוקן
59 והנשאר הוצא את יסודו | והוא המיתר השני. וכאשר
היה זה כן הנה בידך משלש שלשה קויו ידועים הקו
האחד הוא המיתר השני ושני הקוים הנשארים הם
שאריות חצי קשתות המרחב לאלו הכוכבים כשגרענו
60 אותם משששים. ולזה תודע מזה הזוית המגבלת קשת
המרחק בין שני אלו הכוכבים מעגלת המזלות באורך.
61 או תנהג בזה באופן אחר והוא שתדע שעור עמוד
62 הנופל מראש המשלש אל תושבתו. ונשים תושבת המשולש
אחד מהקוים היוצאים ממרכז עגולת המזלות אל מקום
63 נפילת נכחות קשת המרחב לאחד מהכוכבים. וכאשר
תדע שעור העמוד ההוא הערך אותו על ששים ותחלק על
הקו היוצא ממרכז עגולת המזלות אל מקום נפילת
הנכחות לקשת מרחב הכוכב השני שלא הושם תושבת
במשלש והעולה הוא נכחות קשת | המרחק בין הכוכבים
64 וממנו תעמוד על קשת המרחק. ואם היו שני הרחבים

52 למרובע קו] נ: לקו.   59 הנשארים] חסר בכ"י פ.
59 שאריות] נ: שארית.

40 אבגד. ונאמר שקו זח נכחי לקו טה וזה כי זרית זחה
היא נצבת כי קו הח יצא ממרכז העגולה וחלק מיתר כפל
קשת זד לחצאין רזורית טהח היא גם כן נצבת כי קו טה

41 הוא עמוד על שטח עגלת אבגד. ולזה יתבאר כי הוא
נצב על כל קו יצא מנקדת ה בשטח עגולת אבגד ולזה
הוא מבואר כי הוא עומד על זרית נצבה על קו הד ולפי
שקו זח עמוד גם כן על זרית נצבה על קו הד והיו שני
קוי זח הט בשטח אחד והוא שטח עגולת דזט הנה שני

42 קוי זח הט הם נכחיים. ולפי שקו הט הוא עמוד על
שטח עגולת אבגד הנה יהיה קו זח הנכחיי לו עמוד על
שטח עגלת אבגד וכזה יתבאר שכל הנכחיות אשר יהיו
לקשתות המרחב מעגלת המזלות הם עמודים על שטח עגולת
אבגד והם נכחיים זה לזה כי כלם הם נכחיים לקו הט.

נ 32א 43 ואחר | שהתישב זה הנה נבאר איך יודע מזה המבט
קשת המרחק שבין שני הכוכבים באורך מגלגל המזלות

44 כשהיה מרחב הכוכבים ידוע. והנה הדרך בזה אם היה
לכוכב האחד לבד מרחב שתגרע מרובע הנכחות לקשת
המרחב ממרובע המיתר המתוקן והנשאר אחר הגרעון
תוציא את שרשו והוא יהיה המיתר השני וככה נקראהו

45 במה שיבא מזה הפרק. וכאשר היה זה כן הנה בידך משלש
שלשת קוים ידועים הקו האחד הוא חצי הקוטר והקו השני
הוא המיתר השני והתושבת הוא שארית חץ קשת מרחב

46 הכוכב. כשגרענו אותו משים מעלות ובהיות העניין כן
כן הנה יהיה העמוד הנופל מראש המשלש אל החושבת ידוע
והוא נכחות קשת המרחק הוצא קשתו והנה המבוקש.

47 המשל שיהיה שטח עגולת המזלות הישר העובר על

48 מרכז ה עגולת אבגד על מרכז נקדת ה. ותהיה נקדת
א מקום הכוכב שאין לו מרחב ותהיה נקדת ב מקום

ק 10א הכוכב השני | באורך מגלגל המזלות אלא שיש לו מרחב

49 והוא קשת בז והנה הכוכב בנקדת ז. ונוציא קוי הא הב
הישרים ונוציא מנקדת ז עמוד על קו הב והוא קו זח
הנה קו זח הוא נכחות קשת בז וקו חב הוא חצר והוא

---

41 ולפי ... על קו] נ: ויהיו שני קוי (סימנים בכ"י
לצייך את המלים האלה שהן מיותרות.)   42 אבגד הנה]
פ: אבגד.   45 בידך] ק: בדרך.   46 המרחק] נוסף בשולי
נ: הארך.

29 והנה אם היו שני אלו הכוכבים בעצם אזור גלגל המזלות
הנה יהיה זה הקשת הוא המרחק אשר ביניהם מעגולת גלגל
המזלות וכן אם היו שניהם במקום אחד באורך מגלגל
המזלות הנה יהיה זה הקשת קשת המרחק ר"ל שזה יהיה
30 קשת המרחב בין שני אלו הכוכבים. והנה אם היה האחד
בעצם אזור גלגל המזלות הנה יהיה מרחב הכוכב השני
מגלגל המזלות זאת הקשת הנזכרת לצד אשר הוא בו
מהמזלות ר"ל שאם היה הכוכב צפוני מגלגל המזלות הנה
יהיה זה המרחב לפאת צפון ואם היה דרומי יהיה זה
31 המרחב לפאת דרום. ואם היה לאחד מאלו הכוכבים מרחב
ידוע והיו במקום אחד באורך מגלגל המזלות הנה תהיה
זאת הקשת הנזכרת יתרון המרחב אשר לכוכב האחר עליו.
32 אם היו שניהם לצד אחד ר"ל אם היה מרחב הכוכב השני
מהכוכב הראשון אשר מרחבו ידוע לצד מרחב הכוכב
הראשון חבר שני הרחבים והנה רוחב הכוכב השני הבלתי
ידוע המרחב או יחוסר המרחב המועט מהרב אם היו לשני
צדדים והנשאר יהיה מרחב הכוכב אשר לא היה ידוע המרחב
33 לצד המרחב הרב. ואם לא היו שניהם במקום אחד מגלגל
המזלות באורך והיה לאחד מהם או לשניהם מרחב באי זה
אופן שיהיה הנה נבאר לך שאם היה המרחב ידוע הנה
יהיה המרחק אשר ביניהם באורך מגלגל המזלות ידוע.
34 ולבאור זה אנחנו מציעים תחלה הצעת אחת והיא
נ 31ב שנכחות קשת | מרחב הכוכב מאופן המזלות הוא עמוד על
שטח גלגל המזלות כשיוצא התחלת הנכחות ממקום הכוכב.
35 המופת שאנחנו נשים שטח עגולת המזלות שטח עגולת אבגד
הישר ויהיה מרכזה נקודת ה והוא מבואר ששטח עגולת
36 הרחב עוברת על קטביה. ונשים קשת דג חלק מעגולת
הרוחב ויהיה נכחותה הבא מנקודת ז קו זח הישר.
פ 16א 37 ונוציא קו החד והוא מבואר שקו החד הוא | קו אחד לפי
שהקו שיצא מהמרכז אל חצי המיתר יחתך הקשת לחצאין.
38 ונשים קשת דזט רובע עגולה ולזה תהיה נקודת ט קוטב
39 עגולת אבגד. ונוציא קו טה והוא מבואר שקו טה הוא
עמוד על שטח עגולת אבגד לפי שעגולת דזט עוברת על קטבי

30 הנה יהיה ... השני מגלגל המזלות] חסר בכ"י נ ;
בשולי נ: נ' יהיה המרחב.   30 ואם היה דרומי ...
דרום] חסר בכ"י פ.   32 לצד האחד] נוסף בשולי נ:
על האחד או לכוכב האחר עליו.   32 יחוסר] בשולי נ:
נ' חסר.   33 הנה נבאר] נ: נבאר.   35 מרכזה] בשולי נ:
המרכז.   35 קטביה] נוסף בשולי נ: נ' קטבי המזלות.

מעלות ומשמונה מעלות ומד׳ מעלות ומב׳ מעלות ויהיה
רחבה בצד האחד מעלה אחת ובצד השני חצי מעלה או רביע
מעלה או פחות כדי שנוכל להביט בו הכוכבים שהם
קרובים מאד זה לזה כמה מרחק אחד מהם מהאחד באורך
18 או ברוחב. ויהיו אלו הלוחות כלם ישרות **השטחים**
ויהיה נצב חשטח האחר מהם על האחר.
19 וכשנרצה לעיין בזה הכלי בשני כוכבים לדעת כמה
מרחק | האחד מהם מהאחר נקח אחת מאלו הלוחות אשר
תאות יותר לדעת בה זה המרחק ר״ל שאם היה המרחק
עשרים וחמש מעלות ולמעלה נקח הלוח היותר גדול.
20 וכן על זה האופן נשתדל תמיד שיקרב הלוח מה שאפשר
21 לסוף המקל. ונכניס המקל בלוח בדרך שיחתכהו על
זויות נצבות ונשים היתדות אשר בראש המקל אצל הראות
ממאקי העין המבטה ונעצום העין השנית כדי שלא תתבלבל
22 המבט. ונקריב אלו היתדות אל הראות כל מה שנוכל
ואם נקריב הלוח אל הראות או נרחיקהו עד שנראה מראש
שטח הלוח העליון מהצד האחד הכוכב האחד ומהצד השני
הכוכב השני באופן שימששו שני הכוכבים זה הלוח
23 משני קצותיו לפי המבט. ולפי שזה לא יתכן שישלם אם
לא יראו קצות הלוח הראות מבואר הנה נשים הנר
מאחורינו בדרך שישים שטח הלוח הנראה ולא ימנע אורו
24 ראותנו מראיית הכוכבים. וכאשר ישלם זה נכתב מאי
זה לוח נהיה המבט ונראה בכמה מעלות יפול הלוח בעת
המבט ממעלות המקל ובכמה דקים באי זה חלק מהדק.
25 וזה נקראהו מרחק כאלו תאמר כי כבר הובטו אלו
הכוכבים מלוח עשר מעלות מעלות מ׳ מעלות בעת הפלוני.
26 וכאשר תרצה לדעת מזה המבט כמה מעלות יש מהכוכב
האחד עד השני ממעלות העגולה הגדולה | העוברת על
שניהם הנה תחבר מרובע המרחק אל מרובע חצי הלוח
ותוציא שרש המחובר והוא יהיה חצי הקוטר המתוקן.
27 כפל מעלות הלוח בששים מעלות וחלק העולה על חצי
הקוטר המתוקן והעולה יהיה המיתר המתוקן וככה |
28 נקרא שמו במה שנזכר מזה הכלי. הוצא קשתו המלא
בלוח הקשתות והמיתרים והוא יהיה מרחק הכוכב האחד
מהכוכב השני מהעגולה הגדולה העוברת על שניהם.

19 ולמעלה] חסר בכ״י נ. 20 לסוף] פ: לראש.
22 היתדות אל] חסר בכ״י נ. 26 המרחק אל מרובע]
חסר בכ״י נ. 26 לוח] נוסף בשולי נ: עם מרובע
מעלות המרחק. 27 והעולה] חסר בכ״י נ.

6 המבט רחוק מהשטח הממשש לשטח הלוח הדבק לעין לפנים
לצד ראש המביט כמו חלק אחד מעשרים חלקים בזרת ברב
האנשים לפי מה שבחננו אחר טורח ועמל רב. ונחלק
7 המקל למעלות גדולות באופן שתהיינה שמנה מעלות בזרת
ונרשמם שם ברחב השטח הישר מקצה אל קצה. ותהיה
8 התחלת מעלות המקל ממקום נקדת מרכז הראות אשר היא
חוץ מהמקל הזה כמו חלק מעשרים בזרת. ולזה תהיה
המעלה הראשונה שאצל הראות באופן שכאשר יחובר לה
חלק מעשרים בזרת תהיה שוה לשאר המעלות ונכתבם שם
9 בזה האופן. עוד נחלק כל מעלה מאלו לששה חלקים שוים
10 מהצד האחד ומהצד השני נחלקם לי״ב חלקים שוים. עוד
נוציא קו באלכסון מראשית קו המעלה אל סוף החלק
הראשון מהחלקים שחולקה בהם המעלה לי״ב חלקים שוים.
11 ונוציא עוד קו באלכסון מסוף זה החלק הראשון שזכרנו
אל סוף החלק הראשון מהחלקים שחולקה בהם המעלה לששה
חלקים שוים ומסוף זה החלק אל סוף החלק השלישי
מהחלקים שחולקה בהם המעלה לי״ב חלקים ומסוף זה החלק
השלישי אל סוף החלק השני מהחלקים שהם שתות מעלה
ומסוף זה החלק השני אל | סוף החלק החמשי מהחלקים א30 ב
שהם חצי שתות מעלה וכן ען זה האופן בכל מעלה ממעלות
12 המקל עד שתשלם חלוקת על המעלות בזה האופן. והנה
יגביל כל קו מאלו הקוים האלכסוניים חצי שתות המעלה
13 שהוא חמשה דקים. ותוכל לעמוד באומד על החלק הנחתך
ממנו כי אם הוא חמישיתו הוא דק אחד ואם רביעיתו הוא
דק אחד וט״ו שניים ואם הוא שלישיתו הוא דק אחד ומ׳
שניים ואם הוא מחציתו הוא ב׳ דקים וחצי עד שבזה
האופן תוכל לעמוד על הדקים והשניים מהמעלג בקירוב
14 נפלא. ואם תחלק לחמשה חלקים רוחב שטח המקל המחולק
בארכו למעלות הנה יחלקו בזה כל הקוים האלכסוניים
לחמשה חלקי׳ שוים יגביל כל א׳ חלק מהם דק אחד
ממעלות המקל.
15 עוד נעשה לוחות רבות יהיה בכל אחת מהם נקב
עגול באמצעה באופן שיכנס בו המקל בדוחק ונוכל
16 להניע הלוח סביבו לאי זה צד שנרצה. ויהיה שם לוח
יהיה שיעורו כ״ד מעלות משיעור מעלות המקל ויהיה
שטחו העליון גבוה על המקל כגובה מרכז הראות על
17 המקל. ובזאת הצורה בעינה יהיה שם לוח שש עשרה

8 בזה האופן] ב: באופן הזה.   16 משיעור מעלות] ב: ממעלות.

נ 29א 42 כן קו גד על זוית נצבה ולזה הוא מבואר שקו חא הוא
על יושר קו | חל. ולפי שמשלש גדל היה בתוכו קו הז
נכחי לתושבת קו גד הנה יהיה יחס קו הל אל קו גל
43 כיחס קו הז אל קו גד. ובזה יתבאר שיחס קו הט אל
44 קו גח הוא כיחס קו לט אל קו לח. וכאשר המירונו
והבדלנו הוא מבואר שיחס קו הט אל קו לט הוא כיחס
יתרון קו גח על קו חט אל יתרון קו לח על קו לט
45 שהוא קו חט. ואולם יתרון קו גח על קו הט ידוע
ושעור קו חט הוא גם כן ידוע ושעור קו הט ידוע.
46 וישאר שעור קו לט ידוע כי יחס שעור קו הט הידוע
אליו ידוע.
47 ואנחנו כאשר עיננו בזה האופן והפלגנו
החקירה בו בכל הפנים שאפשר מצאנו מקום נקודת ל בזה
48 המשל במרכז העין ר"ל באמצע הלחות הכפורית. ואולם
הוצרכנו לזאת החקירה כי בלעדיה לא נוכל לאמת שעור
זוית המבט בלא טעות כשנביט בזה הכלי בשני כוכבים
49 ונרצה שיודיע לנו כמה קשת מרחק האחד מהשני. וזה
שאם הנחנו הנקודה הזאת במה שבין שתי נקודות ל וט
נחשב המרחק יותר גדול ממה שהוא כי הזוית תהיה שם
פ 15א 50 יותר גדולה ר"ל זוית | המבט. והפך זה יהיה אם
הנחנו הנקודה ההיא חוץ מקו טל כי אז נחשב המרחק
יותר קטן שהוא כי הזוית תהיה שם יותר קטנה.

הפרק השביעי

1 ואחר שכבר הצענו מה שכבר ראוי להציע לבאור זה
הכלי אשר המצאנוהו ללקיחת המבטים בכל עת בתכלית מה
2 שאפשר מהדקדוק. הנה נזכר תחלה אופן מעשה זה הכלי
3 ואחר נזכר אופן החשתמשות בו. וזה מעשהו נקח מקל
ישר ארוך כמו שש זרתות ונעשה בו שטח ישר ורחב כמו
נ 29ב 4 אצבע בכל אורך המקל. ונשים בראש | המקל האחד לוח
ק 9א קטן יהיה בו מקום חקוק באופן | שיצאו ממנו שתי
יתדות רחוקות זו מזו יותר מאצבע מעט באופן שתושם
היתד האחת במאק אחד ממאקי העין המבטת והיתד השנית
5 תפול במאק השני מהמנין ההיא באופן שלא תעיק הראות.
וכאשר יהיה הענין על זה התאר הנה יקרה שיהיה מרכז

49–50 נחשב ... כי אז] חסר בכ"י ב.   4 והיתד
השנית] נ, ק: והיתד השני.

24 וכאשר ישלם לנו זה והפלגנו הדקדוק בו הנה נוכל
לדעת בקלות מקום הנקודה אשר ממנה יהיה זרית המבט.
25 ב14 פ וזה שאלו הלוחות | הם נכחיות ונצבות על המקל
והקוים הנכוחיים יחתכו קוי המשלש על יחסם והקו
26 הנופל מראש המשלש אל תושבתו. ולפי שמרחק אלו
הלוחות קצתם מקצתם ידוע ויחס זה אל זה ידוע הנה מקום
27 נקודת המבט ידוע. וזה כי יחס הקו המגיע ממנו אל
הלוח הקטן אל הקו המגיע ממנו אל הלוח הגדול כיחס
28 הלוח הקטן אל הלוח הגדול. וכאשר המירונו והבדלנו
והפכנו הנה יהיה יחס הלוח הקטן אל הקו המגיע ממרכז
המבט אליו כיחס יתרון הלוח הגדול על הקטן אל יתרון
28ב ג מרחקו ממרכז | המבט על מרחק הלוח הקטן ממרכז המבט.
29 ולפי שזה היחס השני הוא ידוע בהוש הנה ידוע מזה יחס
30 הלוח הקטן אל מרחקו ממרכז המבט. ואולם שעור הלוח
הקטן הוא ידוע ולזה יהיה מרחקו ממרכז המבט ידוע.
31 המשל שיהיה במקל קו ישר בארך בשטה משטחיו אשר
32 עליו יהיה המבט והוא קו אב. ויהיו בלוחות הנכחיות
מצד שטחם אשר אצל הראות שיחתך קו אב המורנח קוי גד
הז ויהיה קו גד יותר גדול מקו הז ויהיה קו הז יותר
33 קרוב אל הראות. ויהיה באופן שיסתיר הלוח הגדול
כמו שקדם ויחתך קו אב בנקודת ט ואולם קו גד יחתך
34 קו אב בנקודת ח. ולפי שמזרית אחת הובטו שני קוי
גד הז בניצוצים ישרים הוא מבואר שימצא שאם נוציא
קוי גה דז יגישו במקום נקודת זרית המבט ויפגשו
35 בנקדת ל דרך משל. ונדביק שתי נקודות אׄ וׄל בקו ישר
והוא מבואר שקו באׄל הוא קו אחד ישר כי כבר הונח
36 מרכז העין על יושר קו בא. ועוד כי מפני שקו גׄח שוה
לקו חד וקו הׄט שוה לקו טׄז וקו חׄט הוא משותף וזוית
37 דׄחט שוה לזוית גׄחט וזוית חׄטז שוה לזוית חׄטה. הוא
מבואר שאם הרכבנו תמונת דׄט על תמונת גׄט ותתדבק
עליה בשווי כי תפול נקודת דׄ על נקודת גׄ ונקודת זׄ
38 על נקודת הׄ. ולזה הוא מבואר שזרית חׄדז שוה לזרית
39 חׄגה. ולפי ששתי הזריות שאצל התושבת ממשלש לׄגד הם
40 שוות הוא מבואר שמשלש לׄגד הוא שוה השוקים. ולזה
יהיה הקו המגיע מנקודת לׄ אל חצי התושבת שהוא בנקודת
41 חׄ חותך קו גד על זרית נצבה. ואולם קו חׄא חותך גם

24 מקום] חסר בכ״י נׄ. 28 המגיע ממרכז] נׄ: המגיע
ממנו. 32 המונח] נוסף בכ״י פׄ: אצל. 34 יגישו]
קׄ: יפגשו. 37 כי] חסר בכ״י נׄ.

9 הנראים בבכח העין באופן הרשם תמונתם במראה המלוטשת.
ולזה יראה שכאשר היה העין באופן בלתי נאות להרשם
10 בו התמונות שהוא לא יראה דבר. וזה ממה שיתבאר ממנו
בלא ספק שמרכז הראות הוא בפנים בעין לא בשטח המקיף
בו מחוץ.
11 וכבר יתאמת גם כן שאין מרכז הראות במקיף העין
החצוני ממה שנראה במלאכת הרפואה מהפסד הראות בסבת
המים אשר בעין כאשר היו באופן שיתחדש כיס סביבם על
12 הלחות הכפורית. וכאשר הגיעו הכיס ההוא לצד אחד
באופן שלא יכסה הלחות הכפורית ישוב הראות כמו שנזכר
13 במלאכת הרפואה. הנה אע"פ שירשמו התמונות בעין ההוא
לא ישיג הדברים הנראים מפני היות מסך מבדיל מהגעת
14 זה הרושם אל המקום שבו הראות. וזה ממה שיראה ממנו
שמרכז הראות הוא במה שבפנים ממקיף הלחות הכפורית.
15 שאם היה במקיפה החיצרוני הנה לא יפסד הראות בסבתה
כי אין למקום ההוא רושם בהשגת המובט לפי זאת ההנחה
16 כמו שהיתבאר מדברינו בזה המקום. והנה יראה שהראות
17 הוא בזה הלחות הכפורית | ושבמרכזו יהיה מרכז המבט.
ובזה האופן יביא הלחות ההוא הרושם המגיע לו מהמובט
במקיף הפנימי מהלחות הכפורית ויובל רושם התמונה לפני
הראש כי שם החוש המשותף כמו שנזכר בטבעיות.
18 וכבר אפשר שנבאר במופת למודי מקום מרכז המבט
באמצעות הכלי אשר המצאנוהו ללקיחת המבטים בכל עת
19 שנרצה אשר נבאר ענינו במה שיבא בג"ה. ואולם בזה
המקום נבאר מענינו מה שיש בו די בבאור זה הדרוש אשר
20 אנחנו בו. והוא שיהיה בידיעו מקל ישר השטחים יהיה
21 בראשו האחד מקום נשים באמצעו העין. ויהיה לנו
לוחות נקובות באמצעם ישרות השטחים ונכניס המקל
בתוכם ויהיו באופן שיהיה גבהם על המקל פחות מעט
22 מגובה הראות עליו. ונשים בו שני לוחות יחד האחד
23 כפל האחד או אי זה שיהיה מהיחסים. ונשים הלוח
היותר קטן יותר קרוב אל הראות ונקריבנו אל הראות
עד שיסתיר הלוח הקטן הגדול בלי תוספת ובלי חסרון.

12 הגיעו] נ: הגיע. 12 שנזכר] נ: שזכר.
13 הרושם] נ: הרשום. 17 הרושם] נ: הרשום.
17 שנזכר] פ: שזכר. 23 נ, ק, ושולי פ: חסרון]
פ: גרעון.

נ 27א 20 יתבאר לנו מזה המקום אם גלגל השמש יוצא ממרכז
העולם אם לא. ונעמד גם כן מזה על שעור היציאה
21 מהמרכז אם היה שיהיה יוצא ממרכז העולם. ומזאת
התמונה יתבאר לך שכבר תעמד מזה בעת הלקויות על
22 שעור האצבעות הנלקות. וכל שכן שיתבאר זה יותר
כשהיה נקב החלון קטן השעור מאד כי אז יבא הענין
בכותל המקבל הניצוץ כדמות תמונה ירחית לפי שעור
23 הלקות. ואם תגרע מקטרי הניצוץ היותר גדול והיותר
קטן שעור החלון תמצא בשאר יחס הנלקה מגשם המאיר כי
הוא כיחס יתרון הקוטר הגדול על הקוטר הקטן אל הקוטר
24 הגדול או כיחס הקוטר הקטן אל הקוטר הגדול. ומזה
גם כן יתבאר שכאשר היה הנלקה בצד מה מהמאיר יבא
25 החסרון בניצוץ בהפך הצד ההוא. וזה שכבר נתבאר
מדברינו בזאת התמונה שהצד העליון מהמאיר ימשך ממנו
תוספת בניצוץ המגיע מהחלון בכותל בצד השפל מניצוץ
26 הכותל. ובזה יתבאר שכל חלק ממנו יחייב תוספת בצד
המקביל לו ורמז מה שזכרנו.

הפרק הששי

1 ולפי הדרך שבו נדע אופן ההשתמשות בזה הכלי
פ 14א 2 שהמצאנוהו ללקיחת | המבטים בתכלית הדקדוק. הוא
שנדע תחלה באי זה מקום מעין הוא מרכז הראות והוא
3 המקום אשר תכלה אליו זוית המובט. הנה ראוי שנחקור
בו בזה המקום ונאמר שלא ימנע משיהיה אם במרכז העין
אם במקיף העין וזה אם במקיף החיצוני או הפנימי או
4 יהיה במה שבין אלו המקומות. וזאת חלוקה הכרחית לפי
שהוא מבואר שהעין הכלי המיוחד לראות.
5 ונאמר שהוא מבואר במעט עיון שאי אפשר היותר
במקיף החיצוני וזה שכבר התבאר בטבעיות שהנראה יבא
נ 27ב אל | הראות באמצעות האויר הספירי האמצעי אשר יקבל
6 רשמו ויובילהו אל הראות. ואלו היה מרכז הראות בשטח
החיצוני המקיף מעין לא היה בכאן דרך אל הכח הרואה
להשיג הדברים הנראים כי מה שיגיע אליו מהדברים
הנראים הוא נקודה ואי אפשר בנקודה שיושג בה מראה
7 ולא גודל ולא תמונה. וכבר ישיג הראות המראה והגודל
והתמונה ולזה הוא מבואר שאין מרכז הראות בשטח
8 המקיף בעין מחוץ. ועוד שכבר יראה בחוש הרשם תמונת

22 בדמות] פ: כדמות. 23 או ... הגדול] חסר בכ"י
נ, וק. 2 מרכז] נ: המרכז. 3 הפנימי או] חסר
בכ"י נ.

פ 13ב 3 יתרחב בחוש הניצוץ שיבא ממנה מפני מרחק החלון
מהכותל שיקבל הניצוץ. כי לעוצם המרחק שיש |
מהגרמים השמימיים עד הארץ לא יהיה שם מוחש כלל
4 מפני זה המרחק המעטי. הלא תראה כי גם חצי קוטר
הארץ איננו דבר מורגש מאד בכוכבים לעוצם המרחק אשר
5 מהם אלינו. כל שכן בזה המרחק המעטי כי הוא אי
אפשר שיורגש בו גם בירח שהוא היותר קרוב אלינו.
6 וזה כלו יתבאר באור שלם בג"ה מכח דברינו במה שיבא.
7 המופת לכל מה שזכרנו תחלה שאנחנו נרשום בכותל
שיהיה בו החלון קו AB ותהיה נקדת H בו ראש החלון.
8 ונרשום בכותל הנכחי לו שיפול בו הניצוץ במקום שהיה
9 ראוי שיפול בו הניצוץ בו GD. ויהיה מרכז עגולת
המאיר נקודת H ויהיה קוטר המאיר הנכחי לקו AB קו
נ 26ב 10 ZHT. ויהיה הקו המגיע ממרכז המאיר | אל נקודת H
11 חותך GD על נקדת L ונגיע קו HL הישר. והוא מבואר
שזה הניצוץ היה ראוי שיפול על נקדת L מצד מה שיגיע
12 מהמאיר אם היה נקודה. אך מפני שכל חלקי המאיר הם
מאירים הנה יצא ניצוץ מנקודת T מקוטר המאיר אל נקדת
13 H ביושר ויכלה בקו GD בנקודת M. ולפי שקוי חהל
טהמ הם מתחתכים הנה תהיה זוית מהל שוה לזוית חהט
14 שהיא זוית חצי קוטר המאיר במקום החלון. ובזה יתבאר
זה הענין בכל צדדי החלון כי המאיר הוא מאיר מכל
קצורת הקטרים הרשומים בו מהצד המביט אלינו והוא מה
שרצינו לבאר.
15 ומזאת התמונה התבאר שכאשר יגיע הניצוץ מחלון
בעל זויות ישרות הקוים הנה לא יהיה הניצוץ בעל
זויות ישרות הקוים לפי שהוא יתרחב לכל צדדי הזויות
16 בשיעור הזוית חצי קוטר המאיר. ויבא הענין בזוית
בדמות רובע עגול מרכזו נקדת הזוית וזה דבר נראה בחוש
מענין הניצוצים המגיעים מהשמש והירח מן החלונות
17 בעלי זויות ישרות הקוים. ולפי שאין הבדל שיורגש
בין הגודל הנראה לשמש מנקדת החלון ובין הגודל שהיה
ראוי שיראה לו ממרכז הארץ לקוטן שעור חצי קוטר
ק 8א הארץ אצל המרחק העצום שיש | בינינו ובין השמש כמו
18 שיתבאר במה שיבא בג"ה. הנה נעמד מזה על שעור קוטר
19 השמש מהעגולה הגדולה אשר יסבב בה בעת המבט. וזה
שער מועיל מאד למה שנרצהו בזאת החקירה וזה שכבר

4 לעוצם] ק: להעדר. 9 נוסף בשולי פ: הנכחי] חסר
בכ"י נ, וק. 11 ל] נוסף בכ"י נ: ויגיע.
14 הקטרים] נ: הגשרים; ק: הקשרים. 15 הזויות
בשיעור] חסר בכ"י נ.

187 זכרנו לך זה בזה המקום שלא תתבלבל בבאורים שנעשה בזה האופן. ומזאת התמונה התבאר שכל משולש ישר הקוים הנה יחס קוויו קצתם אל קצת הוא כיחס נכחויות
188 הזויות שהם מיתרים להם קצתם אל קצת. ומזה יתבאר במעט עיון שאם היו זויות משלש ישר הקוים ידועים ויהיה קו אחד ממנו ידוע השעור ששאר הקוים הם ידועי השיעור כי יחסם אל הקו הידוע הוא ידוע.
189 כאשר היו שני קוים ממשלש מה ידועים והיתה הזוית אשר יקיפו בה הקוים הידועים ידועה הנה השאר
190 מהזויות והקוים ידועים. ויהיה המשלש משלש אבג ויהיו שני קוי אב בג ממנו ידועים והיתה זוית אבג ידועה ואומר שקו אג ידוע ושאר הזויות ידועות.
191 וזה שזוית אבג אם היתה נצבת הנה זה מבואר ממה שקדם.
192 ותהיה גם כן חדה כמו הענין בצורה הראשונה או נרחבת
193 כמו הענין בצורה השנית הנה יהיה קו אג ידוע. וזה שכבר נוציא מנקדת א קו יהיה עמוד על קו בג המונח
194 בלי תכלית והוא קו אד. ואיך שיהיה הוא מבואר שזוית אבד היא ידועה כי היא אם זוית אבג הידועה
195 אם שלמות שתי זויות נצבות מזוית אבג. ותשאר זוית
196 דאב ידועה כי היא שלימות הזוית הנצבה. ולזה יהיה משלש אבד ידוע הזויות כלם וקו אחד ממנו ידוע השעור
197 ויהיה השאר ממנו ידוע. וגם כן הנה שני קוי אד דג
198 הם ידועים בשתי הצורות. ולזה יודע שעור קו אג ממשלש אדג נצב הזוית ותהיינה מפני זה זויות משלש
199 אדג ידעות ולזה תודע זוית אגד. וכבר הונחה זוית אבג ידועה ותשאר זוית באג הנשארת מהמשולש ידועה כי היא שלמות שתי זויות נצבות והוא מה שרצינו לבאר.

הפרק החמשי

1 ניצוץ השמש או הירח או אי זה מאיר שיהיה מן הכוכבים הנכנס בחלון הוא יותר רחב משעור החלון לכל הצדדין כשעור זוית חצי קוטר המאיר במקום החלון.
2 ונניח בזה המקום על דרך השרש המונח עד שיתבאר מדברנו במה שיבא שאם היה המאיר נקדה אחת לבד לא

---

189 השאר] פ: הנשאר.   189 ידועים] ק: ידוע.
194 מזוית אבג] חסר בכ"י נ.   1 זוית] נוסף בכ"י פ (למעלה): תושבת.

171 תהיה זוית באג ידועה בשתי הצורות. ותודע זוית
גבא הנשארת מהמשלש לפי שזוית גבד ידועה וזוית דבא
ידועה וימשך מזה במעט עיון שזוית גבא בשתי אלו
172 הצורות ידועה. ולזה הוא | מבואר שמשלש אבג שהוא
ידוע הקוים הוא ידוע הזויות והוא מה שרצינו לבאר.
173 כשנודעו שני קוים ממשלש מה ותודע זוית אחת ממנו
ויהיה אחד מהקוים הידועים מיתר לה הנה הנה שאר הזויות
174 ידועות והקו השלישי ידוע. ויהיו שני קוי אב בג
175 ממשולש אבג ידועים ותהיה זוית באג ידועה. ואומר
176 שקו אג הוא ידוע והזויות הנשארות הם ידועות. וזה
שאנחנו נקיף עגולה תקיף במשלש באג והיא עגולת באג
177 ונניח שיהיה קוטר העגול קו אד. הנה מפני שזוית
באג היא ידועה והיא אצל המקיף שתהיינה בו שתי זויות
178 נצבות ש״ס חלק הנה תודע קשת בג. ולזה יודע מלוח
הקשתות | והנכחוריות שיעור מיתר זה הקשת בשעור שבו
יהיה שעור קו אד ק״כ חלק ולזה יהיה יחס קו בג אל
179 קו אד ידוע. ולפי שיחס קו בג אל קו אב הוא גם כן
180 ידוע הנה יהיה יחס קו אב אל קו אד ידוע. ולזה הוא
מבואר שכבר יודע מזה שעור קו אב בשעור שבו קו אד
ק״כ מעלה ומזה תודע קשת אב בלוח הקשתות והמיתרים.
181 ולפי ששתי קשתות בג אב ידועות הנה תודע הקשת הנשארת
שהיא קשת גא וממנה תודע זוית גבא ויודע ממנה קו גא
182 בהנהגה הקודמת. ולזה הוא מבואר שמשלש אבג הוא ידוע
הקוים והזויות והוא מה שרצינו לבאר.
183 וראוי שתבין מזה הביאור שאם הונח קוטר העגול ס׳
מעלה ומקיף העגול ק״פ מעלה לא יצטרך לחשב מיתרי אלו
הקשתות בלוח הקשתות והמיתרים | כי אם בדמיון נכחוריות
184 כי הנכחות הוא חצי מיתר כפל הקשת. והנה יחס
הנכחות אשר לקשת מה אל חצי קוטר העגול הוא כיחס
185 מיתר כפל הקשת אל קוטר העגול. ולזה בחרנו אנחנו
כשנשתמש בזה המורפת שיהיו דברינו בו לפי ההנחה הזאת
186 השנית כי יהיו הדברים בביאורן יותר קצרים. והנה

177 ש״ס] חסר בכ״י נ.   177 והיא ... בג] נוסף
בשולי נ: נ״ל יודע שיעורה אצל שעור שתהיינה בו שתי
זויות נצבו׳ ק״פ חלק ולפי שזוית באג היא אצל המקיף
והיא חצי אשר על המרכז תהיה ידועה אשר על המרכז זה
וממנה נודע קשת בג שהיא כמוה   מק״פ שבחצי המקיף.
178 ולזה יהיה] נוסף בכ״י נ, פ (וסימן להשמיטה
בשניהם) : חלק.

156 ידועים הנה הקו השלישי ידוע לפי שהקו שיעשה מיתר
לזוית הנצבה ממנה כחיי על מרובעי שני הקוים
הנשארים. ולזה אם יודעו הם יודע הוא ואם יודע הוא
ואחד מהנשארים יודע הקו השלישי לפי שהוא כחיי על
הנשאר ממרובע הקו שיעשה מיתר לזוית הנצבת כשגרענו
ממנו מרובע הקו השני הידוע.

157 ונניח שתהיה זוית אבג נצבת ויהיו שני קוי אג
בג ידועים ואומר שזוית באג ידועה וזה שאנחנו נשים
נקדת A מרכז ונקוה במרחק אג קשת גז ונדביק קו אבז
הישר והוא מבואר ממה שקדם שקו בג הוא נכחות קשת גז.

158 ולפי ששני קוי אג בג ידועים הוא מבואר ששעור קו בג
ידוע בשעור שבו יהיה שעור קו אג שהוא חצי הקוטר ס׳

159 מעלה. ובזה השעור יודע מלוח הקשתות והנכחוריות הקשת
160 הראויה לנכחות קו בג. וכאשר נודעה בזה האופן קשת
161 גז נודעה זוית באג כמו שהתבאר מאקלידיס. וממנה
תודע זוית בגא כי היא שלמות הזוית הנצבה לפי ששתי
זויות באג | בגא הם יחד צ׳ מעלה.

162 כשהיו קוי משולש מה ידועים הנה זויותיו הם
163 ידועים. ויהיה משולש אבג ידוע הקוים ואומר שזויותיו
164 ידועות. וזה שאנחנו נפיל מנקודת B עמוד על קו אג
המונח בלי תכלית והוא קו בד וחפול בצורה הראשונה
נקודת ד תוך המשלש ובצורה השנית תפול חוץ מהשולש

165 ואומר ששעור קו גד הוא ידוע. וזה כי כשלקחנו יתרון
שני מרובעי קוי גב גא על מרובע קו אב בצורה הראשונה
או יתרון מרובע קו אב על מרובעי שני קוי גב גא
בצורה השנית וחלקנו אותו על כפל שעור קו גא היה

166 העולה שוה לקו גד כמו שהתבאר במעט עיון במאמר השני
מאקלידיס ולזה יהיה שעור קו גד ידוע. ולפי שקו גב

167 שהוא ידוע מוסיף עליו בכח כמו מרובע קו בד הנה יהיה
שעור קו בד ידוע. ולזה יתבאר ממה שקדם שמשלש בדג

168 שהוא נצב הזוית הוא ידוע הזויות. ומזה יודע בצורה
הראשונה זוית בגא וחלק אחד מזוית גבא והוא זוית
גבד ובצורה השנית תודע זוית בגא לפי שזוית בגד
שהיא שלמות שתי זויות נצבות ממנה היא ידועה ותודע

169 גם כן זוית גבד. וגם כן הנה מפני שקוי אג גד ידועים
170 הנה יהיה שעור קו אד ידוע. ויתבאר מזה שמשלש בדא
נצב הזוית שהוא ידוע הצלעות הוא ידוע הזויות ולזה

160 התבאר] נוסף בשולי נ: מסוף הששי.

140 ורבהפך זאת ההנהגה תדע מהחץ הקשת הראויה לו ר"ל
שאם היה פחות מס׳ מעלה תגרעהו מס׳ מעלה והנשאר בידך
תחקור עליו באלו הלוחות לאי זה קשת הוא נכחות ותגרע
הקשת ההיא מצ׳ מעלה והנשאר בידך היא הקשת הראויה
141 לאותו החץ. ואם היה החץ יותר מס׳ מעלה תחקור על
היותר לאי זה קשת הוא נכחות ותוסיף על הקשת ההיא
צ׳ מעלות ויגיע בידך הקשת הראויה לאותו החץ וזה
142 כלו מבואר ממה שזכרנו בדבור שני מזה הפרק. ואם
היה בידך קשת ידוע ורצינו לדעת נכחות הקשת | ההוא
תחקור באלו הלוחות על הקשת ההיא ותמצא כנגצו הנכחות
143 הראויה לו. ובזה האופן גם כן תדע מהנכחות הקשת.
144 ואם לא מצאת הקשת ההיא באלו הלוחות תחקור על שתי
קשתות היותר קרובות לו ותקח מהחלוף אשר בין שתי
145 הנכחויות אשר כנגדם לפי היחס. וכן תעשה אם לא
מצאת הנכחות כרצונך לדעת ממנו הקשת הראויה לו.
146 ולפי שכל מה שיקרב שעור הנכחות להיות בשעור חצי
הקוטר יגדל יותר הטעות כשלקחנו הקשת הראויה לו לפי
היחס משתי נכחויות קרובות לו כמו שיתבאר ממה שזכרנו
במעט עיון הנה ראוי שתבחר תמיד לחקור מהנכחות
היותר קטן אי זה קשת ראויה לו כשהיה אחד מאלו
147 הקשתות מודיע השני. כאלו תאמר שאם היה שעור שתי
קשתות אלו הנכחיות צ׳ מעלה הנה מידיעת הקשת האחת
תודע הקשת השנית ואז תבחר הנכחות היותר קטן.
148 והנה צורת אלו הלוחות הוא לפי מה שאספר נחלק
149 הלוח ברוחב לשלשה טורים. הטור הראשון נכחוב בו
150 הקשתות מרביע מעלה לרביע מעלה עד צ׳ מעלה. הטור
השני נכחוב בו שלמות ק"פ מעלות מהקשת אשר בטור
151 הראשון בכל שטה ושטה. הטור השלישי נכתב בו הנכחות
הראוי לכל קשת וקשת בשטח הקשתות ההם וזה ציור
הלוחות.
152 הדבור החמשי נודיע בו איך נדע זויות המשולש
153 וצלעותיו מפני ידיעתנו בקצת מזה. כשהיו שני קוי
משלש נצב הזוית ידועים הנה הוא ידוע הצלעות
154 והזויות. יהיה המשלש נצב הזוית משלש אבג והיו שני
קוים ממנו ידועים ואומר שהשאר ידוע מהזויות
155 והצלעות. וזה כי איך שהיה שיהיו שני קוים ממנו

140 תגרעהו מס׳ מעלה] חסר בכ"י נ. 151 וקשת] חסר
בכ"י נ. 153 הזוית] פ: הזויות. 154-155 ואומר
... ידועים] חסר בכ"י פ.

	המסודרות ממעלה למעלה תהיה הקשת הראויה לזה אם
	היתה קטנה מצ׳ מעלות פ"ט מעלות ומ"ה דקים וכ"ז
	שניים ואם היא יותר מצ׳ מעלה תהיה הקשת הראויה
128	לזה צ׳ מעלות וי"ד דקים ול"ג שניים. והוא מבואר
	ממה שקדם שאם היה זה צורדק היה ח׳ שניים חץ קשת
129	י"ד דקים ול"ג שניים. וכבר יתבאר ממה שקדם שכאשר
	היה שעור החץ ח׳ שניים היה מרובע נכחות הקשת ההיא
	ט"ו דקים רנ"ט שניים רנ"ח שלישיים רנ"ו רביעיים
130	כי כן תעלה הכאת ח׳ שניים בשלמות הקוטר. ולזה
	יהיה שעור נכחות הקשת ההיא ששעורה י"ד דקים ול"ג
131	שניים ל׳ דקים רנ"ט שניים רנ"ג רביעיים. וזה
	שקר כי זה הנכחות יסכים	לקשת כ"ט דקים ל"ה
132	שניים. ולזה הוא מבואר שהקשת הראויה	לנכחות
	שזכרנו היא מוסיפה על שעור צ׳ מעלה או מחסירה
133	ממנו כמו כ"ט דקים ול"ה שניים. ולזה היה מהטעות
	בזה המקום לפי הדרך האחרת יותר מט"ו דקים ומפני זה
	הסכמנו לסדר אלו הלוחות מרביע מעלה לרביע מעלה כי
	בזה לא יקרה מהטעות דבר מורגש מאד אם ילקח לפי
	היחס.
134	הדבור הרביעי נעשה בו אלו הלוחות לדעת הנכחות
	והחץ לאי זה קשת שיהיה וההפך ונודיע בו דרך
	ההשתמשות באלו הלוחות.
135	ראוי שתדע שכבר חקקנו באלו הלוחות הנכחויות כלם
	וכל הקשתות מרביע מעלה לרביע מעלה ולא חקקנו בהם
136	החצים לאלו הקשתות ולא המיתרים. והנה יודעו מאלו
	הלוחות המיתרים לאי זה קשת שתהיה כשתדע מהם תחלה
137	נכחות חצי הקשת ההיא וכאשר תכפלהו יגיע בידך המיתר
	לקשת ההיא. והנה יודעו מאלו הלוחות החצים לכל
138	הקשתות לפי מה שאומר. והוא שאם היה שעור הקשת אשר
	רצינו לדעת חצה פחות מצ׳ מעלה נחקר באלו הלוחות על
	נכחות הקשת שהיא שלמות צ׳ מעלה מהקשת ההיא ויהיה
139	החץ המבוקש שלמות חצי הקוטר מהנכחות ההוא. ואם
	היה שעור הקשת אשר רצינו לדעת חצה יותר מצ׳ מעלות
	הנה נחקור באלו הלוחות על נכחות קשת יתרונה על צ׳
	מעלות ונחבר אותו עם חצי הקוטר ויגיע לנו החץ
	המבוקש.

130 נ"ג] נ: י"ג.   131 כ"ט] נ: נ"ט.   133 ילקח] נ: לא ילקח.   133 היחס] נ: הירח.

קשת מעלה אחת וחצי וחציהם נעמד על נכחות קשת י׳
מעלות פחות רביע ובכמו זאת ההנהגה נעמד על השאר.
וכבר נעמד על נכחות קשת רביע מעלה בהקש תחבולי 116 נ 21א
לפי מה שאומר. והוא שכבר נעמד מנכחות קשת ח׳ 117
מעלות ורביע מעלה על נכחות קשת ד׳ מעלות ושמינית
מעלה ונרד בזאת ההנהגה עד שנדע מזה נכחות קשת 118
רביע מעלה וחצי שמינית שמינית המעלה. וכן נעמד
בכמו זאת ההנהגה מנכחות קשת ד׳ מעלות פחות רביע
אל שנדע מזה נכחות קשת רביע מעלה פחות שמינית
שמינית. והנה כאשר חקרנו על זה בזה האופן מצאנו 119
כי יחס נכחות קשת רביע מעלה וחצי שמינית שמינית
אל נכחות קשת רביע מעלה פחות שמינית שמינית הוא
בקירוב נפלא כיחס הקשת אל הקשת לא יתבלבל זה היחס
גם ברביעיים. ולזה קיימנו לפי היחס שנכחות קשת 120
רביע מעלה הוא ט״ו דקים ומ״ב שניים וכ״ח שלישיים
ול״ב רביעיים וז׳ חמשיים. וכאשר התישב זה הנה 121
נתישר לדעת מזה שאר הנכחויות כלם איש על מקומו.
וזה כי מנכחות קשת רביע מעלה ונכחות קשת ג׳ 122
רביעיות מעלה נעמוד על נכחות קשת חצי מעלה ונכחות
קשת מעלה אחת ונכחות קשת ב׳ מעלות ונכחות קשת ד׳
מעלות ונכחות קשת ח׳ מעלות ונכחות קשת י״ו מעלות
ונכחות קשת ל״ב מעלות. ומנכחות קשת חצי מעלה 123
ונכחות קשת ב׳ מעלות נעמד על נכחות קשת מעלה אחת
ורביע מעלה ועל נכחות קשת ב׳ מעלות וחצי ונכחות
קשת ה׳ מעלות ונכחות קשת י׳ מעלות וכ׳ מעלות ומ׳
מעלות. ובזאת ההנהגה יקל לעמד על הנכחויות כלם 124
והחצים בקירוב מרביע מעלה לרביע מעלה עד מ״ה
מעלות ומהם נעמד על השאר כמו שקדם.
והנה הסכמנו לחקור בזה מרביע מעלה לרביע 125
מעלה | ולא הספיקה לנו ידיעת הנכחויות ממעלה למעלה נ 21ב
כמו שנמשך המנהג בהרבה מהלוחות המסודרות לזה לפי
שכבר ראינו כי בקצת מקומות יגיע הטעות לכמו ט״ו
דקים מהקשת כשנבקש לדעת מהגבהות הקשת הראויה לו
ובפרט כשהיתה הקשת ההיא מוסיפה מעט על צ׳ מעלה או
מחסירה מעט מצ׳ מעלה. והמשל שיהיה הנכחות אשר בידינו 126
נ״ט מעלות נ״ט דקים רנ״ב שניים. והנה לפי הלוחות 127

115 ובכמו] נ: ובכמות.   122 ונכחות קשת ג׳ ברייעיות
מעלה נעמוד על] נ: נדע.   125 המסודרות] ק: המחוברות.

106 ביאר זה אקלידס ולזה יהיה נכחות קשת ל׳ מעלה שוה
לרביע הקוטר. ואולם חצו יתבאר ממה שקדם שהוא ח׳
מעלות וב׳ ראשונים י"ח שניים ל׳ שלישיים מ"ו
107 רביעיים בקרוב מעט. המיתר השלישי הוא צלע המעושר
והוא מיתר קשת ל"ו מעלות וזה שכבר התבאר מאקלידס
שכאשר חובר צלע המעושר עם רביע קוטר העגול היה
המחובר שוה בכח לשני אלו המרובעים הנזכרים תכף והם
108 מרובע חצי הקוטר ומרובע רביע הקוטר. ומזה יתבאר
שמיתר קשת ל"ו מעלה היא ל"ז מעלות וד׳ דקים ונ"ה
שניים וכ׳ שלישיים ול׳ רביעיים בקירוב מעט ולזה
יהיה נכחות שקשת י"ח מעלות י"ח מעלות ול"ב דקים
וכ"ז שניים ומ׳ שלישיים וט"ו רביעיים בקירוב מעט.
109 ורצו יתבאר ממה שקדם מנכחותה שהוא ב׳ מעלות ונ"ו
110 דקים רי"א שניים ומ"ה שלשיים רנ"ח רביעיים. והנה
יתבאר ממה שקדם כי מנכחות קשת צ׳ מעלות נעמד על
נכחות קשת מ"ה מעלות ונכחות קשת כ"ב מעלות וחצי
111 ונכחות קשת י"א מעלות ורביע מעלה וחציהם. ומנכחות
קשת ל׳ מעלה נעמד על נכחות קשת ט"ו מעלות ונכחות
קשת ז׳ מעלות וחצי ונכחות קשת ד׳ מעלות פחות רביע
112 וחציהם. ומנכחות קשת י"ח מעלה נעמד על נכחות
קשת ל"ו מעלות ועל נכחות קשת ט׳ מעלות ועל נכחות
קשת ד׳ מעלות וחצי ועל נכחות קשת ב׳ מעלות ורביע
וחציהם.
113 ויתבאר עוד ממה שקדם כי מנכחות קשת ל׳ מעלה
ונכחות קשת י"ח מעלה וחציהם יודע נכחות חצי הקשתות
האלו שהוא כ"ד מעלה וממנו יודע נכחות קשת י"ב מעלות
ונכחות קשת ו׳ מעלות ונכחות קשת ג׳ מעלות ונכחות
קשת מעלה וחצי ונכחות קשת ג׳ רביעיות מעלה וחציהם.
114 ובזאת ההנהגה יקל לעמוד מאלו הנכחיות שזכרנו על כל
הנכחיות לכל הקשתות מג׳ רביעיות מעלה לג׳ רביעיות
115 מעלה. כאלו תאמר שמנכחות קשת ט"ו מעלות ונכחות
קשת מעלה אחת וחצי וחציהם ונעמד על נכחות קשת ח׳
מעלות ורביע מעלה ומנכחות קשת י"ח מעלות ונכחות

106 י"ח] ק: רח׳.
107 התבאר] נוסף בשולי כ"י נ: מט׳ ומג׳ מי"ג יוד׳
מב׳. 108 שמיתר] פ: שמעלה מיתרו. 109 ומ"ה]
נוסף בשולי כ"י נ: נ׳ נ"ג. 114 לג׳ רביעיות מעלה]
חסר בכ"י פ. 115 וחצי$_1$ ... אחת$_2$] חסר בכ"י נ.

קו דה במשלינו זה וחץ הקשת הגדולה מרובע עגולה הוא
מוסיף על חצי הקוטר בשעור קו דה במשלינו זה.
92 ולפי שכבר ידע מהחץ מיתר הקשת ההיא כמו שקדם הנה
יתבאר מזאת התמונה שכאשר היה מיתר כפל קשת מה ידוע
93 הנה מיתר הקשת ההיא ידוע. וזה כי כשנודע נכחות
הקשת נודע מיתר כפל הקשת ההיא כי הנכחות הוא חצי
המיתר ההוא וידוע מהנכחות מיתר הקשת ההיא וזה מה
שרצינו לבאר.
94 הדבור השלישי נישר בו לעשיית אלו הלוחות יודעו
מהם הנכחיות והחצים לאי זה קשת שיהיה והפך.
95 כבר יתבאר ממה שקדם כי מה שידע הנכחיות והחצים
לאי זה קשת שיהיה עד מ"ה מעלות ידע הנכחיות והחצים
96 כלם. והמשל שיהיה ידוע נכחות קשת רביע מעלה וחצו
נ 20א ואומר שחץ | קשת צ׳ מעלה פחות רביע ונכחותו ידועים.
97 וזה שחץ זה הקשת השני עם נכחות הקשת הראשון הוא שוה
לחצי הקוטר כמו שהתבאר בדבור הקודם וכן הענין בנכחות
98 זה הקשת השני עם חץ הקשת הראשון. ומחץ זה הקשת
השני ונכחותו יודע חץ קשת צ׳ מעלה ורביע מעלה
99 ונכחותו. וזה כי נכחותו היא נכחות הקשת השני בעינו
וחצו כשחובר עם חץ הקשת השני הוא שוה לקוטר העגול.
100 ומחץ קשת הראשון ונכחותו יודע חץ ק"פ מעלה פחות
רביע ונכחותו וזה כי נכחותו הוא נכחות הקשת הראשון
בעינו וחצו כשחובר עם חץ הקשת הראשון הוא שוה לקטר
101 העגול. ובזה האופן יתבאר שכאשר נודעו נכחיות הקשתות
וחציהם עד מ"ה מעלה יודעו נכחיות שאר הקשתות וחציהם.
102 ולפי שכבר יתבאר בדבור הקודם איך יודע מהידוע
הנעלם בכל הפנים שאפשר בענין הנכחיות והחצים ראינו
בזה המקום להניח שלשה מיתרים ידועים למעיין בזה
103 הספר. המיתר האחד הוא מיתר קשת ק"פ מעלה כי הוא
כשיעור קוטר העגול ומזה יודע נכחות צ׳ מעלה וחצו
כי כל אחד מאלו שוה לחצי הקוטר להיותו בא מהמרכז
104 אל המקיף. המיתר השני הוא מיתר ס׳ מעלה כי הוא
105 כשעור חצי קוטר העגול וזה מבואר במעט עיון. וכבר

92 יודע] נ: ידוע. 92 ההיא כמו ... הנה מיתר
הקשת] חסר בכ"י נ. 93 כפל] נ: כפול. 94 נישר]
נוסף בשולי פ: עשית [?] הלוחו׳ בידיעת מיתר [?]
מקשתו ובהפך. 94 לעשיית] פ: לעשות. 100 ומחץ
קשת ... לקטר העגול] חסר בכ"י נ. 103 להיותו]
נ: להיות.

79 וכזה יתבאר מזאת התמונה שמיתר קשת יתרון הקשת
הגדולה על הקשת הקטנה הוא שוה בכח לשני אלו
המרובעים ר״ל למרובע יתרון הנכחות האחד על הנכחות
80 האחד ולמרובע יתרון החץ האחד על החץ האחר. וזה
שהוא מבואר בזה המשל שקו גד שוה בכח מרובעי קוי גח
חד ואולם קו חד הוא מבואר שהוא שוה ליתרון הנכחות
האחד על האחר וקו גח הוא שוה ליתרון החץ האחד על
האחר והוא מה שרצינו לבאר.
81 נרצה לבאר שכאשר היה נכחות קשת מה ידוע שחץ
הקשת ההיא ידוע לפי שמרחק הנכחות מהמרכז ידוע ומזה
82 יודע החץ. המשל שיהיה נכחות קשת אב ידוע הוא קו
83 אה ויהיה מרכז העגולה | נקודת ד. ונוציא קו בה
ביושר קוטר העגולה באופן שיהיה מרכז העגולה בזה
הקו אם במה שבין שתי נקודות ב וה אם היתה אב גדולה
מרובע עגולה ויהיה זה הקו בדה כמו הענין בתמונה
הראשונה אם במה שאחר נקודת ה ויהיה זה הקו בהד
כמו הענין בתמונה השנית וזה אמנם יקרה כאשר היתה
84 קשת אב קטנה מרובע עגולה. או תהיה נקודת המרכז
85 היא נקודת ה כשהיתה קשת אב רובע עגולה. ואולם
כשהונח הענין כן הנה יהיה החץ ידוע והוא קו הב כי
הוא שוה לחצי קוטר | העגול.
86 ותהיה גם כן נקודת ה על זולת המרכז ואומר ששיעור
קו הד הוא ידוע וזה שאנחנו נקרה קו אד והוא ידוע
כי הוא חצי הקוטר וכאשר חוסר ממרובע קו אד הידוע
מרובע קו אה הידוע היה יסוד הנשאר שוה לקו דה ולזה
87 יהיה קו דה ידוע. ולפי שקו בד ידוע הנה ישאר שעור
88 קו בה ידוע. והנה יתבאר שקשת אב בתמונה הראשונה
היא גדולה מרובע עגולה לפי שזורית בדא היא יותר |
89 גדולה מזורית ה הפנימית ממשולש אהד. ואולם זורית ה
היא נצבת ולזה תהיה זורית בדא נרחבת ויחוייב מזה
90 שתהיה קשת בא גדולה מרובע עגולה. ובדמיון זה יתבאר
בתמונה השנית שזורית בדא היא חדה ולזה יחוייב שתהיה
91 קשת בא קטנה מרובע עגולה. ומזאת התמונה התבאר שחץ
הקשת הקטנה מרובע עגולה הוא פחות מחצי קוטר בשעור

79 יתרון הקשת] נ: יתרון. 79 על החץ] פ: על.
81 יודע] נ: ידוע. 86 נקוה] נ: הוצאנו; חסר בכ״י
ק. 88 יותר] חסר בכ״י פ. 88 נרחבת] ק: נרוחת.

פ 9ב ביושר עד נקודת ד והוא מבואר שהוא | יגיע לנקדת ד
נ 18ב ביושר כי מיתר קשת אב̄ד̄ חולק בנקודת ה לחצאים ויהיה
קו אה נכחות קשת אב ולזה גם הוא מבואר שקו הד שוה
68 לקו אה. ונוציא מנקדת ג קו יהיה עמוד על קו אד
המונח בלי תכלית והוא קו גח ואם לא תפול נקדת ח במה
שבין שתי נקדות א ו ד כמו הענין בצורה השנית הנה
69 יגיע קו אד̄ הישר אל נקודת ח. ונקוה קוי אג גד
הישרים ויהיה קו אג מיתר שתי קשתות אב בג כשחוברו
ויהיה קו גד מיתר קשת קשת יתרון קשת אב על קשת בג וזה
70 כי קשת בגד הונחה שוה לקשת אב. ואומר שכאשר היו
נכחיות שתי קשתות אב בג וחציהם ידועים והם בזה
המשל קוי אה גז בה בז הנה מיתר קשת אבג ידוע והוא
קו אג בזה המשל ומיתר קשת יתרון קשת אב על קשת אב
71 על קשת בג ידוע והוא קו גד בזה המשל. וזה שמיתר
המקובץ משתי הקשתות הוא שוה בכח אל מרובע הקו השוה
לשתי הנכחיות לשתי הקשתות המונחות ואל מרובע יתרון
72 החץ הגדול על החץ הקטן. ר״ל שקו אג בזה המשל הוא
שוה בכח למרובע ההוה משני קוי אה זג כשחוברו בקו
73 אחד ולמרובע ההוה מקו זה יחד. והנה אלו המרובעים
ידועים לפי שהקוים האלו שהם יסודותיהם הם ידועים.
74 ומיתר קשת גד הוא שוה בכח ליתרון הנכחות האחד על
האחר בכח וליתרון החץ האחד על החץ האחר בכח.
75 וזה כי לפי שזויות ז ה ח הם נצבות הנה קוי גז חה
הם נכחיים וכן הענין לקוי זה גח ולזה הוא מבואר
שהקוים הנכחיים הם גם כן שוים אם קו חה לקו גז ואם
76 קו גח לקו זה. ולפי שזוית גחא נצבת הוא מבואר שקו
נ 19א 77 גא שוה בכח למרובעי קוי גח חא. ואולם | קו חא שוה
לשתי נכחיות קשתות אב בג וקו גח שוה לקו זה שהוא
יתרון החץ האחד על האחר.
78 ולזה הוא מבואר שמיתר המקובץ משתי הקשתות הוא
שוה בכח אל מרובע הקו השוה לשתי נכחיות שתי הקשתות
המונחות ואל מרובע יתרון החץ הגדול על החץ הקטן.

67 חולק] נ: חלוק. 68 לא תפול] נ: תפול.
71 וזה] נוסף בשולי פ: המופת; נוסף בשולי נ: הקשת.
72 יחד] חסר בכ״י נ. 75 לפי שזויות] פ: כשזויות.
75 ח] חסר בכ״י נ. 75 לקוי] ק: בקוי. 76 הוא
מבואר שקו] נ: יהיה קו.

51 הוא נכחות קשת אב. ונוציא קו ביושר עד שיגיע
לנקדת ז שהיא מרכז העגול כי חץ הקשת הוא תמיד
52 עובר על קוטר העגול. ונוציא גם כן קו גהז היושר
ויתבאר שקו גהז הוא קו אחד ישר לפי שקו גה הוא
מבואר שהוא חץ קשת בג והחץ הוא בהכרח על יושר קוטר
53 העגול כמו שהתבאר מאקלידיס. והוא מבואר שזורית אזג
54 היא נצבת לפי שהיה קשת אבג רובע עגולה. ולפי
שזוריות אדב גהב הם גם כן נצבות הנה הוא מבואר
מאקלידיס שקוי בד הז הם נכחיים וכן הענין בקוי בה
דז ולזה הוא מבואר שאלו הקוים הנכחיים הם שוים ומזה
55 יתבאר שקו בה שוה לקו דז. ולפי שהיו שני קוי אד
דז חצי הקוטר הוא מבואר כי שני קוי אד בה שוים לחצי
56 הקוטר | והוא מה שרצינו לבאר. ומזה יתבאר שאם היה
חץ קשת מה | ידוע נכחות שארית צ' מעלה ידוע והפך
זה גם כן ר"ל שאם היה נכחות קשת מה ידוע חץ שארית
צ' מעלה ממנו ידוע והוא מה שרצינו לבאר.
57 נרצה לבאר שחץ קשת גדולה מצ' מעלה שוה לנכחות
קשת צ' מעלה שהוא חצי הקוטר ולנכחות קשת היתרון על
58 צ' מעלה יחד. המשל שתהיה קשת אבג יותר גדולה
מרובע עגולה והיא על מרכז נקודת ז ותהיה קשת אב
59 ממנה צ' מעלה. ויהיה חץ קשת אבג קו אזד ויהיה
60 נכחות קשת בג קו גה. ואומר שקו אזד שוה לחצי
הקוטר ולקו גה יחד.
61 המופת שאנחנו נקוה קו גד שהוא נכחות קשת אבג
62 ונקוה קו בז. והוא מבואר שזוריות גדז בזד בהג הם
כלם נצבות ולזה יתבאר שקוי הג זד הנכחיים הם גם
63 כן שוים. ולזה הוא מבואר שקו אזד הוא שוה לשני
קוי אז הג וזה מה שרצינו לבאר.
64 כשהיו נכחיות שתי קשתות מתחלפות וחציהם ידועים
הנה מיתר הקשת השוה לשתי הקשתות ההם ידוע ומיתר
65 קשת יתרון הקשת האחת על שנית ידוע. המשל שיהיו
שתי הקשתות המתחלפות קשתות אב ובג ותהיה קשת אב
הקשת היותר גדולה ויהיה נכחות קשת אב קו אה וחצה
66 קו בה. ויהיה נכחות קשת בג קו גז וחצה קו בז והוא
מבואר שקו בז יפול על יושר קו בה אי אפשר זולת זה
לפי ששני החצים באים מנקדת ב אל מרכז העגולה.
67 ונשים קשת בגד שוה לקשת אב ונוציא קו אה הישר

51 הישר ... הוא] נ: והוא. 51 מאקלידיס] נוסף
בשולי נ: מא' מג'.

35 והוא מבואר כי קו בד הוא נכחות קשת בא. ולזה הוא
מבואר שאם גרענו מרובע קו אד הידוע ממרובע קו אב
36 היה הנשאר ידוע וכאשר לקחנו יסודו יגיע לנו בד.
37 וכבר יתבאר זה בדרך אחר יותר קל. והוא שאנחנו
נאמר שאם היה החץ ידוע הנה הנכחות ידוע ר"ל שאם
היה בזאת התמונה הקודמת שיעור קו אד ידוע הנה
38 שיעור קו דב ידוע. וזה שכבר התבאר שמשולשי אדב
בדג מתדמים לפי שזויות אדב בדג הם נצבות וזוית
אבד שוה לזוית ג כמו שקדם ותשאר זוית א שוה לזוית
39 גבד. ולזה הוא מבואר שצלעות אלו המשולשים שהם
40 מיתרים לזויות השרות הם מתיחסות. ולזה יהיה יחס
קו אד ממשלש בדא אל קו בד ממשלש בדג כיחס קו בד
41 ממשלש בדא אל קו גד ממשלש גדב. ובהיות הענין כן
הנה הוא מבואר שהכאת הראשון ברביעי הוא כהכאת השני
בשלישי | ולזה יתבאר שהכאת קו אד בקו דג שוה להכאת
42 דב בעצמו. והוא מבואר כי קו גד ידוע לפי שקו אד
ידוע וקו אג שהוא קוטר העגול הוא ידוע ולזה הוא
43 מבואר שקו גד הוא ידוע. ובהיות הענין כן הוא
מבואר שהכאת קו אד בקו דג ידוע וכאשר הוצאנו יסודו
יגיע לנו קו בד ולזה יהיה שיעור קו בד ידוע והוא
מה שרצינו לבאר.

44 ורמזה התבאר שכאשר היה מיתר קשת מה ידוע הנה
מיתר כפל הקשת ההיא ידוע כי ידוע ממנו נכחות הקשת
ההיא וכאשר כפלנו אותו יעלה בידינו מיתר | כפל
45 הקשת. ואומר עוד שכאשר היה חץ קשת מה ידוע הנה
46 מיתר שלימות ק"פ מעלה מהקשת ההיא ידוע. וזה שכאשר
נודע קו אד בזה המשל בתמונה הקודמת נודע קו גד
שהוא שלמות הקטר וכאשר נודע שיעור קו גד נודע ממנו
שעור מיתר בג כמו שקדם.

47 וכשחוברו חץ קשת מה עם נכחות שלמות צ' מעלה מהקשת
48 ההיא יהיה המחובר שוה לחצי הקוטר. המשל שתונח קשת
אבג צ' מעלה על מרכז נקדה ז והיה חץ קשת אב קו אד
49 ויהיה נכחות קשת בג קו בה. ואומר כי כשחוברו שני
50 קוי אד בה היה המחובר שוה לחצי קוטר. המופת שאנחנו
נקוה קו בד בין שתי נקודות ב וד והוא מבואר שקו בד

38 אבד שוה] פ, ק: אבג שוה.   38 גבד] פ: ב.
46 ממנו שעור] נוסף בשולי נ: קו בד ומשניהם נדע.

20 קו אד בקו אג. המופת שאנחנו נוציא קשת אב עד ג
   ולזה תהיה קשת אבג חצי עגולה ונקוה קוי בג בד
21 הישרים. והוא מבואר שזוית אבג היא נצבת וכן זוית
   בדא לפי שקו בד הוא נכחות קשת בא והנכחות הוא מקיף
22 בזוית נצבת עם הקוטר. | ואומר שמשולשי אדב אבג הם ב16 נ
   מתדמים וזה כי זויות אדב אבג הם נצבות וזוית א
23 משותפת לשני המשולשים. ותשאר זוית אבד הנשארת
   ממשולש אדב שוה לזוית אגב הנשארת ממשולש אבג.
24 ולזה הוא מבואר שמשולשי אדב אבג הם מתדמים
   וצלעותיהם אשר הם מיתרים לזויות השרות הם |
25 מתיחסות. ולזה יהיה יחס קו אד ממשולש אדב אל קו ק 3ב
   אב ממשולש אבג כיחס קו אב ממשולש אדב אל קו אג
26 ממשולש אבג. ובהיות הענין כן הוא מבואר שהכאת
   השני בשלישי שהוא הכאת קו אב בעצמו היא שוה להכאת
   הראשון ברביעי שהיא הכאת קו אד בקו אג והוא מה
   שרצינו לבאר.
27 ומזאת התמונה התבאר שאם היה שיעור קו בא ידוע
28 שיעור קו אד הנה ידוע. וזה כי מפני שיעור קו בא
   ידוע הנה מרובעו ידוע וכאשר חלקנו אותו על שיעור
   קו גא הידוע כי הוא קוטר העגולה אשר הונח שיעורו
29 ק"כ מעלה היה היוצא מהחלוקה קו אד. וכבר התבאר
   מזאת התמונה שאם היה שיעור קו אד ידוע הנה שעור
30 קו בא ידוע. וזה כי כשהיה שעור קו אד ידוע הנה
   הכאת קו אד בקו אג הוא ידוע לפי ששיעור קו אג הוא
31 ידוע כמו שקדם. ולזה יהיה יסוד העולה מההכאה
   הזאת ידוע והוא שיעור קו אב.
32 ואומר עוד שכאשר היה המיתר לקשת מה ידוע או
33 החץ הנה נכחות הקשת ההוא ידוע. וזה כי מאחד מאלו
   ידוע השני ומידיעת שניהם תגיע ידיעת הנכחות לפי
   שהמיתר שוה בכח למרובע החץ ולמרובע | הנכחות.
34 וזה מבואר מזאת הצורה הקודמת כי מפני שזוית בדא א17 נ
   היא נצבת היה מרובע קו בא שוה למרובעי קוי בד דא

21 שזוית] נוסף בשולי נ: מל' מג' לאקלי'.
26 הכאת קו אב] פ: קו אב. 27 ומזאת] נוסף בשולי
נ: שאם המיתר ידוע החץ יהיה ידוע. 27 הנה] ק
הוא. 29 נוסף בשולי נ: שאם החץ ידוע המיתר יהיה
ידוע. 29 הנה שיעור קו בא] נ: ששיעור קו אב.
29 בא] ק: בא הוא. 33 יודע] נ: ידוע. 33 לפי
שהמיתר] פ: כי המיתר.

לאין קץ לפי מה שירצו מהדקדוק.

5 ואולם גלגל המזלות יחלקוהו ראשונה לשנים עשר
6 חלקים שוים כל חלק מהם יקרא מזל. והמזל יחלק עוד
לשלשים מעלות בדרך שיהיו מעלות הגלגל שלש מאות
7 וששים. והנה יחלקו קוטר העגולה לק״ך מעלות ואע״פ
ששיעור אלו המעלות מתחלף לשיעור המעלות אשר יהיו
8 במקיף. והנה יקרא חלק מה מהמקיף קשת והקו הישר
המגיע בין שתי קצות הקשת יקרא מיתר לקשת ההוא או
9 לשארית העגול. והקו המגיע מחצי המיתר ההוא לחצי
הקשת ההיא יקרא חץ לחצי הקשת ההיא או מיתר נזור.
10 וחצי המיתר ההוא יקרא נכחות לחצי הקשת ההיא או
11 מיתר מחוצה. וכבר התבאר מאוקלידס כי החץ אשר בזה
התואר יקיף עם הנכחות בזוית נצבה והוא גם כן
מונח על יושר קוטר העגול ר״ל שכאשר נגיעהו ביושר
אל הצד השני ממקיף העגולה יעבור על מרכז העגולה.
12 והתנועה אשר ייחסה בטלמיוס לגלגל ההקפה נקראה
אנחנו תנועת החלוף והמקום אשר יקראהו בטלמיוס גובה
בכוכבים נקראהו אנחנו גם כן גובה על צד ההסכמה
בשמות לא בענין כי אפשר שיהיה המקום הזה בלתי רחוק
מן הארץ יותר משאר חלקי הגלגל.
13 הדבור השני נביא בו קצת מופתים גימטריים נתישר
מהם לידיעת מה שיצטרך מענין הקשתות והמיתרים.
14 מיתר קשת אחת הוא בעינו המיתר לקשת הנשארת
15 מהעגול. וזה מבואר כשתשלים עגולת הקשת ההיא כי
אז יתבאר לך שזה המיתר בעינו הוא מיתר לכל אחת מאלו
16 הקשתות. ומזה יתבאר כי המיתר המחוצה לקשת מה הוא
בעינו המיתר המחוצה לשלמות ק״פ מעלה מהקשת ההיא.
17 המרובע ההווה ממיתר קשת מה הוא שוה אל השטח
נצב הזויות שיקיפו בו חץ הקשת ההיא וקוטר העגולה.
18 המשל שיונח הקשת קשת אב ומיתרו קו אב הישר ויהיה
קוטר העגול קו אג הישר ויהיה קו אד מזה הקו חץ
19 קשת אב. ואומר שהכאת קו אב בעצמו הוא שוה להכאת

---

8 מה] חסר בכ״י פ. 11 התבאר] בשולי נ: מכ״ט מג׳.
11 קוטר] נוסף בשולי נ: מא׳ מג׳ לאקלי׳.
15 כשתשלים] נוסף בשולי נ: מכ״ד מג׳ לאקלי׳.
16 מה] פ: אחת.

52 במבט מתנועות הכוכב באורך וברוחב.   וזה מספיק
למעיין כי אותן החקירות הקשות אשר יגיע מהם לעשיית
ההקשים התחבוליים הצריכים לזה הדרוש אין בהם תועלת
53 רב | אם לא קודם הגעת האמת בזאת החקירה.   אך אחר
זה אין בהם תועלת רב אם לא לבקשת הכבוד.
54   ואם ירצה השם ויגזור בחיים וישלמו לנו המבטים
הראויים בזה לבאר אלו הדרושים בקלות יותר נשוב
לבאר זה בג"ה מאותם המבטים בספר נפרד או בזה הספר
בעצמו אם ישלם לנו זה קודם החפשט זה הספר בארצות.
55 כי אז יתכן לנו להרחיב הביאור בדרכים אשר ילקחו
מהם אותם ההקשים התחבוליים בזולת עומק בדברים
ובזולת אורך רב בהם.
56   והנה קודם שנתחיל זכרון בזה הכלי נציע קצת
הצעות מועילות לכל שנחקור בו מעניין זה הכלי ומה
שיתחייב מהתנועה לכל אחת מהתכונות אשר אפשר שירונחו
באופן יחשב שימשך מהם מה שיראה מעניין התנועות
57 לכוכבים מהמהירות והאיחור והיושר והנזורות.   ועם
זה נחקור מעניין הקשתות והמיתרים כי בזולתם אין
58 דרך להכנס בזאת החקירה.   ואם היה שיספיק בזה מה
שזכרו בטלמיוס בספר המגסטי הנה לא הסכמנו שיהיה
זה הספר חסר דבר ממה שיצטרך בזאת החקירה כי אולי
יפול זה הספר ביד מי שלא יהיה אצלו ספר המגסטי
59 רימנע ממנו תועלת זה הספר.   ולזאת הסבה בעינה
הסכמנו לדבר בזה הספר דרושים | אחדים יספיק בהם
גם כן מה שזכרו בטלמיוס כמו העניין במצעדי המזלות
באופן הישר ובאופן הנוטה ושאר הדברים הדומים לזה.

הפרק הרביעי

1 והוא יחלק להמישה דבורים.
2 הדבור הראשון בביאור קצת שמות נשתמש בהם בזאת
המלאכה.
3   ראוי שתדע כי חכמי המזלות הסכימו לחלק הקשת
המקיף בעגולה לשלש מאות וששים חלק יקראו כל חלק
מהם מעלה וכל מעלה יחלקו לששים חלקים יקראו דקים
או רשאונים וכל אחד מאלו הדקים יחלקו לששים חלקים
4 יקראו שניים.   ובזה האופן יחלקו השניים לשלישיים
והשלישיים לרביעיים ובזה האופן תמשך זאת החלוקה

52 אותן] נ, ק: אותם.   56 זכרון בזה] פ: בזכרון זה.

38 עשר מעלות מסרטן לא הוכר שם נטייה בכמו אלו הכלים.
    ולזה יוביל הטעות המעטי בזה המקום בגובה הכוכב לטעות
    רב במקום הכוכב באורך.
39 וראוי שתדע כי נטיית הכוכבים לא נודעת עדיין
    בשלימות | לפי מה שהתבאר לנו ממבטים רבים. ואם
40 נחשוב בגובה הכוכב הנמצא במבט הנטייה המתחייבת לו
41 לפי חשבון בטלמיוס יקרה מזה טעות רב. ועוד כי לא
    יתאמת לנו מגבהם מקומם באורך אם לא ידענו תחלה
42 מקומם באורך שיעור נטייתם לצפון או לדרום. ולזה
    יחויב שנדע תחלה מקומם באורך קודם שנוכל לחקור בכמו
43 אלו הכלים על מקומם באורך וזה מגונה. ולקושי ההשגה
    בזה כמו שבארנו והיה שכבר התבאר לנו בחוש פעמים
    רבות שאין העניין במהלכות הכוכבים ולא בנטיותיהם בכמו
44 האופן שיתחייב מכונת בטלמיוס. השתדלנו להמציא כלי
45 לא יקרה טעות במעשהו ולא במבטו. והחלנו לקחת בו קצת
    מבטים מועילים מאד בזאת החקירה והוא יישירנו בג״ה אל
    האמת בתכונה אשר תחוייב לכוכבים שימשך ממנה מה שיראה
46 לחוש מעניני תנועותיהם ונטיותיהם. ולא רצינו להמתין
    עד שישלמו לנו כל המבטים שיצטרכו להשלמת זאת החקירה.
47 אבל הספיקו לנו קצת המבטים נוכל להוציא מהם אחר
    העמל הרב מה שיעמידנו על אמתת זאת התכונה בקירוב
    נפלא ליראתנו פן תפקד זאת החכמה בהפקדנו ולקושי
    אותן הדרכים אשר העמידנו על אמתת זה ועמקם עד
    שנצטרך אם באנו להזכירם למאמר ארוך אורך נפלא.
48 לא רצינו לזכור כל אלו הדרכים ההם אשר הגענו מהם אל
    ההקשים התחבוליים אשר במה שבארנו בזה בכוכב כוכב כי
49 יארך המאמר מאד. ויהיה עם זה רב העומק באופן שלא
    יקבל בו המעיין תועלת אבל יקוץ בדברים לאריכותם
50 ועמקם. והנה זכרנו מהם קצת יוכל להתיישר האיש שלם
    העיון ממה שזכרנו אל דרך עשיית אלו ההקשים התחבוליים
51 אשר יגיעו מהם למה שבארנו מזה בכוכב כוכב. והנה |
    בארנו עם זה במה שאין ספק בו אמתת התכונה שהנחנוה
    לכוכב כוכב ושאין שם תכונה אחרת יאות עמה מה שיראה

41-42 אם לא ... באורך] חסר בכ״י נ.   42 נוכל]
נ, פ: יוכל.   42 פ (למעלה): על] חסר בכ״י נ, ק.
45 נ: והחלנו] נ (למעלה) : והתחלנו ; פ, ק: והחלונו.
45 יישירנו] פ, ק: הישירנו.   46 עד] חסר בכ״י נ,
פ.   47 ולקושי אותן] נ: ולקושי אותם ; ק: ולקושי
אותם.   47 שנצטרך] נ: שצטרך.   48 אורך] חסר
בכ״י נ.   48 אלו] חסר בכ״י ק.   48 אשר] חסר
בכ״י פ, ק.

שלא יתכן שתשלם ההבטה בו בכוכב הקיים בזמן קצר
מאד אחר ההבטה בכוכב האחר באופן שלא יתנועע הכוכב
25 הקיים שעור מורגש במה שבין שני המבטים. ולזה הוא
מבואר שלא יתכן שיודע מזה הכלי באמתות מרחק קצת
הכוכבים המובטים בו מקצת.
26 ועוד כי אם הודינו שנוכל לקחת המבטים בזה האופן
או | באופן הראשון שזכרנו הנה נצטרך תחלה לדעת מקום
27 הכוכב אשר ידיענו מקום הכוכב הרץ הנראה אמר. וזה
ממה שיש בו ספק היום כמו הספק שיש במקום הכוכב הרץ.
28 וזה כי בתנועת הכוכבים הקיימים ספק עצום עד היום
והתחלפו בה הקודמים חלוף רב וזה ממה שיפיל ספק
29 במקום הכוכבים הקיימים. ואם חשבנו לעמוד על מקום
הכוכבים באצטורלאב או ברביע העגול מצד הגובה הנמצא
להם בעברם על קו חצי היום כמו שנזכר בספר באור
30 האצטורלאב. הנה יהיה בזה המבט קירוב מצדדים רבים.
31 הצד האחד מצד שבוש הכלי במעשהו והטעות המעטי שיקרה
32 בלקיחת הגובה יוביל לטעות גדול. וכבר עמדנו על זה
בכל הכלים שהגיעו | אלינו ומצאנו בהם טעות רב
בעשייתם עד שכבר יגיע זה הטעות לכמו מעלה אחת.
33 וזה יקרה אם מצד חלוקת המעלות כשלא תהיה על נכון
אם מצד קוטר הכלי שיהיה נוטה במצבר לפאת שמאל או
לפאת ימין או מצד הלוח בעל שני דפין הנקרובים.
34 כשלא היו הנקבים באופן שיהיה הקו המגיע מזה אל זה
נכחי לקו הלוח אשר יורח על גובה הכוכב לפי המקום
35 שיחתך בו מעלות האצטורלאב. כי בכל אחד מאלו הפנים
יתכן שיפול טעות רב וכבר יתקבצו כלם או קצתם ויוביל
זה לטעות רב בהודעת מקום הכוכב.
36 והצד השני כי מצד קוטן המעלות לא יתכן לעמוד
על החלק מהמעלה כי אם באופן יהיה בו קירוב.
37 והצד השלישי הוא מצד הקירוב אשר ימצא בדרך שבה
ילקחו המבטים בכמו זה הכלי עד שכבר יתכן שיהיה
הקירוב יותר מעשר מעלות ובפרט כשהיה לכוכב מרחב
ויהיה בתאומים או בסרטן או בקשת או בגדי שלא תוכר
הנטייה שם לגלגל המזלות בכמו אלו הכלים בפחות עשר
מעלות או ביותר עשר מעלות ר"ל שמעשרים מתאומים עד

31 המעטי] ק: המעט. 32 שהגיעו] נ: שיגיעו.
33 יקרה אם] נ: אם יקרה. 33 תהיה] נ: יהיו.
34 מעלות] נ, פ: מעלת. 37 הכלי עד] חסר בכ"י
נ. 37 מרחב] נ: מרחק. 37 הכלים] נ: הכוכבים.

11 מהטעות לסבה סבה מאלו.   ובזה קושי נפלא וצורך
למבטים רבים יתכן שתשלם בהם זאת הבחינה כמו שיתבאר
12 מדברינו במה שיבא.   ולרבוי המבטים אשר יצטרכו לזה
ידמה שלא תשלם השגתה כי אם בזמן ארוך אורך נפלא
יעבור קץ החיים האנושיים כפלים רבים.
13   וזה כי גם השגת מבט אחד מהם לא תקרה כי אם
בקושי לפי שהמבט האמתי בזה הוא שיראה במבט הכוכב
ההולך דבק עם אחד מהכוכבים הקיימים אשר נודע מקומם.
14 כי בזה המבט לא יפול ספק ולא מבוכה כי לא יקרה בו
דבר מהטעות מצד הענבים או האידים אשר באויר שיהיה
15 המבט באמצעותם.   ולא נצטרך במבט הזה לשום כלי מכלי
16 ההבטה ולזה לא יקרה בו טעות מזה הצד.   אלא שזה המבט
לא יקרה כי אם מעט ולא יתכן שישלם בזה האופן הרבוי
שיצטרך שיגיע מהמבטים עד שתשלם הבחינה מאי זה צד
נפל החלוף הזה אשר יגיע במקום הכוכבים הנראה במבט
עם מה שהיה ראוי שיראה מזה לפי חשבון בטלמיוס.
17   וכבר תמצא שהקודמים אשר לפני בטלמיוס כלם בחנו
כמו אלו המבטים שזכרנו מפני ראותם שבאלו המבטים לא
18 יקרה טעות.   ואם דמינו לקחת המבטים בכלי שהמציא
בטלמיוס לעמד על מקום אי זה כוכב שיהיה באי זה עת
שיהיה והוא כלי הטבעות והנה יהיה במה שיראה בכמו
נ 13ב 19 הכלי הזה מן הספק מה שלא יעלם.   וזה כי כבר יקרה
מהטעות במעשהו בסבת הקושי הנופל בעשייתו עד שלא
ימצא אומן בארצנו ידע לעשות בדקדוק והוא מבואר
שהטעות הנופל במעשה הכלי הזה יהיה סבה לטעות במבטים
20 הלקוחים בו.   ואם הודינו שכבר ימצא זה הנה יקרה
שיתערות הכלי אחרי הפרדו מיד האומן וזה שכבר
21 תתערתנה קצת העגולות או כלן.   ולזה לא נבטח בכמו
22 זה הכלי שיתן האמת במה שנשפט בו.   עם שכבר יקשה
להביט בו מבין הנקבים קושי נפלא כמעט שלא ישלם זה
23 בשום פנים.   וכבר יתבאר לך זה אם תשתדל להביט כוכב
אחד מהכוכבים אשר שעורם בלתי גדול מאד מבין שני
24 נקבי לוח האצטורלאב.   ובהיות הענין כן הוא מבואר

11 וצורך] נוסף בשולי פ: נ"א יצטרך.   12 ארוך] חסר בכ"י נ.   13 וזה ... שיראה] חסר בכ"י נ.
15 הזה] פ: זה ה.   16 מעט] פ: כמעט.   16 שישלם] נ: ששלם.   16 החלוף הזה] פ: חלוף זה.   16 שיראה] חסר בכ"י פ.   17 בחנו] ק: בחרו.   18 יהיה] פ: יהיה זה.

נב 12 24 ‏ והנה לעוצם מעלת זאת החקירה היה החלק המעטי
המגיע לנו מההשגה בה יותר חשוק ויותר נכבד אצלנו
מהחקירה השלמה שתהיה בדברים אשר למטה ממנה בכבוד
ק 2א ובמעלה רזה מבואר | מאד מעניין התשוקות האנושיות.
25 ובהיות העניין כן הוא מבואר שאם היה אפשר שנשלים
החקירה בזה בשלם שבפנים עד שלא ישאר בעניינה ספק
ומבוכה שזה יהיה יותר נכבד ויותר חשוק לאין שעור
ולזה ראוי שנפליג בהשלמתה בכל הפנים שאפשר לנו זה.

הפרק השלישי

1 וראוי שלא יעלם ממנו מה שבהשגת הדרוש הזה מן
2 הקושי לקושי ההגעה מהחוש מה שיצטרך למי שירצה
להשלים החקירה בו. כל שכן בזמנינו זה אשר סרו
3 האנשים מחקור בזאת החכמה בכלי ההבטה אבל הספיק להם
מה שאמרו בה הקודמים. עד שהיה זה סבה שלא מצאנו
למי שקדמנו מבטלמיוס עד היום מבטים נעזר בהם בזאת
החקירה זולת מה שראינו שהסכים עליו אלבתני מצד
מבטיו וראוהם המבטים בלתי ידועים לנו.
4 ועוד שנביט אנחנו בכוכבים ונמצא העניין בהם בלתי
5 הולך על הסדר המתחייב מתכונת בטלמיוס. הנה תפול
6 המבוכה מאי זה צד. נפל בהם זה הטעות כי אפשר היותו
מפני שאין הגובה מהם במקום המתחייב מהחשבון ההוא
אשר הניחו בטלמיוס או אלבתאני אחריו או מפני ששעור
תקון תנועת המרכז הוא על זולת מה שיתחייב מהחשבון
ההוא או מפני ששעור תקון תנועת החלוף הוא על זולת |
נ 13א 7 מה שיתחייב מהחשבון ההוא. או מפני ששעור תקון
נטיית קטר תנועת החלוף על זולת מה שיתחייב מהחשבון
8 ההוא. או מפני שמקום הכוכב באורך הוא על זולת
9 המקום המתחייב מהחשבון ההוא. או מצד היות מקום
פ 7א הכוכב מתנועת | החלוף על זולת המקום המתחייב
10 מהחשבון ההוא. או יהיה זה להתקבץ קצת אלו הסבות
או כלם ואז תרבה המבוכה יותר כי יצטרך לבחון מה נפל

11 מצד מה שתראו מעוצם החכמה ביצירתם ומעוצם היכולת
לקרוא לכלם שמות ייוחדו בהם קצתם מקצתם במה שהשפיע
מהם פעולות מתחלפות בזה העולם השפל. ולא נעדר איש
12 מהם שלא יהיה רב אונים ואמיץ כח לעשות אשר גזר השם
ית' להם ביום הבראם. ויהיה הרצון באמרו המוציא
במספר צבאם שהוצאתו צבא השמים מהאופק המזרחי הוא
במספר משוער באופן שיורה שאין שם עיפות ויגיעה
13 בתנועתם ולזה תמצא תמיד הולכת על סדר אחד. או
יהיה הרצון באמרו המוציא במספר צבאם שהוא מוציא
14 צבאם במספר הראוי אל שישלם ממנו מה שיצטרך. והנה
נמשך מיתרון גבורת השם ית' אל שלא נעדר | איש מהם
15 יצטרך להשלמת המציאות. וזה כמו התחלת מופת לבאר
מזה חדוש אלו הגרמים השמימיים כמו שיתבאר במה
שיבא בג"ה.
16 ובזה אמר דוד מספר עוצם מדרגת השם ית' כי אראה
שמיך מעשה אצבעותיך ירח וכוכבים אשר כוננת מה אנוש
17 כי תזכרנו וגו'. ר"ל כי אראה שמיך אשר יתפרסם מהם
שהם מעשה אצבעותיך למה שנראה מהפלא והחכמה ביצירתם
ובפרט מה שיראה מענין הירח וכוכבים אשר כוננת כמו
שיתבאר מדברינו במה שיבא בג"ה.
18 וזה דבר יותר מבואר בירח ממה שהוא בשאר הכוכבים.
19 וזה שעם **היות** הירח וכוכבים כלם מטבע הגרם השמימי
הנה יראה חלוף רב בין הירח לשאר הכוכבים וזה כי
הירח גוף חשוך יקבל אורו מזולתו מה שאין כן בשאר
20 הכוכבים כמו שיתבאר. ועוד כי חלקי הירח בעצמם
מתחלפים קצתם לקצת כמו שיראה מהצל הנראה בגרם הירח.
21 ועוד יראה דבר נפלא בכלם כי עם היותם מטבע אחד היה
מראה ניצוציהם וצבעו מתחלף בכוכב כוכב וזה כולו
היה בלתי אפשר אם לא הונחו אלו הגרמים השמימיים
22 מחודשים כמו שיתבאר במה שיבא. ובהיות הענין כן
הנה הוא דבר נפלא מאד שעם עוצם מדרגתך הנפלאה
המתפרסמת מהעיון באלו הגרמים השמימים תזכר ותפקד
23 בריאה שפלה וקטנה מאד והיא האדם. וזה כלו ממה שהוא
ראוי שיעירנו להפליג בעיון שזאת החקירה הנפלאה
שיושג ממנה זה הפרי הנפלא.

12 אחד] נ: אחת. 17 וגו'] חסר בכ"י פ.
18 וזה דבר] נ: והנה דבר. 18 ק, פ (למעלה):
ממה] נ, פ: מפני.

הפרק השני

1 וראחר שהתבאר ששלמות זאת החקירה יאות להיות בזה הספר ראינו להפליג בה להיות זאת החכמה יקרה מאד אם מצד עצמה אם מצד מה שתיישיר אליו משאר החכמות.

2 ואולם היותה יקרה מצד עצמה הוא מבואר וזה כי מעלת הדרוש תהיה | לפי מעלת הנושא אשר בו החקירה והוא מבואר שהנושא לזאת החקירה והוא הגרם השמימי הוא היותר נכבד שבכל הגמשים הטבעיים והצורה המניעה אותו היא היותר נכבדת שבכל הצורות הטבעיות עד שלא יפול ביניהם יחס כי שם הנושא והצורה יאמר על אלו ועל שאר הדברים הטבעיים בשתוף השם לבד כבר התבאר זה כלו בחכמה הטבעית.

3 ואולם היותה מישרת לשאר החכמות כי היא מישרה לחכמת הטבע ולחכמת האלהות הישרה נפלאה עם שהעיון בצורותיה ומדרגתם מהצורה הראשונה הוא פרי החכמה האלהית ותכליתה ובעבור זאת החקירה תיוחד זאת החכמה שתקרא אלהית.

4 והיא גם כן מישרת אל הפילוסופיא המדינית באופן מה כבר באר זה בטלמיוס בראשון מספר המגסטי.

5 והנה העירונו הנביאים והמדברים ברוח בקדש אל שהוא ראוי שנפליג העיון בזאת החקירה כי בזה נתיישר להשגת השם ית׳ למה שיתפרסם מהעיון הזה.

6 כי אלו הגלגלים והכוכבים הם ברואים בדבר השם ית׳ כמו שיתבאר מדברינו בזה המאמר בג״ה עם מה שיתפרסם בזה מעוצם חכמת השם ית׳ ועוצם יכלתו בהמציאו אלו

7 הגרמים הנכבדים בזה האופן הנפלא מהחכמה. והשפיעו מהם ענינים מתחלפים ישלם מהם זה המציאות השפל עם היותם כלם מטבע אחד מופשט מהאיכויות אשר ישפעו מהם.

8 והנה שמם באופן שלא ייעפו ולא ייגעו מעשות אשר גזר השם ית׳ מפעולותיהם בירם הבארם.

9 ולזה אמר הנביא ישעיה שאי מרום עיניכם וראו מי ברא אלה המוציא במספר צבאם לכלם בשם יקרא מרוב

10 אונים ואמיץ כח איש לא נעדר. רצה בזה שאו מרום עיניכם לראות הכוכבים והשכילו מזה הענין מי הוא שברא אלה כי זה העיון יישירכם לעמוד על שהם נבראים ויישירכם עם זה להשיג מי הוא שברא אלה

---

9 ישעיה] פ: ישעיה ע״ה.   10 כי זה ... אלה] חסר בכ״י נ.

נ 10ב   אם לא תהיה זאת החקירה על אמתתה ללומדי ולא לטבעי
9   ולא לפילוסוף למי תהיה.   האם בכאן חקירה לא תהיה
10   לאחד מבעלי העיון.   ואנחנו אומרים בהתר זה הספק כי
         אע"פ שאין זאת החקירה על אמתתה לחכמה הכוללת
         החוקרת בנמצא במה שהוא נמצא לפי שאין המאמר בזאת
         החקירה על זה האופן אבל היא חקירה מנמצא מה מתנועע
11   מצד שהוא מתנועע.   ואינה גם כן לחכמה הטבעית ואם
         היה שתחקור בנמצא המתנועע מצד שהוא מתנועע לפי שרוב
         מופתי זאת החקירה או כלם הם לקוחים מחכמת הנדסה
         ובכלל במה שתבאר מכמות התנועות המתחייבות מהתכונה
         ההיא אשר בעבורם יוכרח להניח התכונה באופן אשר יניח
12   אותה.   ואינה גם כן על שלמותה לחכמה הלמודית כי בכאן
         תכונות יחשב שיסכימו לתנועות הנראות ולא יאותו לפי
13   הטבע או לפי הפילוסופיא.   הנה היא על שלמותה לכלל
         החכמות האלו כאלו תאמר שלחכמה הלמודית העיון במה
         שיתבאר בה במופתים הלמודיים ולחכמת הטבע והפילוסופיא
         העיון במה שיתבאר בה במופתים הטבעיים והפילוסופיים.
14   ובהיות הענין כן והיה בלתי אפשר החקירה האחת במה
         שהיא אחת שתהיה קצתה לבעל חכמה אחת וקצתה הנשאר
         לבעל חכמה אחרת כי לא ידע השני מה שיחסר בהשלמת
         החקירה אשר חקר בה הראשון אם לא שידע תחלה מה שבארו
פ 6א   הראשון בזה הדרוש ויבחן בזה האופן מה שנשלמה | בו
15   החקירה לראשון ממה שלא נשלמה החקירה לו בו.   ואז
         תהיה החקירה הזאת לשני מצד מה שהוא בעל שתי החכמות
נ 11א   וזה מבואר מאד למעיין | בזה הספר עד שהאריכות
         בבאורו הוא מותר אין צורך לו.
16   הנה יחוייב שתהיה זאת החקירה על שלמותה למי
         שהוא למודי וטבעי ופילוסוף כי מי שזה דרכו אפשר
         שישלים זאת החקירה מפני העזרו בכל אחת מאלו החכמות
17   ולקחו מהם מה שיצטרך לו בהשלמותה.   ולפי שהמאמר בזה
         הספר הוא בזה האופן רצוני שאין מתנאי זה הספר שיהיה
         המאמר בו למודי או טבעי או פילוסופי הנה הוא מבואר
         כי החקירה הזאת על שלמותה היא נאותה בזה הספר.

8 על אמתתה] בשולי פ: באמות.   10 מצד] פ:
במה; פ (למעלה): מזה.   13 בה$_1$] חסר בכ"י נ.
17 או טבעי] חסר בכ"י נ.   כי החקירה] פ, ק:
שהחקירה.

החלק הראשון והוא נחלק למאה ושלשים וששה פרקים כמו שזכרנו

הפרק הראשון

1. בעבור שראינו שלא השתדל אחד מהקודמים אשר הגיעו אלינו דבריהם לחקור בחכמת התכונה חקירה שלימה והיתה נשארת מפני זה במה שהיא מהחסרון.
2. התעוררנו אנחנו לדבר בה בזה המקום. וזה שכבר
3. מצאנו שהחוקרים בה מחקר ראוי מהלמודיים הספיק להם שיעמדו על תכונה יתכן שימשך ממנה בקירוב מה שיראה לחוש ולא השתדלו שיבארו התכונה אשר תחוייב לפי האמת.
4. והנה בתכונה אשר המציאו ימצאו ספקות רבות עד שלא יתכן בשום פנים שתהיה בצד אשר הניחו אותו.
5. וכבד הרגישו בקצת אלו הספקות קצת מהטבעיים והפילוסופים ומפני זה החליטו המאמר שאי אפשר שתהיה תכונת הגלגלים בזה האופן אלא שהם עם זה השתדלו לבאר הצד אשר בו יתכן שימשך מה שיראה לחוש מענין תנועת הכוכבים. אבל סלקו המשא מעליהם ואמרו שהלמודי
6. הוא אשר ראוי שיחקר בזה כי זה בלתי אפשר להם במה שהם טבעיים או פילוסופיים. והנה הלמודי יאמר גם כן
7. שאין לו במה שהוא למודי שישלים זאת החקירה אבל *יספיק* לו שיסדר תכונה יתכן שימשך ממנה מה שיראה לחוש מעניני תנועת הכוכבים באורך וברחוב והעמידה והיושר והנזורות מזולת שישגיח אם התכונה ההיא נאותה לפי הטבע אם לא כי אין ללמודי מבוא לחקור בזה במה שהוא למודי.
8. ובהיות הענין כן הנה למספק שיספק ויאמר |

3 וזה] חסר בכ"י פ.   5 אלו] פ: אלה.

	מהדקדוק לעמוד על זה מהשעה מהיום או מהלילה לעמוד על זה מרחק הכוכב מאופן המישור כשילקח זה הגובה לכוכב בעברו על קו חצי היום.
התשיעי.	נבאר בו שכבר נעזר בזה לעמוד על שיעור קוטר הכוכב ביחס אל העגולה אשר יסוב בה.
העשירי.	נישיר בו לעמוד מזה הכלי על שיעור מרחק הירח מהשמש מגלגל המזלות כדי שנתישר מזה קצת הישרה לעמוד על מקום הכוכבים הקיימים.
הי"א.	נישיר בו לדעת לדרך ההשתמשות בזה הכלי באופן שלא יקרה טעות במבטו.
הי"ב.	נישיר בו לעשיית הכלי האצטורלאב באופן שלא יקרה טעות במבטו כדי שיגיעו ממנה ללקיחת גובה הכוכב בתכלית מה שאפשר מהדקדוק והיישרנו עם זה בו לחלק מעלותיו לדקים ואם הם קטנות.
הי"ג.	נישיר בו לדעת קו חצי היום באי זה מקום שיהיה בתכלית מה שאפשר מהדקדוק.
הי"ד.	נבאר בו מה שיש בו מהקושי בעמידה על האמת במקום הכוכבים הקיימים מאורך וברוחב.
הט"ו.	נישיר בו לידיעת מקום השמש האמתי באי זה עת שנרצה.
הי"ו.	נישיר בו לידיעת מקום הכוכבים הקיימים.
הי"ז.	יתבאר בו מה שיראה מחלוף שיעורי הכוכבים שאין הענין בתכונתם על האופן שהניח בטלמיוס.
הי"ח.	נבאר בו שאין לו דרך בזה המקום לבאר איך זה הענין בתכונת הכוכבים על האופן שהניח בטלמיוס מצד מה שיראה מתנועותיהם שאינו הולך על הסדר ההוא.
הי"ט.	נציע בו על צד השרש וההתחלה בזאת החקירה מה שהושג בחוש מסגולות תנועות הכוכבים באופן שלם לנו או לזולתנו ממה שהוא הכרחי בהשלמת זאת החקירה.
הכ׳.	החקור בו בחלקי הסותר אשר אפשר שידומה שימשך מהם חלוף לתנועת האורך בהנחת זאת התנועה באופן פשוט ונזכור סגולותיהם.

השמיני: על זה מרחק] נוסח רינן: על זה מהשעה מרחק (אבל בכ"י יש סימן לציין את המלה "מהשעה" שהיא מיותרת).

השני.	נבאר בו שהוא ראוי שנפליג בו החקירה בזה הדרוש לגודל מעלתו.
השלישי.	נבאר בו קצת מה שבהשגת הדרוש הזה מהקושי ויתבאר עם זה שם מה שהכריחנו להשתדל בהמצאת הכלי שהמצאנוהו ללקיחת המבטים בתכלית הדקדוק והקלות.
הרביעי.	נבאר בו קצת הצעות מועילות לכל מה שנחקור בו מזאת החקירה במבטים ובזויות התקונים בתכונה אשר יביא אליה העיון והוא יחלק לחמשה דבורים.
הדבור הראשון.	נבאר בו המכוון בקצת שמות נשתמש בהם בזאת החכמה.
השני.	נביא בו מופתים גימטריים יישירו לדעת מה שיצטרך בזאת החכמה מענין הקשתות והמיתרים והחצים כי זהו מהדברים היותר הכרחיים בזאת החכמה.
השלישי.	ניישיר בו לעשיית לוחות יודעו בו המיתרים כלם החצים מהקשתות והפך.
הרביעי.	נעשה בו לוחות לזה ונודיע דרך ההשתמשות בהם.
החמישי.	נודיע בו איך נדע הזויות והצלעות במשולש מונח בעד ידיעתנו בקצת מזה.
הפרק החמישי.	נבאר בו הצעה מועילה לעמוד על שיעור חצי קוטר השמש או הירח מן העגולה אשר יסוב בה מפני שיעור ניצוץ אורו הנכנס בחלון.
הששי.	נחקור בו על מקום מרכז זרית המבט מהעין המבטח כדי שנתישר מזה לעמוד על מרחק הכוכבים קצתם מקצת מפני מה שיראה מהמרחק ביניהם בזה הכלי אשר המצאנוהו ללקיחת המבטים.
השביעי. פ 2א	נזכור בו אופן עשיית זה הכלי והדרך אשר יניעו בה ממנו │ לעמוד על מרחק הכוכבים קצתם מקצת מעגולת גלגל המזלות.
השמיני	נישיר בו לעמוד מזה הכלי על גובה השמש או איזה כוכב שיהיה בתכלית מה שאפשר

הפרק החמשי: בחלון] נוסח רינן: בחלונות.
הששי: מהעין] נוסח רינן: מעין.
השמיני: גובה השמש] נוסח רינן: עבה השמש.

המאמר החמישי ממ"י

1. אמר לוי בן גרשום אחר שהצענו במאמרים הקודמים באור מה שראוי שנציעהו לקצת מה שיתבאר בזה המאמר.
2. הנה כוננתנו בזה המאמר לחקור איך יתכן שתונח תכונת הגרמים השמימיים ומספרם בדרך שתשלים ממנה התנועה הנראית להם.
3. ויאות אל מה שיראה לכוכבים מחלוף הגודל בעצמם ויאות אל השרשים הטבעיים.
4. ונחקור אחר זה למה היו אלו התנועות הנמצאות לגרמים השמימיים באופן מה שהם בו ממהירות והאיחור הנזרות והיושר ומהנטיה לצפון ולדרום ולמה היו שאר המשיגים הנמצאים לגרמים השמימיים בזה האופן אשר הם בו.
5. ונחקור אחר זה איך מדרגת מניעי הגרמים השמימיים קצתם עם קצת ואיך מדרגם הש׳ית׳ מהם לפי השגת יד שכלנו.
6. ומפני זה יחלק המאמר לשלשה חלקים החלק הראשון בחקירה בתכונת הגרמים השמימיים ומספרים.
7. החלק השני בנתינת הסבות בכל משיגי הגרמים השמימיים לפי
8. מה שאפשר לנו. החלק השלישי בחקירה איך מדרגת מניעי הגרמים השמימיים קצתם עם קצת ואיך מדרגת הש׳ית׳ מהם לפי מה שאפשר לנו.

החלק הראשון. בחקירה בתכונת הגרמים השמימיים ומספרם והוא יחלק למאה ושלשים וששה פרקים.

הפרק הראשון. נבאר בו שזה הדרוש הוא נאות שיחקר בו בזה הספר.

7 משיגי] נתקן בכ"י מ"מניעי".   7-8 לפי מה ... השמימיים] חסר בנוסח רינן.

ספר התכונה

לר' לוי בן גרשום

Sources in the History of Mathematics and Physical Sciences

Vol. 1: G.J. Toomer (Ed.), **Diocles on Burning Mirrors: The Arabic Translation of the Lost Greek Original,** Edited, with English Translation and Commentary by G.J. Toomer

Vol. 2: A. Hermann, K.V. Meyenn, V.F. Weisskopf (Eds.), **Wolfgang Pauli: Scientific Correspondence I: 1919–1929**

Vol. 3: J. Sesiano, **Books IV to VII of Diophantus'** *Arithmetica:* **In the Arabic Translation of Qusṭā ibn Lūqā**

Vol. 4: P.J. Federisco, **Descartes on Polyhedra: A Study of the** *De Solidorum Elementis*

Vol. 5: O. Neugebauer, **Astronomical Cuneiform Texts**

Vol. 6: K. von Meyenn, A. Hermann, V.F. Weisskopf (Eds.), **Wolfgang Pauli: Scientific Correspondence II: 1930–1939**

Vol. 7: J.P. Hogendijk, **Ibn Al-Haytham's** *Completion of the Comics*

Studies in the History of Mathematics and Physical Sciences

Vol. 1: O. Neugebauer, **A History of Ancient Mathematical Astronomy**

Vol. 2: H. Goldstine, **A History of Numerical Analysis from the 16th through the 19th Century**

Vol. 3: C.C. Heyde/E. Seneta, **I.J. Bienaymé: Statistical Theory Anticipated**

Vol. 4: C. Truesdell, **The Tragicomical History of Thermodynamics, 1822–1854**

Vol. 5: H.H. Goldstein, **A History of the Calculus of Variations from the 17th through the 19th Century**

Vol. 6: J. Cannon/S. Dostrovsky, **The Evolution of Dynamics: Vibration Theory from 1687 to 1742**

Vol. 7: J. Lützen, **The Prehistory of the Theory of Distributions**

Vol. 8: G.H. Moore, **Zermelo's Axium of Choice**

Vol. 9: B. Chandler/W. Magnus, **The History of Combinatorial Group Theory**

Vol. 10: N.M. Swerdlow/O. Neugebauer, **Mathematical Astronomy in Copernicus's De Revolutionibus**

Vol. 11: B.R. Goldstein, **The Astronomy of Levi ben Gerson (1288–1344)**